WITHDRAWN
University of
Illinois Library
at Urbana-Champaign

GUIDEBOOK

TO THE GEOLOGY OF

NORTHWEST COLORADO

1955

INTERMOUNTAIN ASSOCIATION OF PETROLEUM GEOLOGISTS

ROCKY MOUNTAIN ASSOCIATION OF GEOLOGISTS

UNIVERSITY OF
ILLINOIS LIBRARY
AT URBANA - CHAMPAIGN
GEOLOGY

GUIDEBOOK TO THE
GEOLOGY OF
NORTHWEST COLORADO
1955

INTERMOUNTAIN ASSOCIATION OF PETROLEUM GEOLOGISTS

Sixth Annual Field Conference

held jointly with

ROCKY MOUNTAIN ASSOCIATION OF GEOLOGISTS

Annual Field Conference

INTERMOUNTAIN ASSOCIATION OF PETROLEUM GEOLOGISTS
SALT LAKE CITY, UTAH

ROCKY MOUNTAIN ASSOCIATION OF GEOLOGISTS
DENVER, COLORADO

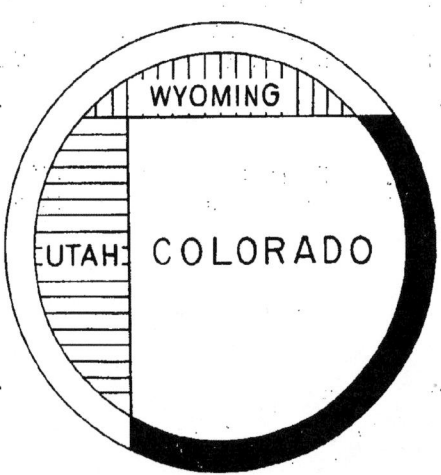

HOWARD R. RITZMA
STEVEN S. ORIEL
Editors

SOCIETY OFFICERS

I. A. P. G.

President:
  KEITH M. HEBERTSON,
  El Paso Natural Gas Company

Vice President:
  MILTON ZENI,
  Standard Oil of California

Secretary:
  JOHN W. REID,
  Standard Oil of California

Treasurer:
  KENNETH W. LARSON,
  Sun Oil Company

Executive Committee Member:
  JOHN W. COOKE, JR.,
  Continental Oil Company

R. M. A. G.

President:
  JOHN MAXSON,
  Aerial Exploration Company

1st Vice President:
  JACK W. KNIGHT,
  Petroleum Research Company

2nd Vice President:
  ROBERT L. KRETZ,
  Petroleum Consultants

Secretary-Treasurer:
  JOE W. JACKSON,
  Sinclair Oil and Gas Company

Councilors:
  B. PETE HARDER,
  Geophoto Services, Inc.

  ROBERT J. "SCOTTY" KNOX,
  Knox, Bergman and Shearer

PAST PRESIDENTS

R. M. A. G.

1922	MAX BALL	1937	H. W. OSBORNE	1946	J. W. VANDERWILT
1928	P. WHITNEY	1938	C. S. LAVINGTON	1947	C. A. HEILAND
1929	J. S. IRWIN	1939	W. O. THOMPSON	1948	C. E. MANION
1930-31	J. H. JOHNSON	1940	H. N. HICKEY	1949	DART WANTLAND
1932	C. E. DOBBIN	1941	NINETTA DAVIS	1950	R. L. SEILAFF
1933	A. E. BRAINERD	1942	H. E. CHRISTENSEN	1951	ROBERT MCMILLAN
1934	HERMAN DAVIES	1943	R. D. COPLEY	1952	G. R. DOWNS
1935	W. A. WALDSCHMIDT	1944	BEN H. PARKER	1953	N. W. BASS
1936	ROSS HEATON	1945	J. J. ZORICHAK	1954	ROBERT MUNOZ

I. A. P. G.

1949	FRANK NEIGHBOR	1951	M. DARWIN QUIGLEY	1953	LEWIS F. WELLS
1950	ORLO E. CHILDS	1952	GRAHAM S. CAMPBELL	1954	ROGER G. ALEXANDER

FIELD TRIP COMMITTEES

I. A. P. G.

Co-Chairman:
  PATRIC W. GAINES,
  The Texas Company

Co-Treasurer:
  KEN LARSON,
  Sun Oil Company

Reservations:
  JOHN W. REID,
  Standard Oil of California

Advertising:
  JOHN C. OSMOND,
  Gulf Oil Corporation

Caravan:
  JAMES B. LA FEVERS,
  Shell Oil Company

FIELD TRIP COMMITTEES (Continued)

R. M. A. G.

Co-Chairman:
GEORGE H. FENTRESS,
Lion Oil Company

Co-Treasurer:
THAD R. CARPEN,
Continental Oil Company

Reservations:
FLOYD H. MILLER,
British-American Oil Producing Company

Housing:
T. G. LARSON,
Phillips Petroleum Company, Steamboat Springs

Advertising:
CHARLES J. MCGINNIS,
The California Company

G. ALLAN NELSON,
Colorado Oil and Gas Corporation

Public Relations:
MAURY GOODIN,
Petroleum Information

EDITORIAL STAFFS

I. A. P. G.

Editor:
HOWARD R. RITZMA
General Petroleum Corporation

Staff Members:
JOHN W. DAHM
El Paso Natural Gas Company

JOHN C. OSMOND
Gulf Oil Company

ARTHUR E. OWEN
Sinclair Oil and Gas Company

M. D. "DUKE" PICARD
Shell Oil Company

SEYMOUR L. SHARPS
Shell Oil Company

R. M. A. G.

Co-Editor:
STEVEN S. ORIEL
U. S. Geological Survey

Staff Members:
JOHN CHRONIC
University of Colorado

ALVIN E. DUFFORD
The California Company

GEORGE FENTRESS
Lion Oil Company

WELCOME

The Rocky Mountain Association of Geologists and the Intermountain Association of Petroleum Geologists extend a welcome to all our visiting colleagues in geology and to the members of our companion organizations. This field trip in northwestern Colorado affords the opportunity for cooperative effort of our two organizations in an area of great mutual interest. Not only is the region one where much of the geology is spectacular and well-exposed but it is also highly important in oil production. It includes Rangely, the largest oil producer in the Rocky Mountains, and other significant fields discovered some years ago; yet, it is an area of current exploratory interest.

We thank the many individuals in both participating organizations who will make this guidebook and joint field trip an outstanding success.

JOHN H. MAXSON, *President, R.M.A.G.*
KEITH M. HEBERTSON, *President, I.A.P.G.*

DEDICATION

To the geologists of the early territorial surveys whose analysis of the geology of Northwest Colorado provided a stratigraphic nomenclature, a tectonic framework and a foundation for subsequent economic exploration.

TABLE OF CONTENTS

	Page
Society Officers — Field Trip Committees — Past Presidents — Editorial Staffs	II
Welcome	III
Dedication	IV
Table of Contents	V

GENERAL GEOLOGY

Paleozoic and Mesozoic Correlation Chart of Northwestern Colorado and Adjacent Areas	Quey Hebrew and M. Dane Picard	2
Lower Paleozoic Rocks of the White River Uplift, Colorado	N. Wood Bass and Stuart A. Northrop	3
Stratigraphy of the Steamboat Springs Area, Colorado	T. G. Larson	10
Pennsylvanian of Northwest Colorado	R. E. Landon and F. A. Thurman	12
Correlation of pre-Mancos, post-Weber Formations, Northwestern Colorado	S. L. Sharps	16
Geology of the Eastern End of the Uinta Mountains, Utah-Colorado	G. E. and B. R. Untermann	18
Jurassic and pre-Mancos Cretaceous Stratigraphy of the Eastern Uinta Mountains, Utah-Colorado	Whitney A. Bradley	21
Pre-Cretaceous Stratigraphic Cross Section, Northwest Colorado	M. Dane Picard — facing	26
Microfossils of the Curtis Formation, Eastern Uinta Mountains, Utah-Colorado	Don L. Eicher	27
The Cretaceous Rocks of Northwest Colorado	Charles C. O'Boyle	32
Early Cenozoic History of the Sand Wash Basin	Howard R. Ritzma	36
Earliest Eocene Vertebrates from the Sand Wash Basin, Northwest Colorado	Malcolm C. McKenna	41
The Elkhead Mountains Volcanic Field, Northwestern Colorado	Byrl D. Carey, Jr.	44
Review of the Browns Park Formation	Byrl D. Carey, Jr.	47
Correlation of Cenozoic Deposits of Northwestern Colorado	M. Dane Picard and Paul O. McGrew	50
Tectonic Map of Northwest Colorado	Editorial Staff — facing	52
A Structural History of Northwestern Colorado and Parts of Northeastern Utah	A. J. Crowley	53
Geomorphology of Northwestern Colorado	William Lee Stokes	56
Geology of Cross Mountain, Moffat County, Colorado	S. P. Kanizay	60
Juniper Mountain Area, Colorado	C. P. Abrassart and G. A. Clough	63
Geology of the North Hahns Peak Area, Routt County, Colorado	James M. Hunter	71
The Geology of the South Hahns Peak District, Routt County, Colorado	William W. Barnwell	73

ECONOMIC GEOLOGY

OIL AND GAS

		Page
Map of Oil and Gas Fields and Pipelines	W. A. Gillespie	77
Vermilion Creek Basin Area	Victor B. Gras	78
Bell Rock Dome	John C. Wyeth	84
Craig Dome	John C. Wyeth	86
Elk Springs Oil Field	William D. Fenex	88
Iles Dome	Erik Nelson	90
Moffat Dome	Don Vieaux and E. R. Haymaker	92
Powder Wash - Ace Field	L. W. Folsom	94
Weber Pool of Rangely Field, Colorado	Graham S. Campbell	99
Fracture Production from the Mancos Shale, Rangely Field, Rio Blanco County, Colorado	V. E. Peterson	101
Temple Canyon Oil Field	W. A. Clough	106
Thornburg Dome	Contributed	108
Tow Creek Oil Field	A. Saterdal	110
White River Dome	Galen Helmke	113
Geology of the Williams Park - Fish Creek Anticlines, Routt County, Colorado	C. L. Severy	116
Selected Oil and Gas Fields of Northwestern Colorado and Southwestern Wyoming	Daniel S. Turner	119
Penetration Chart of Oil and Gas Fields of Northwestern Colorado and Adjacent Wyoming	J. N. Dahm and C. E. White — facing	122

URANIUM

The Uranium Deposits of the Fish Creek District, Colorado	E. P. Beroni and R. C. Derzay	123
Uranium Deposits in the Skull Creek and Uranium Peak Districts, Northwest Colorado	Y. William Isachsen	124
Uranium in Northern Colorado and Southern Wyoming	Eugene W. Grutt, Jr., and Jerry F. Whalen	126

ROAD LOGS — SIDE TRIPS 130

Master Road Map 131

ROAD LOGS — Introduction 132

First Day 133
    The Uinta Fault from Clay Basin Eastward 138
    Geology in Vicinity of First Night's Camp 139

Second Day 140
    Irish Canyon Section 140
    Skull Creek Section 143
    Piceance Creek Gas Field 146

Third Day 147
    Oil Fields Along Danforth Hills Anticline 149
    Pagoda Gas Field 150

	Page
Additional Logs	151
Log, U.S. 40, Craig to Maybell	151
Log, Colo. 789-13, Junction with Moffat 4 to Baggs, Wyoming	151
Log, East end of Browns Park to Clay Basin	152
Log, Colo. 789-13, Rio Blanco to Rifle	153

SIDE TRIPS 154

Dinosaur National Monument	155
Sand Wash Basin Roads	157
Baggs to Slater	158
Elkhead Loop Trip	159

SELECTED BIBLIOGRAPHY, Compiled by John Chronic 161

HISTORICAL SKETCHES .... throughout guidebook

Sketches by:

    Colorado State Historical Society (Maurice Frink) — Leland H. Creer, Department of History, University of Utah — Howard R. Ritzma.

Quotation from:

    Gregory C. Crampton, Department of History, University of Utah.

ADVERTISING 166

    Alphabetical Index to Advertisers 166

IN THE POCKET

ACCOMPANY PAPERS IN GUIDEBOOK

Plate   I — Correlation of pre-Mancos, post-Weber Formations, Northwest Colorado ...... *S. L. Sharps*

Plate II — Geologic Map of Cross Mountain Anticline ...... *S. P. Kanizay*

Plate III — Geologic Map of Juniper Mountain ...... *C. P. Abrassart and G. A. Clough*

Plate IV — Geologic Map of North Hahns Peak Area ...... *James M. Hunter*

Plate V — Geologic Map and Structure Sections of South Hahns Peak District ...... *William W. Barnwell*

Plate VI — Structure Contour Map of Rangely Field ...... accompanies paper by *Graham S. Campbell*

SEPARATE PLATES

Plate VII — Generalized Areal Geologic Map, Northwest Colorado and adjacent Southern Wyoming ...... *A. E. Owen*

Plate VIII — Correlation Chart, pre-Cretaceous, Rangely to Tow Creek ...... *Contributed*

Plate   IX — Photogeologic Interpretation of the Danforth Hills Anticline ...... *Richard C. Oburn*

Plate   X — Geologic Map and Section of Piceance Creek Dome ...... *after U.S.G.S.*

GENERAL GEOLOGY

Northwest Colorado is, in a way, a geologic crossroads in the Rocky Mountains. Its complexity has made it a fascinating area for the geologist. Documented here in outcrops are a Proterozoic geosyncline, two Paleozoic uplifts, the orogenic pulses of the Laramide Revolution, the unique east-west Uinta Arch and extensive later Tertiary warping, faulting and vulcanism. Repeated invasion and ebb of seaways and the waste of mountain ranges are recorded in a colorful sequence of marine and continental formations.

This is not a new area of geologic investigation; Powell, King, Hayden and other pioneers of Western geology knew these hills and rivers well. Classic studies of stratigraphy, structure and geomorphology have had their inception here. With new tools for exploration and the time and capability for detailed investigation, the modern day geologist, eager to resolve the complexity into orderly knowledge often finds truth in the old maxim: "The more you know, the less you know."

There are contradictions and inconsistencies in the pages that follow, subjects for controversy and argument. These are presented in the spirit of scientific investigation. May controversy provoke thought; argument burgeon into ideas.

PALEOZOIC AND MESOZOIC CORRELATION CHART OF NORTHWESTERN COLORADO AND ADJACENT AREAS

Era	System	Series	NORTHEASTERN UTAH	SOUTH CENTRAL WYOMING	NORTHWESTERN COLORADO	WEST CENTRAL COLORADO	NORTH PARK COLORADO
MESOZOIC ERA	CRETACEOUS	UPPER	MANCOS SH: UPPER MANCOS SH. / MESAVERDE FM	LANCE FM / LEWIS SH / MESAVERDE GROUP: ALMOND FM, ERICSON FM, ROCK SPRINGS FM, BLAIR FM, BAXTER (STEELE) SH	LANCE FM / LEWIS SH / MESA VERDE GROUP: WILLIAMS FORK FM (TWENTYMILE SS), ILES FM (TROUT CREEK SS) / MANCOS SH: UPPER MANCOS SH	MESA VERDE GROUP: HUNTER CANYON FM, MT GARFIELD FM (ROLLINS SS, SEGO SS) / MANCOS SH: UPPER MANCOS SH	PIERRE SH / NIOBRARA / BENTON SH
			FRONTIER FM / MOWRY SH	FRONTIER FM / ASPEN (MOWRY) SH	FRONTIER FM / MOWRY SH	FRONTIER FM / LOWER MANCOS SH	FRONTIER FM (WALL CREEK SS) / MOWRY SH
		LOWER	DAKOTA SS / CEDAR MTN FM / BUCKHORN CGL	CLOVERLY GROUP: DAKOTA SS, FUSON SH, LAKOTA SS	DAKOTA SS / FUSON SH / LAKOTA SS	DAKOTA SS / BURRO CANYON FM	THERMOPOLIS SH (MUDDY SS) / COVERLY GROUP: DAKOTA SS, FUSON (?) SH, LAKOTA SS
	JURASSIC	UPPER	MORRISON FM / SAN RAFAEL GROUP: CURTIS FM, ENTRADA FM, CARMEL FM	MORRISON FM / UPPER SUNDANCE (CURTIS) / "SUNDANCE RED" / LOWER SUNDANCE FM	MORRISON FM / CURTIS FM / ENTRADA FM / CARMEL FM	MORRISON FM (BRUSHY BASIN SH, SALT WASH SS) / CURTIS FM – SUMMERVILLE FM / ENTRADA FM	MORRISON FM
		MIDDLE					SUNDANCE (?) FM
		LOWER	NAVAJO SS (NUGGET SS)	NUGGET SS	NUGGET (NAVAJO)	KAYENTA FM / WINGATE SS	
	TRIASSIC	UPPER	CHINLE FM / SHINARUMP CGL	CHUGWATER FM: JELM MBR, ALCOVA LS	CHINLE FM / SHINARUMP CGL	CHINLE FM / SHINARUMP CGL	CHUGWATER FM
		LOWER	MOENKOPI FM	RED PEAK MBR	STATE BRIDGE FM / MOENKOPI FM	MOENKOPI FM	
PALEOZOIC ERA	PERMIAN		PARK CITY – PHOSPHORIA FM	PHOSPHORIA – PARK CITY FM	PHOSPHORIA FM / SO CANYON CREEK DOL / SCHOOL HOUSE TONGUE / WEBER FM / MAROON FM (RESTRICTED)	SO CANYON CREEK DOL / MAROON FM / SCHOOL HOUSE TONGUE / MINTURN FM / BELDEN SH	?
	PENNSYLVANIAN		WEBER SS / MORGAN FM	WEBER SS / AMSDEN FM	MORGAN FM / MOLAS FM	MINTURN FM / BELDEN SH / MOLAS FM	
	MISSISSIPPIAN	U	MANNING CANYON FM / MOLAS FM			MOLAS FM	
		L M	MADISON LS	MADISON LS	MADISON LS / LEADVILLE LS	MADISON LS	
	DEVONIAN	U			CHAFFEE FM (DYER DOL, PARTING QTZ)	DEVONIAN UND.	
	SILURIAN						
	ORDOVICIAN	U			FREMONT LS	ORDOVICIAN – UNNAMED	
		M			HARDING SS		
		L			MANITOU LS		
	CAMBRIAN	U	LADORE FM	DEADWOOD FM	SAWATCH QTZITE (PEERLESS SH)	SAWATCH QTZITE (PEERLESS SH)	

Compiled by: QUEY HEBREW AND M. DANE PICARD

INDEX MAP (Drafted by J.D. MOORE)
- NORTH EASTERN UTAH
- SOUTH CENTRAL WYO.
- NORTHWESTERN COLO.
- WEST CENTRAL COLO.
- NORTH PARK COLO.

LOWER PALEOZOIC ROCKS OF THE WHITE RIVER UPLIFT, COLORADO[1]

By N. WOOD BASS and STUART A. NORTHROP

U. S. Geological Survey, Denver, Colorado and Albuquerque, New Mexico

INTRODUCTION

The White River uplift is in northwestern Colorado and includes parts of Garfield, Rio Blanco, Routt, and Eagle Counties. The uplift lies largely between 107° and 108° W. longitude and 39°30′ and 40°10′ N. latitude. Rocks ranging in age from Precambrian to Recent crop out on the uplift. Figure 1 shows the relationship of the White River uplift to the main physiographic and structural features of Colorado. The stippled areas represent the mountain ranges which are large anticlines elongated northwestward. The White River uplift appears to be a northwestward extension of the Sawatch Range and is slightly en echelon to it.

UPPER CAMBRIAN

Overlying a Precambrian complex of schists and granites, a sequence of quartzite, sandstone, dolomite, and limestone flat-pebble conglomerate interbedded with very thin beds of shale is exposed widely in the mapped area. The sequence is assigned tentatively a Late Cambrian age. The total thickness of the sequence is about 600 feet; the Sawatch quartzite includes the lowermost 500 feet and the Dotsero formation includes the uppermost 100 feet.

Sawatch Quartzite

The Sawatch quartzite consists chiefly of regularly bedded quartzitic sandstone, sandstone, and quartzite, in beds commonly ranging from 2 to 5 feet thick. In addition, the formation contains a few units of thin-bedded dolomite. All these are interbedded with beds of light greenish-gray shale that range from a fraction of an inch to several inches in thickness. The regular bedding of the formation is one of its chief characteristics.

A unit, 75 feet or more in thickness, composed of dull-brown, thin-bedded dolomite and sandy dolomite, containing much glauconite, is present in the upper part of the formation. Such a unit is present at all exposures examined throughout the White River uplift; presumably it is the same unit at all places. Its position below the top of the formation varies considerably, however. On upper Grizzly Creek the dolomite unit lies 220 feet below the top of the Sawatch quartzite; in the eastern part of Glenwood Canyon it is 150 feet below the top; on Main Elk Creek it is 65 feet below the top; on the east slope of Blair Mountain, in the northwestern part of the area, it was estimated to be 100 feet below the top of the formation. The only fossils found in the Sawatch were a very few phosphatic shelled brachiopods in this dolomite unit.

Other units of dolomite are present in the Sawatch at some places. One of these, 45 feet below the top of the formation, is 11 feet thick on the upper part of Grizzly Creek. On Main Elk Creek, about 2½ miles upstream from the Clinetop road bridge, a sandy dolomite, 3 feet thick, lies 10 feet below the top of the formation, and a sandy dolomite, 17 feet thick, lies 185 feet below the top, which is 50 feet below the "75-foot dolomite unit" described previously. Near the east end of Glenwood Canyon beds of dolomite or sandy dolomite, ranging in thickness from 4 to 11 feet, lie at intervals 85 feet, 92 feet, 126 feet and 288 feet below the top of the Sawatch quartzite.

The thickness of the Sawatch is 520 feet in sec. 16, T. 5 S., R. 87 W., near the mouth of French Creek in Glenwood Canyon; at least 470 feet (the base is concealed) in sec. 8, T. 4 S., R. 88 W., near the head of Grizzly Creek; and 397 feet in sec. 21, T. 2 S., R. 90 W., on South Fork of White River.

The Sawatch forms sheer cliffs, 400 to 500 feet high, in Glenwood Canyon and cliffs nearly as high in the canyons of Deep, Grizzly, and Canyon Creeks and South Fork of White River. The contact of the formation with the underlying Precambrian rocks is sharp at the few places where it is exposed; it is generally marked by brush-covered slopes, on the Precambrian rocks, that extend outward from the base of the cliffs formed by the Sawatch quartzite. The contact of the top of the formation with the Dotsero formation is defined by relatively thick beds of quartzite below, with shale and thin beds of dolomite above. Inasmuch as the Dotsero forms a slope that recedes from the edge of the cliff of Sawatch strata, the contact is readily distinguishable in mapping.

The Sawatch is cleanly exposed at many places in Glenwood Canyon, and in many other canyons in the area. Two of these in Glenwood Canyon are 1.4 miles and 4.2 miles northeast of the bridge over Colorado River at Glenwood Springs. At the first place, essentially the entire formation is exposed across the river

---

[1] Publication authorized by the Director, U. S. Geological Survey.

from the highway, dipping southward and forming the lower part of the cliff. At the second place, also, the exposed rocks are across the river from the highway. This place in directly in front of the house and garage of the Colorado State Highway Department. Here the upper part of the cliff, which is set back a small amount, is formed by the Dotsero and Manitou formations and the main cliff is formed by the Sawatch quartzite. Most of the large springs on the White River Plateau issue from beds of the Sawatch.

UPPER CAMBRIAN AND LOWER ORDOVICIAN

Dotsero and Manitou Formations

Overlying the Sawatch quartzite is a sequence of rocks, 185 to 250 feet thick, consisting of interbedded flat-pebble limestone conglomerate, dolomite, and shale of Late Cambrian and Early Ordovician age. The sequence is essentially divisible into three main lithologic units: (1) A lowermost unit, whose thickness is about one-fourth of the total thickness, consisting of thin-bedded tan dolomite and a few beds of flat-pebble dolomite conglomerate interbedded with greenish-tan dolomitic shale; (2) a middle unit whose thickness is about one-half of the total, consisting mainly of thin beds of gray to grayish-tan flat-pebble limestone conglomerate that weathers to a reddish cast, interbedded with greenish-gray limy shale; and (3) an uppermost unit, whose thickness is about one-fourth of the total, consisting of thin beds of regularly bedded tan dolomite that commonly forms a cliff. These lithologic units, however, do not coincide with the named stratigraphic units. The sequence is divided into two formations of about equal thickness—the Dotsero formation of Late Cambrian age and the Manitou formation of Early Ordovician age.

Dotsero Formation

The Dotsero formation comprises a sequence, 96 to 106 feet thick, consisting of tannish gray thin-bedded dolomite, flat-pebble limestone conglimerate, a few beds of flat-pebble dolomite conglomerate, all interbedded with very thin beds of greenish-gray dolomitic shale. The lowermost beds of the formation are commonly sandy. The top beds are composed of algal limestone and conglomerate that forms a distinctive ledge. Many beds contain grains of glauconite. The formation comprises the lower half of the Dotsero dolomite of Bassett (1939, pp. 1855-1858). Six collections of Late Cambrian fossils were made from the formation. Trilobites of Late Cambrian age were identified by Palmer (1951) in four collections and graptolites of Late Cambrian age in two collections. The Dotsero formation represents most of the Trempealeau stage—the uppermost stage of the Cambrian, according to Palmer. The formation is divisible into two members—the Glenwood Canyon member including all but the top few feet of the formation, and the Clinetop algal limestone member, which embraces the top few feet of beds.

The base of the Dotsero formation is readily distinguishable by the presence of interbedded shale and thin beds of dolomite, overlying beds of quartzite of the Sawatch. The top of the formation is drawn at the top of a thin algal limestone unit that is readily recognized throughout the White River uplift. Moreover, a Late Cambrian fauna was collected from the algal limestone unit and beds below it and an Early Ordovician fauna was collected from beds only 3 feet above it.

The formation commonly forms a steep slope between an underlying cliff formed by the Sawatch quartzite and an overlying slope and cliff formed by the Manitou formation. The Dotsero formation is cleanly exposed at many places in Glenwood Canyon and in many other canyons in the area. It is exposed and readily accessible north of U. S. Highway 6, 14.8 miles northeast of the bridge over Colorado River at Glenwood Springs, which place is 0.55 miles west of the Garfield-Eagle County boundary sign.

Glenwood Canyon Member

All beds in the Dotsero formation lying below a ledge-forming algal limestone that forms the top unit of the formation are included in the Glenwood Canyon member. The lower one-half of the member consists of thin beds of light-gray to tannish-gray dolomite and a few thin beds of flat-pebble dolomite conglomerate interbedded with thin beds of light greenish-gray dolomitic shale. The upper one-half of the member consists of thin beds of flat-pebble limestone conglomerate and interbedded light greenish-gray limy shale.

The beds of flat-pebble dolomite conglomerate consist of gray to grayish-tan dolomite matrix and flattish pebbles with rounded edges of gray to tannish-gray dolomite. The pebbles lie at varying angles with the bedding planes. The beds of conglomerate range in thickness from about 2 inches to 1½ feet. Most beds are 5 to 8 inches thick. The beds of shale are commonly thinner than the beds of conglomerate. The beds of flat-pebble limestone conglomerate are composed of flat pebbles of dense medium-gray limestone in a gray limestone matrix. The pebbles range from ¼ inch or less to 5 inches in length and are rarely as much as 8 inches. Many are 1 to 2 inches long; most are less than ½ inch thick. Essentially all have rounded edges. Most pebbles lie at small but varying angles with the bedding planes. A few beds are characterized by pebbles standing at very steep angles to the bedding planes. The

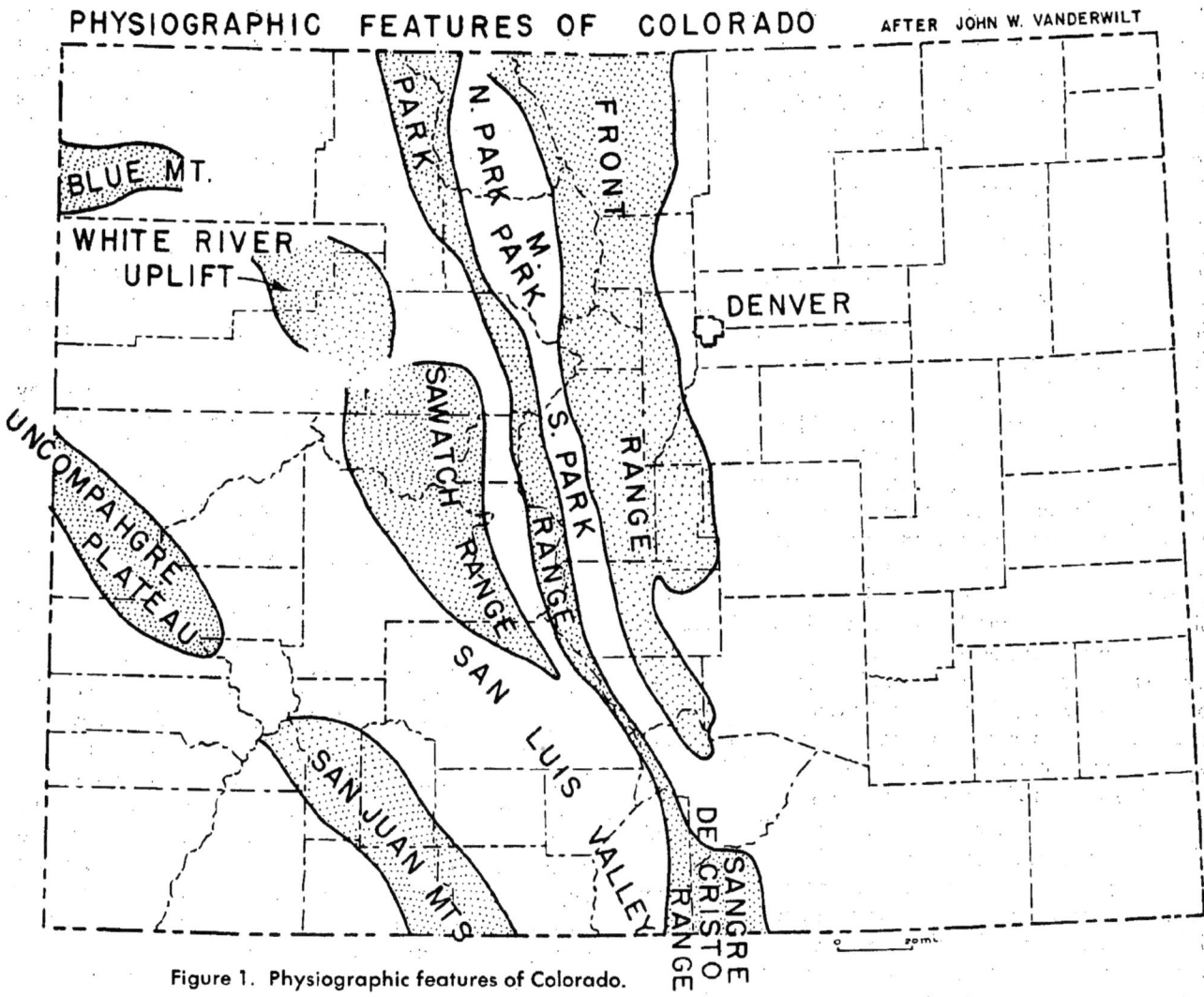

Figure 1. Physiographic features of Colorado.

matrix of both the limestone and dolomite beds of conglomerate weathers more readily than the enclosed pebbles, which condition imparts irregular surfaces to the beds.

Grains of glauconite are common throughout the member, but glauconite is not as abundant as in the dolomite beds of the underlying Sawatch quartzite. Most beds of limestone, dolomite, and conglomerate are fine-grained and dense; a few beds and lenses of limestone, however, are coarsely crystalline and contain fossils. A bed of dense, light tannish-gray dolomite, having a hummocky or wavy upper surface, which suggests an algal origin, is 25 to 35 feet above the base of the member. It forms a relatively distinct ledge. Nearly all beds contain a lattice work of stem-like casts at the bases of the beds that some geologists refer to as burrows and others as fucoid markings. Many of the beds of conglomerate weather reddish; hence the term "red cast" beds, used in early reports.

Clinetop Algal Limestone Member

The most distinctive unit in the Dotsero and Manitou formations is the Clinetop algal limestone member, a ledge-forming algal limestone and conglomerate unit, 3 to 5 feet or more thick, at the top of the Dotsero formation. It contains Late Cambrian fossils.

At most places the lower half of the Clinetop algal limestone member consists of coarse flat-pebble limestone conglomerate, and the upper half consists of crystalline to dense algal limestone with a crinkly to wavy structure, and some conglomerate. Essentially everywhere that the member crops out on the White River Plateau it weathers to a bone-white ledge having a distinct lavender tint, particularly on its top surface; the lavender color appears to impregnate the light-gray to white rock. Aside from the lavender tint, the most distinctive features of the rock are the circular or disc-shaped swirl patterns developed on the top surface and

the crinkled structure shown in vertical sections of the rock. The swirls range in diameter from about 3 to 12 inches; many are 4 to 8 inches. In the center of each swirl is a crudely circular or elliptical disk, 1 to 4 inches in diameter, which is a cross section of a cone-shaped core that is normal to the bedding and tapers downward through the rock; the cone is composed of limestone conglomerate and presumably is the material that was deposited between the closely spaced algal columns or stromatolites. Many cystoid columnals, about 1/16 inch in diameter, and ring-shaped cross sections of an undescribed genus of sponge, three-fourths to 1 inch in diameter, are commonly weathered into relief on the top surface of the bed.

J. Harlan Johnson (in Bass and Northrop, 1953, p. 900) reported that the limestone is unquestionably algal and that the "algae represent an undescribed species of *Collenia* which is similar in appearance and growth habits to *Cryptozoon undulatum* Bassler described from the Late Cambrian or Early Ordovician (Ozarkian) strata of Maryland (Bassler, 1919, p. 190, pls. 29-30). In growth habit, it is similar to *Crytozoon rosemontensis* Stauffer (1945, p. 379, pl. 58) except for developing much longer and wider colonies. In general appearance, it suggests *Collenia spissa* Fenton and Fenton from the Cambrian of the Teton Range, Wyoming (Fenton and Fenton, 1939, pp. 101-102, fig. 4, pl. 6, fig. 2) and probably is closely related to it.

Unusual conditions must have prevailed in the White River Plateau region toward the close of Cambrian time to permit the development of such an extensive algal biostrome, for it was seen at many places throughout an area of 400 square miles. Clearly, the occurrence of flat-pebble conglomerates both below and above the biostrome suggests considerable agitation of the water in Late Cambrian time and again in Early Ordovician time. The wide-spread algal biostrome, however, may indicate a period of quiet water just prior to the close of the Cambrian.

Manitou Formation

The Manitou formation consists chiefly of thin beds of gray flat-pebble limestone conglomerate interbedded with greenish-gray limy shale in the lower half to three-fifths—the Dead Horse conglomerate member. The upper portion of the formation—the Tie Gulch dolomite member — consists of regularly bedded, thin-bedded, medium-brown somewhat siliceous dolomite that commonly forms a cliff. Such cliffs include the lower part of the formation at many places and at other places the lower part of the formation forms a steep slope. The thickness of the formation ranges from 80 to 155 feet in the four stratigraphic sections that were measured.

A total of 17 collections of Early Ordovician fossils, which include trilobites, brachiopods, gastropods, a cephalopod, conodonts, sponge spicules, pelmatozoans, and a graptolite, were made from the lower three-fifths of the formation. Eight of these collections were made from Glenwood Canyon on the east side of a prominent spur in the cliffs north of U. S. Highway 6, half a mile northeast of the bridge over French Creek. There the fossil-bearing beds are distributed through the sequence that extends from 3 to 93 feet above the base of the formation. The fossil collections made at the other places are from beds in the lower 35 feet of the formation.

The Manitou formation is cleanly exposed at many places in Glenwood Canyon and in many other canyons in the area. It is readily accessible north of U. S. Highway 6, near the east end of Glenwood Canyon, half a mile west of the Garfield-Eagle County boundary.

Dead Horse Conglomerate Member

The Dead Horse conglomerate member of the Manitou formation consists largely of thin beds of gray flat-pebble limestone conglomerate. The beds of conglomerate in the member are composed of flat pebbles of dense medium-gray limestone embedded in a gray limestone matrix; the pebbles range in length from one-fourth of an inch or less to 5 inches, and rarely 8 inches; many are 1 to 2 inches long; most are less than half an inch thick; essentially all have rounded edges. In most beds the flat pebbles lie at small but varying angles with the bedding planes; however, the steep inclination of some of the pebbles gives the rock a complex structure. Locally a few beds are composed of flat pebbles, most of which are oriented at very steep angles to the bedding planes. The limestone matrix between the pebbles weathers somewhat more readily than the pebbles, and this differential weathering produces a slightly irregular surface on the tops and particularly on the edges of the beds. A few beds in the lowermost part of the member contain some glauconite, and a very few beds higher in the member contain traces of glauconite.

The beds of conglomerate range in thickness from 3 inches to a foot or more; most beds are 5 to 8 inches thick. The beds of shale are commonly thinner than the beds of conglomerate. The conglomerate weathers dull mottled gray to brown, and many beds and slabs weather mottled red and brown; hence the term "redcast" has often been applied to these beds, as well as to those in the underlying Dotsero formation. Sequences 5 to 20 feet thick of the interbedded conglomerate and shale commonly form alternating slopes and low ledges, rising above the shoulder of the underlying Clinetop algal limestone member of the Dotsero formation.

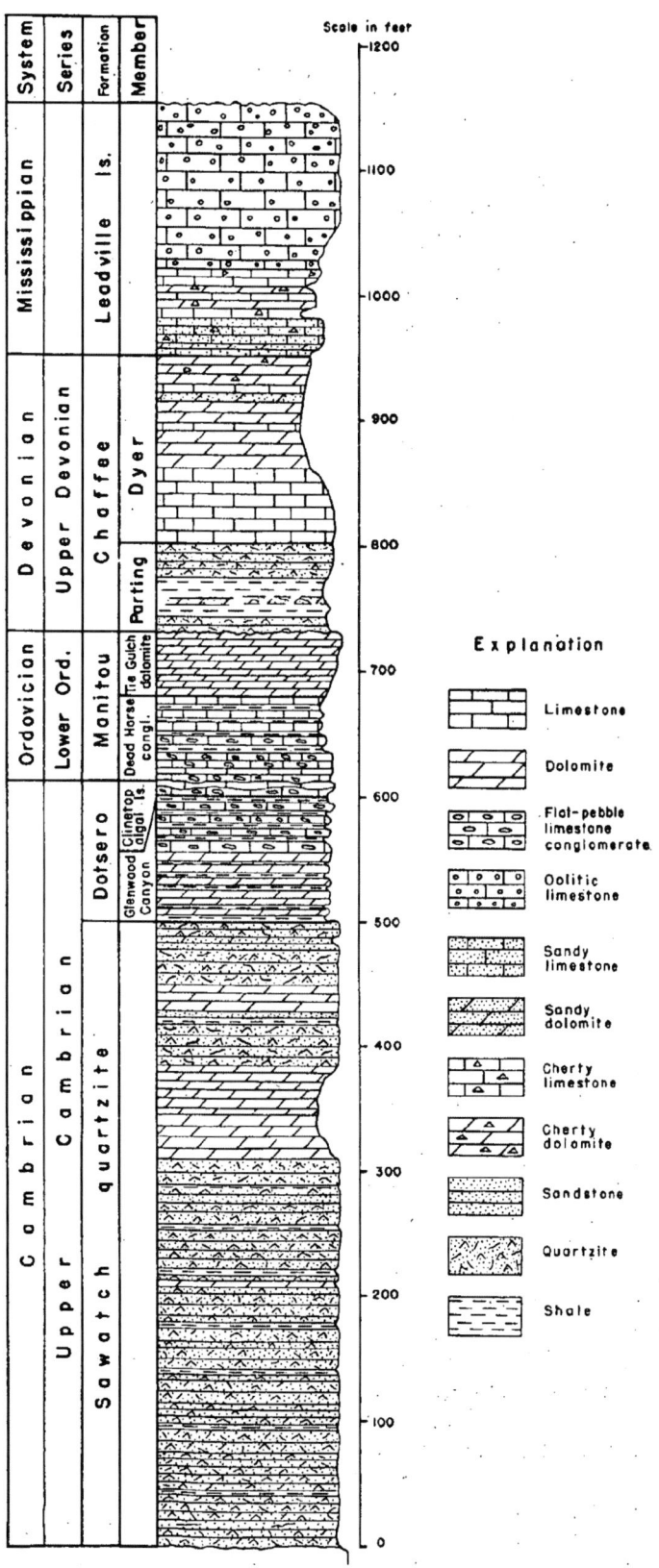

Generalized columnar section of lower Paleozoic rocks.
Figure 2. Generalized columnar section, lower Paleozoic, White River Uplift.

Tie Gulch Dolomite Member

The uppermost 50 feet, more or less, of the Manitou formation consists of regularly bedded, thin-bedded medium-brown dolomite that commonly forms a cliff. This portion of the formation constitutes the Tie Gulch member. Most of the dolomite is fine-grained, but some beds are medium-grained. Many beds are slightly siliceous and a few beds are sandy. Thin stringers of light-yellow chert are present on weathered surfaces of beds in many places. A few beds contain traces of glauconite. No fossils were found in the member.

The Tie Gulch dolomite member forms one of the key units for mapping. In most canyons it forms a light-brown cliff that caps steep slopes, and its top forms a prominent bench where the basal part of the overlying Parting member of the Chaffee formation is eroded back from the edge of the cliff. At many places, particularly in Glenwood Canyon, the cliff includes the underlying Dead Horse conglomerate member.

UPPER DEVONIAN

Chaffee Formation

The Chaffee formation of Late Devonian age unconformably overlies the Manitou formation. The bedding in the Chaffee formation is essentially parallel to that in the underlying Manitou. The Chaffee consists of two members that are widely different in composition. The Parting member occupies about the lower one-third of the formation and consists of interbedded shale and quartzite. The Dyer member occupies the upper two-thirds of the formation and consists of interbedded limestone and dolomite.

Parting Member

The thickness of the Parting member ranges from 63 to 95 feet in 10 stratigraphic sections that were measured. It consists of interbedded light-green shale, black shale, light-tan glassy quartzite, and some dolomite and sandy dolomite. The member is the most variable unit in the Paleozoic sequence of rocks. Stratigraphic sections only a few hundred feet apart show considerable variation, bed for bed. The beds of shale are commonly micaceous; many are sandy; some contain salt-crystal casts. Fish plates and scales of fresh-water placoderms (fishes) are locally common in the quartzite and shale beds. The beds of quartzite are lenticular; cross-bedding at steep angles is common. The quartzite is glassy and composed mostly of quartz. The grain size commonly is variable, ranging from very fine to coarse and, locally, to pebbles. The quartzite can be distinguished from quartzite of the Sawatch quartzite by its denser, glassier appearance, by generally poorer sorting of the grains, and by the association of the quartzite beds with the bright-green shale. The most conspicuous features of the Parting member on the outcrop are the bright-green shale and the blocks of quartzite which range from less than one to three feet thick.

At most places the member forms a slope above the cliff formed by the Manitou formation. Generally the slope is covered by vegetation, but locally it contains a few blocks of quartzite.

Dyer Member

The Dyer member of the Chaffee formation consists of fairly thick units of limestone and dolomite. A few beds are sandy, and others are cherty. The thickness of the member ranges from 140 to 175 feet in 5 sections that were measured. Its thickness is 152 feet near the east end of Glenwood Canyon, 165 feet on Main Elk Creek, and 158 feet on the South Fork of White River. The basal unit is a cliff-forming dull dark-gray limestone ranging from 48 to 75 feet in thickness; it constitutes one of the main key units in the Paleozoic sequence. It can be identified readily in canyon outcrops by the nearly sheer dull gray wall that it forms, rising above the slope formed by the Parting member, and by the presence of solution cavities. The cliff face formed by the unit commonly contains solution cavities ranging from 4 inches to 1½ feet in diameter. The rock weathers nodular; hence, was referred to by the field party as the Knobbly bed. It is abundantly fossiliferous at many places.

A persistent unit, 10 to 30 or more feet thick, of relatively dense light-brown weathering dolomite overlies the so-called Knobbly bed. The rock breaks down into conchoidal-shaped fragments, commonly about 1 to 4 inches thick and 4 to 12 inches wide.

The upper one-half of the Dyer member contains interbedded dense gray dolomite and limestone. A few beds of dolomite contain rounded grains of quartz scattered through the rock. A few beds contain nodules of dark-gray chert. At most places, fossils are scarce in the portion of the member lying above the so-called Knobbly bed. On the other hand, at some places, the uppermost beds of the member contain many fossils. The Dyer member has yielded abundant fossils of about 50 species (in descending order of number of species): brachiopods, gastropods, bryozoans, corals, cephalopods, pelecypods, etc.

MISSISSIPPIAN

Leadville Limestone

The Leadville limestone consists chiefly of limestone, although the lower third of the formation contains interbedded dolomite and limestone, many beds of

which contain dark-gray chert. Furthermore, the basal 20 to 30 feet of the formation consists of beds of very sandy dolomite and locally some beds of sandy limestone.

This lowermost portion of the formation is believed to be equivalent to the Gilman sandstone member, which is at the base of the Leadville limestone in the region of Leadville and Minturn, Colo. (Tweto, 1949), 50 and 35 miles, respectively, southeast of the White River uplift. The bedding planes in this lower portion are wavy to irregular and probably represent minor disconformities. This unit served to identify the boundary between the Leadville limestone and the underlying Chaffee formation, for the general lithology of the beds above and below it is similar and they are only sparingly fossiliferous.

The upper half of the Leadville consists of massive gray coarsely oolitic limestone. Oolites are present in the limestone throughout the mapped area and far beyond it. They were observed in the Leadville limestone on Lime Creek, 25 miles southeast; at Minturn, 30 miles east; and at Juniper Mountain, 40 miles northwest of the area. The oolitic character of the rock is a particularly important key to the identification of the formation at isolated exposures on the White River Plateau for, except for the oolites, weathered rock of the Leadville and parts of the Dyer member of the Chaffee formation are strikingly similar.

Determinable macrofossils are generally scarce and only a few small lots were collected in the mapped area. A composite list of determinations of collections from this region by other workers includes 39 species: algae, brachiopods, foraminifers, bryozoans, gastropods, and other groups.

The Leadville limestone forms the most striking outcrops of all the formations. Its light-gray cliff capping steep slopes and cliffs is conspicuous throughout much of the area. It is readily accessible for examination half a mile west of the east end of Glenwood Canyon above and a few hundred feet west of the Garfield-Eagle County boundary. It is accessible also one-quarter mile northeast and one-half mile northwest of the bridge over Colorado River at Glenwood Springs, where it forms a prominent prong-shaped ridge north of the highway.

REFERENCES

Bassett, C. F. (1939), "Paleozoic Section in the Vicinity of Dotsero, Colorado," Geol. Soc. America Bull., v. 50, no. 12, p. 1855-1858.

Bassler, R. S. (1919), "The Cambrian and Ordovician Deposits of Maryland," Maryland Geol. Survey, p. 190.

Fenton, C. L., and Fenton, M. A. (1939), "Pre-Cambrian and Paleozoic Algae," Geol. Soc. America Bull., v. 50, no. 1, pp. 101-102.

Johnson, J. H., in Bass, N. H., and Northrop, S. A. (1953), "Dotsero and Manitou Formations, White River Plateau, Colorado," Am. Assoc. Petroleum Geologists Bull., v. 37, no. 5, p. 900.

Palmer, A. R., 1951 in Bass, N. W., and Northrop, S. A. (1953), "Dotsero and Manitou Formations, White River Plateau, Colorado," Am. Assoc. Petroleum Geologists Bull., v. 37, no. 5, pp. 908-911.

Stauffer, C. R. (1945), "Cryptozoons of the Shakopee Dolomite," Jour. Paleon., v. 19, no. 4, p. 379.

Tweto, O. L. (1949), "Stratigraphy of the Pando Area, Eagle County, Colorado," Colo. Sci. Soc. Proc., v. 15, no. 4, pp. 177-184.

CANON PINTADO

"The Dominguez-Escalante expedition crossed the Colorado River at about the railroad point of Una and ascended Roan Creek through the Book and Roan Cliffs, escarpments of the East Tavaputs Plateau. At the divide they passed over to the watershed of the Green River and descended the long northern slope of the plateau by Douglas Creek . . . named Canon Pintado from . . . Indian paintings . . . seen at two different places."

"At the mouth of Douglas Creek the expedition crossed the White River, which the explorers named the San Clemente . . ." (September 1776)

C. Gregory Crampton
"The Discovery of the Green River",
Utah Historical Quarterly, vol. xx, no. 4, 1952

STRATIGRAPHY OF THE STEAMBOAT SPRINGS AREA, COLORADO

By T. G. LARSON

Phillips Petroleum Company, Steamboat Springs, Colorado

The sediments described belong to the Permo-Pennsylvanian, Triassic, Jurassic and Cretaceous systems. These sediments total 6740 feet including an estimated thickness of 5000 feet for the Mancos shale. In the vicinity of Steamboat Springs the Cretaceous formations are well exposed but the older formations are poorly exposed. For this reason, much of the information on the pre-Dakota section is derived from the Texas Company's deep test at Tow Creek twelve miles west of Steamboat Springs. The best surface section available is that on the east slope of Emerald Mountain about one mile south of Steamboat Springs where Permo-Pennsylvanian sediments may not be present.

PERMO-PENNSYLVANIAN

The pre-Cambrian granites, gneisses and schists of the Park Range near Steamboat Springs are overlain by about 300 feet of redbeds belonging to the Maroon formation of Pennsylvanian age. This formation is typically arkosic and is similar to the Fountain formation of the east flank of the Front Range. Steamboat Springs is near the northeastern wedge edge of the Maroon formation.

TRIASSIC

The Triassic interval found at the Tow Creek oil field is represented by the Moenkopi, Shinarump and Chinle formations.

The Moenkopi is logged as 463 feet thick. It is made up of dull red sandy shale, sandstone, siltstone and some greenish gray and yellow shales. The Moenkopi is usually non-calcareous. Stringers of gypsum occur in the lower part of the formation.

The Shinarump is logged as 73 feet thick. This is somewhat thicker than normal for this region. The formation is typically conglomeratic, composed chiefly of varicolored quartz grains ranging from medium sand to pebbles one-half inch or more in diameter.

The Chinle is logged at 234 feet in thickness. This formation consists of shale, mudstone and fine-grained sandstone which may be calcareous. The dominant color of the rock is red with minor amounts of brown, yellow, gray and purple.

Total Triassic found at Tow Creek measures 770 feet.

JURASSIC

The Jurassic section at Tow Creek including the Entrada sandstone and Morrison formation is about 520 feet thick. It is locally unconformable with the underlying Chinle.

The Entrada consists of massive beds of fine-grained, sugary, light gray sandstone which usually crops out in high sheer cliffs.

The Morrison formation is about 300 feet thick at Tow Creek and elsewhere. The formation consists predominately of variegated green, greenish-gray and maroon shale. Some thin beds of limestone, lenses of chert and white sandstone are sometimes found near the base.

CRETACEOUS

The Cretaceous system at Steamboat Springs is represented by the Dakota sandstone and Mancos shale.

The Dakota sandstone is about 150 feet thick; it consists of massive non-calcareous fine-grained sandstone containing lenses of coarse sand and conglomerate. The sandstone is usually hard and often quartzitic with siliceous cement. The conglomerates are usually confined to the basal part of the formation. Carbonaceous material is often disseminated through the sand.

The Mancos shale in the Steamboat Springs area is here defined as including all sediments from the top of the Dakota to the base of the Tow Creek sandstone which is taken as the basal bed of the Mesaverde group. So defined, the formation is about 5000 feet thick.

The lower five hundred feet is mainly thin-bedded black carbonaceous shale which correlates with the Benton of the eastern slope. Immediately above the Dakota is a slightly siliceous zone containing fish scales which is assumed to be equivalent to the Mowry.

At the top of the Benton zone is a fossiliferous sandy limestone which correlates with the Frontier formation of Wyoming and with the Codell member of the Carlile formation of the eastern slope. Above the Frontier is a calcareous zone 800 to 900 feet thick which correlates with the Niobrara. This zone is characterized by hard, brittle, thin-bedded limy shales which weather white on outcrop.

The Mancos sediments above the calcareous zone are characterized by sandy gray shales. Toward the top the formation becomes more sandy with three or more well-defined sandstone beds. The more prominent sandstones occur at about 300, 600, and 1200 feet below the base of the Tow Creek sandstone. The lowest of these may be the Morapos sandstone.

Traveling west toward Craig, the Mesaverde group is encountered near Milner; the Lewis shale just beyond Mt. Harris; and the Lance formation west of Hayden.

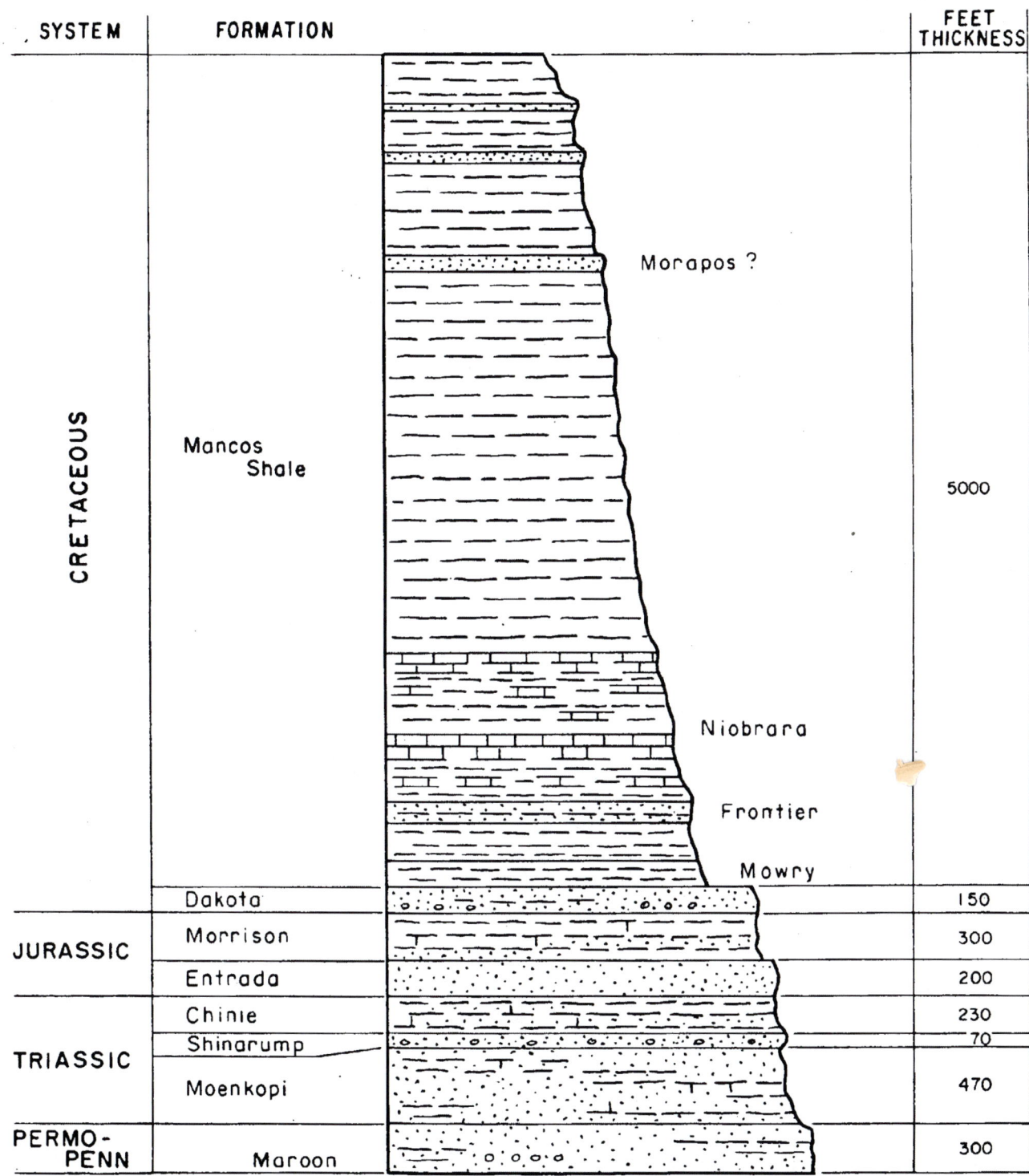

COLUMNAR SECTION OF ROCKS
VICINITY OF STEAMBOAT SPRINGS

PENNSYLVANIAN OF NORTHWEST COLORADO

By R. E. LANDON
General Petroleum Corporation, Denver, Colorado

and

F. A. THURMAN
Thurman Consultants, Denver, Colorado

Pennsylvanian stratigraphy of northwestern Colorado has long been considered very complex and results derived from studies are generally inconclusive. The red-beds lithology of much of the section and the resultant "downgrading" as probable oil producing horizons, have deprived these sediments of the attention given to other, more prosperous rock units.

When individual local sections are compared for purposes of correlation, the stratigraphy appears extremely variable, and because of lack of data and wide spacing of subsurface control, fully reliable correlations cannot be made. However, when the Pennsylvanian stratigraphic units of the region are considered as facies of a sedimentary epoch in a structural-depositional unit, these complexities appear to fall into place.

The Pennsylvanian history of northwestern Colorado is a good example of the structural-depositional history of a basin formed during a major deformational orogeny and subsequently filled by marine and later by non-marine sediments derived from surrounding uplifts. Thus, these complexities probably are more apparent than real when considered in the light of Pennsylvanian paleogeography. The following is an attempt to outline those major tectonic elements which must be considered in applying a paleogeographical approach toward reconciliation of stratigraphic vagaries.

During the Mississippian period there was widespread carbonate deposition throughout the Rocky Mountain province as well as over most of the North American continent. This period marks the last great carbonate deposition in northwestern Colorado. Though minor thicknesses of younger limestones and dolomites are present, the area of interest throughout post-Mississippian time has been close enough to major tectonic uplifts to receive predominantly clastic sediments.

At the close of Mississippian time, gently regional uplift brought the sediments above sea level and a long period of gentle erosion followed. As a result, Mississippian sediments were deeply weathered and a topography of low relief developed.

Pennsylvanian seas, advancing over the deeply weathered carbonate-derived soils and rubble, reworked and stratified the material over which they advanced. Sediments from this ancient soil horizon, plus slight amounts of sediment from the incipiently developing highlands, form the Molas formation. As is to be expected under these conditions of sedimentation, the Molas is not everywhere present; its absence in local areas is due either to non-deposition over areas of slight topographic relief or to complete removal of the weathered zone by erosion prior to the deposition of sediments younger than Molas.

The formation is predominantly a variegated shale sequence with occasional thin limestone and sandstone stringers. It is generally less than 50 feet in thickness, though as much as 200 feet has been reported. In general, the thicker sections of Molas contain a larger percent of carbonates.

By the end of Molas time, the tectonic units which were to dominate the major portion of the Pennsylvanian and Permian were beginning to localize. The relative movement intensified, as indicated by the variety and limited areal distribution of those deposits immediately younger than Molas, namely, the Lower Morgan and Belden and, to the southeast in Central Colorado, the Kerber. Regional studies suggest the presence of four major geographic units which began development at this time and persisted throughout the Pennsylvanian and Permian. To the west of the area of interest was the great shifting orogenic belt of the "Cordilleran geosyncline". Trending southeast from this "deep" an arm of the sea extended into northwestern Colorado following a relatively narrow, trough-like depression which was being down-warped as the positive units of the Uncompahgre and Colorado Range were being elevated to the southwest and northeast. As Pennsylvanian time advanced the highlands continued to develop, ultimately exposing pre-Cambrian rocks to erosion. During middle or upper Pennsylvanian time the trough extended across what is now central Colorado, connecting with the seas advancing from the southeast.

The Morgan formation, as originally defined, included those sediments between the Mississippian and the base of the Weber. Blackwelder considered it to rest unconformably on the Mississippian and to pass transitionally upward into the base of the Weber. At its type locality, in Weber Canyon, Utah, it is predominantly soft, earthy, red sandstone. Some beds are shaly and some thin beds of gray, fossiliferous limestones are

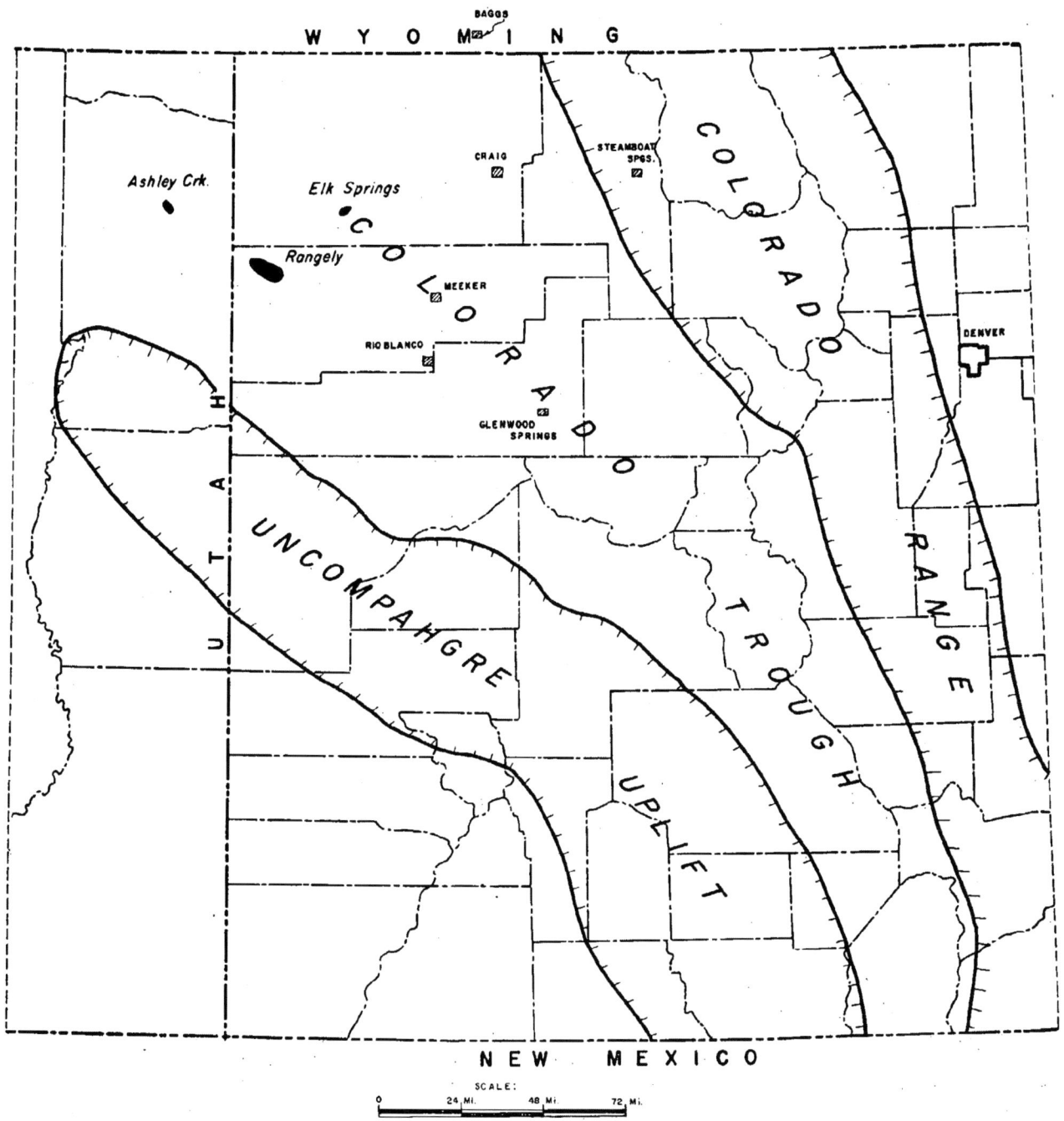

ELEMENTS OF THE ANCESTRAL ROCKIES IN PENNSYLVANIAN

included. It was originally considered to be of probable terrestrial origin. The name Morgan has been retained, though concepts of the origin and lithology of the formation have changed radically with additional control.

The relative carbonate-clastic content of the Morgan changes rapidly with the distance from the uplifts bounding the trough of deposition. Near the center of the trough more than a thousand feet of limestone has been classed as Morgan. Near the uplifts this interval is predominantly clastic.

The lower Morgan, in general, consists of sandstone, sandy and argillaceous limestone and dolomite attaining a thickness of several hundred feet in wells near the axis of the trough. The Belden formation, believed correlative with the lower Morgan, due to its geographic

location of deposition, is composed, principally, of gray and dark shales with occasional limestone and sandstone stringers. This formation may reach a thickness of more than 1200 feet. Southeastward of the present area of interest the Belden grades into the Kerber formation, a contemporaneously deposited terrestrial facies. In view of rapid lateral facies changes, terminology becomes a problem and selection of a regional datum a virtual impossibility.

Middle Morgan time is characterized by a great increase in carbonate deposition in the northwestern part of the area. To the southeast, correlative sediments are red-beds and evaporites. Throughout most of the present area of interest the term Morgan is applicable.

The evaporite-redbed clastic facies of the Middle Morgan, known as the Minturn, attains thicknesses of 6000 feet or more in nearby areas to the southeast. The presence of this great thickness of gypsum, anhydrite and red silt in the middle of an otherwise normal sedimentary series poses intriguing questions bearing on the origin and stratigraphic and paleogeographic relationships of these deposits.

The Minturn consists of arkose, conglomerate, sand, shale and evaporites. Several limestone and dolomite members are present. The Minturn evaporite basin extends from southeastern Moffat County on the north to Fremont County on the south, a distance of approximately 175 miles. Evaporite deposition was not continuous over this entire distance. In most places evaporites are interstratified with clastics, indicating a periodic dessication of the waters in the semi-enclosed basins. Laterally the evaporites interfinger with, and are replaced by, clastics. During this period there was a constriction in the trough, which may have developed periodically into a bridge, in south-central Colorado. It is probable that the floor of the trough was higher in the northwestern part than in the more rapidly downwarping central area. Such conditions would account for observed variations in lithology.

The terrestrial type of deposition of much of the Minturn continued into the Maroon. As its name implies, the Maroon is predominantly dark red in color and consists of rocks of clastic origin—arkose, conglomerate, sandstone, silt and shale. The base of the Maroon is taken as the top of the Jacque Mountain limestone where that member is present. Elsewhere the base is called on a color change. The upper limit of the Maroon is the base of the Weber sandstone where that formation is present. Where the Weber is present as only a tongue it is considered as a member of the Maroon which then continues upward to the Phosphoria.

As a unit, the Maroon is a typically developed basin-filling formation with coarse clastics near the edges and much finer clastics nearer the center of the trough. The formation is in part equivalent to the Upper Morgan and Weber and interfingers with them. By the end of Maroon time the trough was essentially filled and peneplanation of the highlands almost completed. The highlands remained slightly positive into Triassic time, but by the upper part of Chinle time sediments were deposited across the uplifts, with the exception of a few local areas where the remnants of the ranges were not buried until the Jurassic. This type of sedimentation is not unique to this area of Colorado, but is duplicated in the Cutler formation, southwest of the Uncompahgre; the Sangre de Cristo formation, of south-central Colorado; and the Fountain formation, on the east side of the Colorado Range.

It is very probable that the Pennsylvanian-Permian boundary is in the Maroon formation. Conclusive evidence is lacking, but a summation of floral, faunal and paleogeographical indications, plus an attempt to fix the major regional diastrophic events, leads to the conclusion that both ages are represented.

Overlying the Maroon in much of the area of interest is the Weber sandstone. Its base is gradational with the Morgan formation which is of known Pennsylvanian age. The top of the Weber in the northwestern part of the area is marked by the Phosphoria formation of known Permian age. The Weber attains a maximum thickness of some 1300 feet in northwestern Colorado and southeastwardly it thins rapidly into, and is replaced by, redbeds of the Maroon formation.

In its typical development the Weber is a white to yellowish white, clean, quartz sandstone. As it approaches the transition zone into the Maroon formation, stringers of each type lithology are interfingered and the clean quartz sands become red and silty. At or near the top of the formation a thin tongue of typical Weber sand extends southward and southeastward beyond the present area of interest. In places this member, known as the Schoolhouse tongue, is overlain directly by the Phosphoria. To the southeast it is overlain by a thickness of Maroon redbeds which is, in turn, overlain by the Phosphoria or South Canyon Creek dolomite of Phosphoria age. It is believed here that the Schoolhouse tongue is of Permian age.

A preponderance of accumulating evidence indicates that the Weber, as developed in the present area of interest, was deposited from two sources. The clean, well sorted, typical Weber sands probably came from the west and northwest and represent the edge of the great sandstone deposits of the "Utah deep". Near the margins of this deposition the geography was such that the eastern limit of clean sand deposition coincided with the northwestern limit of Maroon deposition from

the Uncompahgre and Colorado Range highlands.

Oil production from Pennsylvanian sediments of the area of interest and the immediately adjacent region has been limited to Weber sand. Three fields, to date, have been discovered. They are Rangely and Elk Springs in Colorado and Ashley Valley in Utah. Rangely field is located, paleogeographically, at about the eastern limit of clean sand deposition and near the southwesterly margin of the trough. There is a notable difference in sand quality between the western and eastern ends of the field. In general, the farther east, the less the effective porosity and permeability and the greater the percentage of redbed, Maroon-type sediments. The Rangely field is ideally located for stratigraphic accumulation and it is possible that a major oil field would have been in the area regardless of the presence of the Rangely structure.

The paleogeographic location of the Elk Springs field is less favorable than that of Rangely, being located apparently in a deeper water facies of more silty texture. Average porosity is approximately three times as great in Rangely as in Elk Springs and average permeability ten times greater.

Many wells drilled in areas surrounding these fields have encountered saturated Weber, but have been non-productive, in most cases due to lack of adequate porosity and permeability. There has been no penetration of the Weber in the central part of the trough southeast of the Rangely field. Possibilities for adequate sand conditions appear better here than higher on the flanks of the trough on which considerable drilling has been done, probably because of the admixture of Maroon-type sedimentation coming from the bounding highlands.

Other Weber fields will be found. To date, the emphasis has been on structural traps. As information accumulates, to the point that detailed stratigraphic and lithologic trends can be mapped, production can be expected from permeability traps as well as from structural.

CORRELATION OF PRE-MANCOS, POST-WEBER FORMATIONS, NORTHWESTERN COLORADO

By S. L. SHARPS

Shell Oil Company, Salt Lake City, Utah

INTRODUCTION

The purpose of this paper is to correlate the upper Paleozoic and Mesozoic strata of Northwestern Colorado from the Wilson Creek oil field to a measured section on King Mountain, Routt County, Colorado. This information, presented as a stratigraphic cross section, is included in the pocket at the back of this guidebook.

ACKNOWLEDGMENTS

The work was done in partial fulfillment of the requirements for the Master of Science Degree at the University of Colorado, Boulder, Colorado, under the direction of Dr. Warren O. Thompson.

The writer is grateful to the Shell Oil Company for permission to publish this paper.

STRATIGRAPHY

Pennsylvanian

The Schoolhouse tongue of the Weber formation, as originally deposited, probably covered the entire area of Northwestern Colorado, thickening regionally westward. However, because of subsequent erosion during the development of a major unconformity (noted by Murray, 1949, in the McCoy area) at the base of the State Bridge formation, local variations in the regional pattern are evident; that is, the unit thickens eastward from 95 feet in the Wilson Creek No. 20 well to 150 feet in the Iles Dome well, and is not present either at the Beaver Creek well or at King Mountain. Its thickness of 200 feet in the Poose Creek well is considered excessive due to steep dips in the subsurface. The depositional environment was marine littoral.

Permian

The Phosphoria (?) formation is present throughout the area studied. It consists primarily of red to Maroon shale and siltstone interbedded with minor but persistent beds of dolomite and limestone. Some gypsum beds are present. The formation thins eastwardly from 105 feet at the Wilson Creek No. 20 well to 80 feet at the outcrop on King Mountain. The upper dolomite (see enclosure) is believed to be the South Canyon Creek Dolomite of Bass and Northrop (1950). The environment of deposition was at least in part, shallow water marine. In the McCoy area this formation is included in the basal State Bridge formation.

Triassic

The Moenkopi formation thickens irregularly eastward from 480 feet at the Wilson Creek No. 20 well to 680 feet at the outcrop on King Mountain. At Wilson Creek No. 20, it is primarily green shale and siltstone and is characterized by abundant pyrite. The basal part is dolomitic and suggestive of "Dinwoody formation" lithology. Eastwardly the green coloration gives way to red; this is accompanied by a marked decrease in the amount of pyrite. A basal gypsum member is present in all sections except the Wilson Creek No. 20 well. The environment of deposition was possibly near-shore marine with reducing conditions off-shore. In the McCoy area this formation constitutes the major part of the State Bridge formation.

The Shinarump (?) conglomerate unconformably overlies the Moenkopi formation and is a persistent unit that varies in thickness from 38 to 240 feet. Primarily it is a pebble conglomerate, but contains considerable amounts of maroon shale. Unusual lithologies such as shale pebble conglomerates and features such as bleached mud-cracked shale occur in minor amounts but are characteristic of the formation in this area. There is a possibility that rocks called Shinarump (?) in this paper may be correlative with the Mossback sandstone of Utah. The depositional environment was probably fluvial and subaerial—subaqueous mudflat.

The Chinle formation conformably overlies the Shinarump and varies in thickness from 260 feet to 550 feet. The lithology is fairly consistent throughout the area studied and is red siltstone and shale interbedded with minor amounts of red sandstone. Limestone and limestone pebble conglomerate are present and may, in local areas, serve as good marker beds. The depositional environment was probably fluvial and lacustrine.

Jurassic

The Entrada sandstone thins from 360 feet at the Wilson Creek No. 20 well to 22 feet at King Mountain, where variation in thickness from 22 feet to 220 feet was noted. This local variation in thickness is attributed to the effects of the major unconformity at the base of the formation. The maximum thickness at King Mountain was plotted to show that the regional thinning is not uniform. The thick section indicated at the Wilson Creek No. 20 well is believed to contain, in the lower

CORRELATION OF PRE-MANCOS, POST-WEBER FORMATIONS
NORTHWESTERN COLORADO

BY
S. L. SHARPS
JUNE 1965

part, remnants of the Navajo sandstone. The unconformity mentioned above is probably within this interval. The formation is composed of sandstone that is very fine to fine grained, very friable and massively cross-bedded. The depositional environment is eolian.

The Curtis formation conformably overlies the Entrada and thins uniformly eastward from 110 feet at the Wilson Creek No. 20 well to 40 feet at the outcrop on King Mountain. It is predominately glauconitic sandstone and limestone with minor beds of grey-green and green shale. Oolites and shell fragments are characteristic of this formation. The limestone and sandstone beds are intergradational. The marine pelecypod *Camptonectes* (sp.) is common in this formation and was collected at King Mountain.

The basal sandstone of the Morrison formation is probably fluvial and thins from 160 feet in the Wilson Creek No. 20 well to 35 feet at King Mountain. The middle part of the formation is composed principally of varicolored greyish-green, green and red shale. Freshwater limestones are consistently present in this middle portion of the formation. The upper part is principally composed of green siliceous shale, chert beds and siliceous sandstones. The contact with the overlying Dakota is unconformable.

Lower Cretaceous

The Dakota interval in the Wilson Creek No. 20 well consists of three distinct units: (1) a lower marine beach pebble conglomerate (Lakota) 30 feet in thickness, (2) a middle green shale and sandstone unit of the "Morrison type" (Fuson) 42 feet in thickness and (3) an upper marine beach sandstone unit, 70 feet in thickness, that probably includes both the Dakota and Muddy sandstones. In the area of outcrop, on King Mountain, only the Lakota conglomerate and the Dakota-Muddy sandstones are present.

Lower Cretaceous (?)

The Mowry formation is present throughout the area. The formation is composed of hard, black, siliceous shale interbedded with black, brittle, non-calcareous shale, sandstone and bentonite. Fish remains are characteristic of the formation. The formation is assigned a lower Cretaceous age by Cobban and Reeside (1951) and an upper Cretaceous age by T. C. Yen (1954).

REFERENCES

Baker, A. A., Dane, C. H., and Reeside, J. B., Jr. (1936), "Correlation of the Jurassic Formations of Portions of Utah, Arizona, New Mexico and Colorado," U. S. Geol. Survey Prof. Paper 183.

Bass, N. W., and Northrop, S. A. (1950), "South Canyon Dolomite Member. A Unit of Phosphoria Age in the Maroon Formation Near Glenwood Springs, Colorado," Bull. Amer. Assoc. Petrol. Geol., vol. 34, no. 7, pp. 1540-51.

Brill, K. G., Jr. (1942), "Late Paleozoic Stratigraphy of the Gore Area, Colorado," Bull. Amer. Assoc. Petrol. Geol., vol. 26, no. 8, pp. 1375-97.

---------- (1944), "Late Paleozoic Stratigraphy, Westcentral and Northwestern Colorado," Bull. Geol. Soc. America, vol. 55, pp. 621-56.

---------- (1952), "Stratigraphy in the Permo-Pennsylvanian Zeugogeosyncline of Colorado and Northern New Mexico," Bull. Geol. Soc. America, vol. 63, no. 8, pp. 809-80.

Cobban, W. A., and Reeside, J. B., Jr. (1951), "Lower Cretaceous Ammonites in Colorado, Wyoming and Montana," Bull. Amer. Assoc. Petrol. Geol., vol. 35, no. 8, pp. 1892-93.

Donner, H. F. (1949), "Geology of the McCoy Area, Eagle and Routt Counties, Colorado," Bull. Geol. Soc. America, vol. 60, no. 8, pp. 1215-48.

Kummel, B. (1950), "Triassic Stratigraphy of the Area Around the Green River Basin, Wyoming," Wyoming Geol. Assoc., Southwest Wyoming Guidebook, August, 1950, pp. 28-36.

Langenheim, R. L., Jr. (1954), "Maroon Formation, Colorado," Bull. Amer. Assoc. Petrol. Geol., vol. 38, no. 8, pp. 1748-79.

Murray, H. F. (1950), "Stratigraphic Study of Pennsylvanian of McCoy Area, Colo.," Masters Thesis, Dept. of Geology, Univ. of Colorado.

Thomas, C. R., McCann, F. T., and Raman, N. D. (1945), "Mesozoic and Paleozoic Stratigraphy in Northwestern Colorado and Northeastern Utah," U. S. Geol. Survey Prelim. Chart 16 (in 2 sheets).

Yen, T. C. (1954), "Age of Bear River Formation," Bull. Amer. Assoc. Petrol. Geol., vol. 38, no. 11, pp. 2412.

GEOLOGY OF THE EASTERN END OF THE UINTA MOUNTAINS, UTAH-COLORADO

By G. E. and B. R. UNTERMANN
Utah Field House of Natural History, Vernal, Utah

GENERAL STRUCTURE

Folds

The principal structural features of the Uinta Mountain and Basin region consist of the main Uinta arch together with its accompanying synclinal and anticlinal flank folds, producing a broad anticlinorium with steeply dipping beds, sharp monoclinal flexures and faults of considerable displacement. These folds represent part of an original geosynclinal trough into which were deposited sediments from pre-Cambrian to late Cretaceous time. Folding and elevation was initiated near the close of the Cretaceous period, resulting in the present Uinta Range which represents the principal east-west trending system in the western hemisphere. Maximum elevation in the western part of the range is 13,498 feet (Kings Peak), while in the eastern part the maximum is 9,006 feet (Zenobia Peak). The range pitches downward at both ends and is somewhat convex northward and asymmetrical, dips being steeper on the north side. The greatest convexity occurs approximately where the axis crosses the Utah-Colorado line. This rather sharp southeastward bending, with a concentration of minor folds along the south slopes, suggests that the compressive stresses, which produced a more intense deformation in this region, probably came from the positive land areas to the southwest which had been supplying much sediment to the geosyncline.

A broadening of the range occurs near the east end where there are a larger number of flank folds. This structural broadening, together with thinning of many of the formations to the east and a generally lower elevation of the range in the eastern section, is indicative of a shallowing of the Uinta trough in that direction.

Some of the smaller structures such as the Split Mountain and Section Ridge anticlines, the latter with its steep south scarp facing U. S. Highway No. 40, pitch at moderate angles in a westerly direction. They duplicate the typical Uinta Mountain structure exhibiting relatively flat-topped anticlines having sharply-flexed flanks that are often steepened by drag-folding along shear zones of accompanying faults.

Faults

Faults are numerous throughout the region and are chiefly of the normal type. They commonly occur as strike-faults, the major systems having a general east-west trend roughly paralleling the flanks of the range. Some of the faults are rotational and may be diagonal in direction. Those along part of the Yampa fault zone are distributive in character, producing step-faulting along the south side of the Yampa Canyon. Maximum displacement of these high-angle faults varies from 4,000 feet along the Yampa Fault in the eastern end of the range (Uinta Mountain graben) in Dinosaur National Monument, to approximately 25,000 along the Uinta or Crest Fault north slope of the range, which diminishes eastward.

The South Flank Fault is chiefly confined to the western half of the range, but its eastern counterpart may be represented by the Yampa and Miners Draw-Wolf Creek faults that bound Blue Mountain on the north and south, respectively. Faulting in Dinosaur National Monument and adjacent areas in the eastern part of the Uinta Range has greatly ruptured the anticlinorium, producing the Uinta Mountain graben of the Browns Park and Yampa Canyon areas, bounded by the Uinta Fault on the north and by the Yampa Fault on the south, and by the Island Park Fault on the west.

Joints

Erosion of some of the formations has been greatly accelerated by the presence of joints, particularly in the softer, more poorly cemented sandstones such as the Weber. Tensional stress is frequently indicated by small offsets along some of these joints. Parallel vertical-walled gorges have been incised along the slopes of Split and Blue mountains and are particularly well developed on the slope east of Split Mountain gorge. This is the same type of weathering which occurs in much of the Pats Hole-Yampa Canyon, Jones Hole and Brush Creek areas. These spectacular scenic features have been carved chiefly in Weber sandstone.

GENERAL STRATIGRAPHY

Formations in the Uinta Mountain and Basin area consist principally of marine and continental (lake, floodplain and eolian) origin. A few minor exposures of igneous rocks occur, which include lava flows on the extreme western end of the range and a small number of pegmatitic and granitic dikes outcropping in the northeastern portion or at one or two localities near the center of the range. The stratigraphic history of the region is generally one of great regularity, with deposi-

tion often being continuous and transitional. Noticeable unconformities are rare, indicating the predominance of vertical crustal movements with little or no orogenic deformation throughout the Paleozoic and most of the Mesozoic, until near the close of the latter. Breaks, for the most part, therefore, are faunal and lithologic rather than stratigraphic. Formations range in age from pre-Cambrian to Cretaceous (Mesaverde), representing more than 30,000 feet of sediments, all of which were involved in the depositional and diastrophic history of the Uinta Mountains. Pre-Cambrian deposits were probably chiefly of stream origin consisting of conglomerate, coarse sandstone and minor amounts of shale, now exhibiting varying degrees of metamorphism and containing infrequent granitic dikes. During Paleozoic time marine conditions predominated with rapid lateral change of facies from west to east, and a general shallowing of the seas in an eastward direction. Absence of Ordovician, Silurian and Devonian strata does not reveal a pronounced erosional unconformity. Most conspicuous erosion surfaces formed prior to Laramide folding are between pre-Cambrian and Cambrian (also slightly angular), between Mississippian and Pennsylvanian and between Moenkopi and Shinarump (Triassic).

Rise of the Uncompahgre Range of the "Ancestral Rockies" to the southeast during Carboniferous time no doubt acted as a source of the later sediments, somewhat affecting lateral facies change. Thicker marine deposition continued to the west, however, during the Carboniferous with intertonguing of shallow shelf deposits to the east.

Conditions of deposition similar to those during the Paleozoic prevailed during Triassic time with the sediments derived from highlands to the east (Ancestral Rockies) grading, in the Uinta region, into marine facies which thicken to the west. These early Mesozoic sediments in the eastern Uinta Mountains indicate near-shore shallow water beds or on-shore subaerial deposits characteristic of deltas, flood plains, evaporating basins and lagoons. The "Nevada Mountains" on the west and southwest may have supplied much of the sediments during the Jurassic, bringing about a change in the source of material from east to west. These orogenic movements were possibly also reflected in the widespread accumulation of coarser clastics of the Upper Jurassic and Lower Cretaceous series.

The Uinta geosyncline, an east-west arm of the Wasatch trough, remained fairly stable until late Cretaceous time, receiving sediments from surrounding highland areas. Until the late Cretaceous the deepest parts of the marine basin were located to the west, with gradual shallowing taking place in an eastward direction, producing a broad foreland shelf area on which sediments of increasing shallow marine character graded into and interfingered with subaerial continental deposits derived from the east or southeast, thus accounting for a persistent change of facies from west to east. However, during Upper Cretaceous time the thickest marine deposition occurred to the southeast and spread westward grading laterally into coarser sediments derived from the west, which are thought to signify the initial impulses of the "Early Laramide" deformation in the Uinta Range.

Generalized columnar section of the rock formations exposed in the vicinity of Dinosaur National Monument

After Untermann, G. E., and Untermann, B. R., Geology of Green and Yampa River Canyons and vicinity, Dinosaur National Monument, Utah and Colorado: Amer. Assoc. Petroleum Geologists Bull., vol. 33, no. 5, table 1, pp. 686-689.

System	Series	Group or formation	Thickness in feet	Character of sediments
Cenozoic	Tertiary Pliocene	Browns Park	1,200	Chalk-white to grayish crossbedded sandstone with some volcanic tuff and with a basal bed of conglomerate. Deposited in lakes and stream channels and by wind.
		Erosion interval		
	Miocene	Bishop (?)	0 to 100	Conglomerate with boulders up to several feet in diameter. Deposited by streams.
		Erosion interval		
Mesozoic	Cretaceous Upper Cretaceous	Mancos	5,000	Mainly dark-gray shale which weathers light gray to buff; contains thin sandstone lenses. Marine deposit.
		Frontier	200	Marine and continental limy sandstone with dark shale and sub-bituminous coal beds.
		Mowry	125	Dark-gray marine shale weathering silver gray and containing abundant fish scales.

Generalized columnar section (continued)

System	Series	Group or formation	Thickness in feet	Character of sediments
Mesozoic	Cretaceous — Lower Cretaceous	Dakota	100	Yellowish crossbedded sandstone and conglomerate with chert and limestone pebbles less than 1 inch in diameter underlain by dark-gray to reddish shale and yellowish quartzitic sandstone. Contains fossil wood.
		Erosion interval		
Mesozoic	Jurassic — Upper Jurassic	Morrison	756+	Varicolored marl and siltstone and brownish crossbedded conglomeratic sandstone with abundant dinosaur remains, some fossil wood, and fresh-water shells. Deposited in lakes and stream channels and on flood plains.
		Curtis	260	Greenish gray, sandy at base and with shale and oolitic limestone in upper part. Marine deposit.
		Entrada	165	Light-gray to buff and pink medium-grained sandstone weathering into domes. Deposited by wind.
		Carmel	125	Red siltstone and shale and some gypsum. At least in part a marine deposit. Occasional tracks of three-toed dinosaur.
		Erosion interval		
Mesozoic	Jurassic — Middle (?) Jurassic	Navajo	700	Buff to red crossbedded sandstone, mainly wind-deposited. Water-laid basal layers contain occasional tracks of four-toed dinosaur.
		Erosion interval		
Mesozoic	Triassic — Upper Triassic	Chinle	230	Red shale, marl, and siltstone with some sandstone and conglomerate. Contains teeth and bones of reptiles.
		Shinarump	50	Light buff to yellow. Stream-carved coarse sandstone and conglomerate.
		Erosion interval		
Mesozoic	Triassic — Lower Triassic	Moenkopi	800	Red, limy, and gypsiferous shale and sandstone with abundant ripple marks. May be in part marine.
		Erosion interval		
Paleozoic	Permian	Park City	50	Light-gray, thin-bedded, fossiliferous limestone, limy sandstone, and gray, red, and yellowish shale with some phosphate nodules. Marine deposit.
Paleozoic	Carboniferous — Pennsylvanian	Weber	1,000	White to light-gray or buff crossbedded sandstone; oil-reservoir rock in Rangely and other fields in the region.
		Morgan	1,280	Interbedded gray limestone and white to red sandstone in upper part underlain by limy shale and limestone. Marine deposit.
		?	185	Black carbonaceous shale with light-colored plant-bearing sandstone underlain by white to yellow sandstone with carbonaceous streaks.
		Erosion interval		
Paleozoic	Carboniferous — Mississippian	Madison	600	Buff to dark-gray limestone and dolomite with occasional layers containing fossils. Marine deposit.
		Erosion interval		
Paleozoic	Cambrian — Upper Cambrian	Lodore	400	White to red, coarse-grained, quartzitic, and in part arkosic sandstone with interbedded silty shale. Marine deposit.
		Erosion interval		
Pre-Cambrian		Uinta Mountain group	12,000	Red to white, coarse-grained, quartzitic, and in part arkosic sandstone and conglomerate with some thin beds of shale.

JURASSIC AND PRE-MANCOS CRETACEOUS STRATIGRAPHY OF THE EASTERN UINTA MOUNTAINS, UTAH-COLORADO

By WHITNEY A. BRADLEY
The Ohio Oil Company, Casper, Wyoming

INTRODUCTION

Location

Sediments included in this study lie stratigraphically between the Triassic Chinle formation and the Cretaceous Mancos formation. These beds outcrop almost continuously from Dinosaur National Monument headquarters, Utah, to Lily Park, Colorado, (Figure 1).

Purpose and Method of Study

This paper is a revised condensation of a thesis written in 1952 in partial fulfillment of requirements for the degree, Master of Science, from the University of Colorado.

Although many geologists have studied the stratigraphy of the Uinta Mountains, little detailed information on the Jurassic and Pre-Mancos Cretaceous sediments has been published. This detailed study of the lateral variation in thickness and lithology of the sediments was conducted to aid interpretation of the geologic history of the Uinta Mountains.

Eight stratigraphic sections were measured in detail and samples of each lithologic unit were collected for examination with a binocular microscope. Heavy mineral separations of the Navajo, Entrada, and Curtis formations isolated minerals which were examined under a petrographic microscope. Formational age determinations are based on fossil identification where possible and on regional correlations.

DISCUSSION OF FORMATIONS

JURASSIC

Navajo Formation

The Navajo formation is pale orange to yellowish-gray, clean, fine- to medium-grained, quartz sandstone. The lower half of the Navajo at Wolf Creek and a band 50 feet below the top at West Draw are reddish brown. A band of limonite and hematite concretions occurs 150 feet above the base at Red Wash, Cocklebur Creek, and Pipe Spring Draw.

More than 95 per cent of the grains are quartz, accessory minerals being garnet, zircon, and magnetite. Grains average between 0.05 mm. and 0.2 mm. in diameter, but grains as large as 0.5 mm. are common. The large scale crossbedded sandstones have thick beds of uniform size grains; however, many of the intricately cross-laminated sands exhibit well sorted laminae one grain thick. The characteristics of the grains and the nature of the crossbedding indicate that the Navajo formation is an eolian deposit.

The Navajo of the eastern Uinta Mountains has been correlated with the Navajo formation in Arizona (Baker, Dane and Reeside, 1936, p. 44). According to Heaton (1950, p. 1683), it is of Lower Jurassic age.

Carmel Formation

Reddish-brown siltstones, sandstones, and shales comprise most of the Carmel formation. A greenish-gray shale, interbedded with thin limestones, occurs near the middle of the formation, but it thins eastward and pinches out before reaching the Colorado-Utah border. As the Carmel formation thins eastward it is more sandy, becoming entirely siltstone and sandstone before it pinches out near Skull Creek, a short distance east of Pipe Spring Draw. In most places the top is marked by interbedded gray, purplish-gray, and red shales. The base is sharply defined, except at Willow Creek where some Navajo-like sandstone occurs interbedded with the lower red shales.

The siltstones, all poorly stratified, contain many grains as large as 0.5 mm. in diameter and are interbedded with dirty, poorly sorted, non-resistant sandstones. The grains, many of which are frosted, range in diameter from 0.1 mm. to 0.6 mm., averaging 0.2 mm. Shaly siltstones near the top contain 10 per cent biotite.

Heaton (1939, p. 1172) traced the Twin Creek formation of north-central Utah into the Carmel formation of the Uinta Mountains, where it occupies the same stratigraphic position as does the Carmel formation in the San Rafael Swell. In southern Utah the Carmel formation changes from thick, fossiliferous limestones in the west to siltstones, shales, and thin limestones in the east. This facies change is similar to the eastward transition from the thick Twin Creek limestone in north central Utah to the red siltstones and shales of the Carmel formation in the eastern Uinta Mountains (Figure 2).

Entrada Formation

Binocular and petrographic examination of the eolian Entrada and Navajo formations revealed no differences

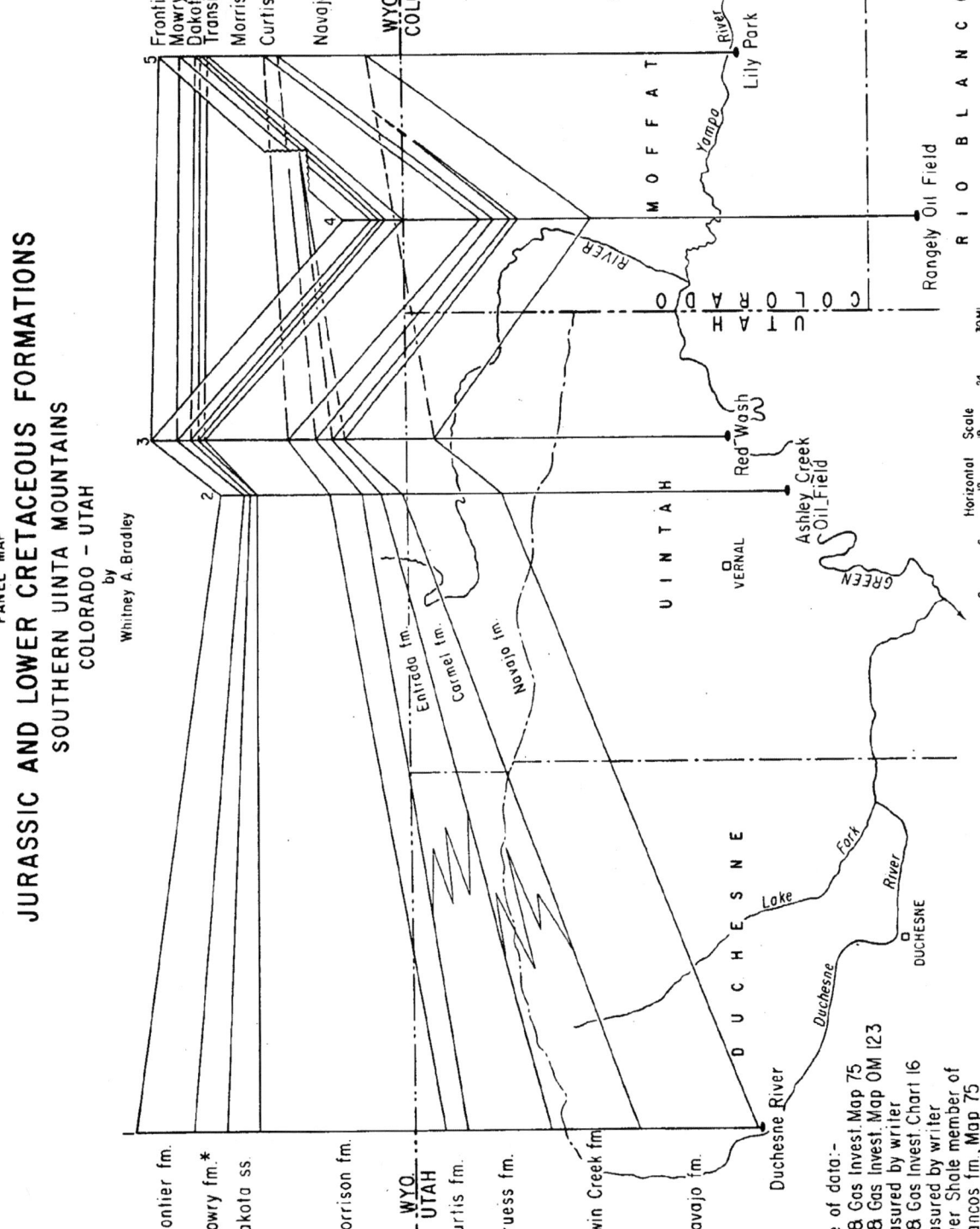

Figure 1.

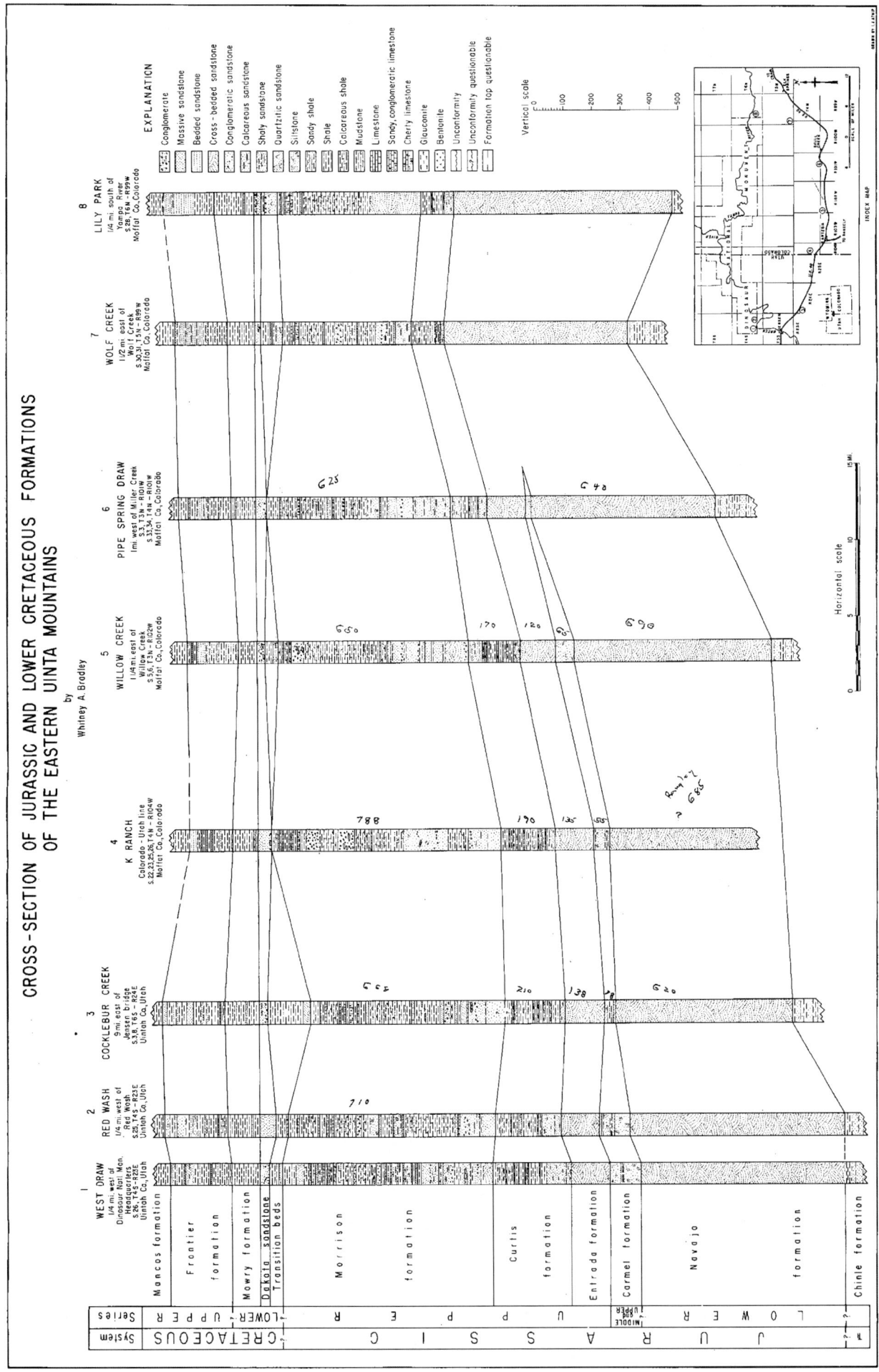

BRADLEY: FIGURE 2. CROSS SECTION, JURASSIC AND CRETACEOUS FORMATIONS OF EASTERN UINTA MOUNTAINS.

in the mineral suites and no difference in the character of the grains or the type of crossbedding. These formations can be differentiated only where they are separated by the Carmel formation. The transitional Navajo-Carmel contact and the conformable Carmel-Entrada contact indicate that the Carmel pinchout is depositional. The undivided eolian sandstones is called Entrada-Navajo in this paper in view of the probability that eolian deposition was continuous east of the pinchout.

Intertonguing of the Entrada with the overlying Curtis formaiton is well exposed at the K Ranch. The Curtis is a resistant, nearly parallel-bedded, glauconitic sandstone; whereas the Entrada tongues are soft, lighter-colored, eolian, crossbedded sandstone.

Thomas and Krueger (1946, p. 1277) report that at Lake Fork, Utah, 65 miles west of Red Wash, the eolian Entrada sandstone interfingers with and is replaced by silty redbeds and sandstones of the Pruess formation (Figure 2). Baker, Dane, and Reeside (1936, p. 7) noted a similar facies change across southern Utah, where the intricately crossbedded, clean sandstone of the east is replaced by red siltstone in the west.

Curtis Formation

The Curtis formation is a series of marine sandstones, shales, and limestones. Thomas, McCann, and Raman (1944) include the basal, parallel-bedded, glauconitic sandstone in the upper Entrada formation; however, this sandstone is here included with the marine Curtis formation because of the occurrence of *Pachyteuthis densus* (Meek) and fragmentary brachiopods near the base at K Ranch and Pipe Spring Draw.

Groundwater circulation in the basal, resistant sandstone has deposited azurite and malachite along bedding planes and has developed perfect secondary quartz crystals. A series of calcareous shales interbedded with thin, gray, ripple-marked sandstones and limestones overlies the basal sandstone. The sandstones, some of which contain 50 per cent glauconite, are fine- to medium-grained. The limestones are dense, sandy, glauconitic, and somewhat oolitic. Most of the shales are greenish-gray, calcareous and silty, and contain thin selenite beds. The eastward thinning of the formation is chiefly due to the thinning of the middle shales, although there is some thinning of the limestones and basal sandstone. Fossils collected from the limestones and upper sandstones are *Rhynchonella myrina* (Meek and Hayden), *Pseudomonotis curta* (Hall) (?), and unidentifiable pelecypods.

The Curtis formation in the eastern Uinta Mountains is correlated with the type Curtis in the San Rafael Swell because of the similarity in lithology, stratigraphic position, and fauna (Baker, Dane, and Reeside, 1936, p. 8). The Curtis formation is also correlated with the Stump formation of North-Central Utah by Heaton (1939, p. 1177) and with the upper Sundance of Wyoming by Thomas and Krueger (1946, p. 1279).

Morrison Formation

A clean, medium-grained, crossbedded, gray sandstone found at the base of the Morrison formation in some localities varies in thickness from 145 feet at Lily Park to 15 feet at Wolf Creek and 60 feet at Willow Creek. The grains average 0.2 mm. in diameter, and approximately 30 per cent of them are frosted.

The sandstones interspersed with the shales in the remainder of the lower third of the formation are poorly sorted, calcareous, medium-grained, and locally crossbedded. Lenticular, coarse-grained, conglomeratic, torrentially cross-bedded chert sandstones are common. The middle third of the formation is a series of mottled and variegated, gray, greenish-gray, and grayish-purple siltstones and bentonitic shales and mudstones.

Thin, gray, dense limestones and lenticular chert conglomerates interspersed in varicolored shales typify the upper third of the Morrison. A ridge-forming, light gray limestone near the top of the formation is characterized by chert and calcite veins, as well as pebbles of limestone and black and gray chert up to 15.0 mm. in diameter. A cliff-forming conglomerate 70 feet thick occurs 120 feet below the top of the Morrison at K Ranch. This conglomerate is composed of dark gray to black, rounded chert pebbles, ranging in size up to four inches in diameter in a coarse-grained sandstone matrix. Another chert conglomerate sixty feet lower contains silicified logs up to 18 inches in diameter. The silicified dinosaur bones at Dinosaur National Monument are quarried from a lenticular conglomerate 150 feet below the ridge-forming limestone.

The top of the Morrison is here designated as the highest greenish-gray or grayish-red shale below the lowest quartzitic, Dakota-like sandstone. Although the age of the Morrison formation may vary throughout the Rocky Mountain Region, vertebrate fossils from the quarry at Dinosaur National Monument are believed to be Upper Jurassic in age (Kay, personal communication).

Stokes (1944, p. 965) believes the upper variegated shales containing the bone-bearing conglomerates should be called Brushy Basin shales, as in the Colorado Plateau. He further states that, inasmuch as Upper Jurassic and Lower Cretaceous faunas have not been clearly differentiated, it is possible that part of the Morrison beds in Utah are of Lower Cretaceous age.

CRETACEOUS (?)

Transition Beds

The beds which overlie the Morrison in most of the area are in part similar to the Morrison formation and in part similar to the Dakota sandstone. These beds are called Transition beds in this paper. A quartzitic, medium-grained, crossbedded sandstone of varying thickness, strongly resembling the Dakota sandstone, marks the base of the Transition beds. The crossbedding, laminae, sorting, and partial frosting are typical of beach deposits; whereas some of the coarse-grained sandstones and conglomerates above the basal sandstone are torrentially crossbedded.

The Morrison-like shales of the Transition beds are grayish-red, greenish-gray, and gray and vary greatly in thickness. The top of the Transition beds is here considered to be at the top of those varicolored shales which are overlain by the persistent, massive, blocky, quartzitic Dakota sandstone.

Stokes (1944, p. 989) applies the name Cedar Mountain Group to all beds between the Morrison and Dakota formations in the Colorado Plateau. He divides the Group into the Buckhorn conglomerate and the overlying Cedar Mountain shales. These names have been applied by some geologists to stratigraphically similar beds in the eastern Uinta Mountains. Thomas, McCann, and Raman (1945) have called this sequence in the Uinta Mountains "Beds of undetermined age." The coarse chert conglomerates of the sequence have been included in the Morrison in this paper, leaving only the interbedded Morrison-like and Dakota-like beds in the Transition beds. The Transition beds, together with the overlying Dakota, can possibly be correlated with the Dakota of western Colorado.

CRETACEOUS

Dakota Sandstone

The Dakota sandstone is a massive, ridge-forming, somewhat conglomeratic sandstone with much red and yellow iron staining. The sandstones are quartzitic, medium-grained, fairly well sorted and exhibit littoral type crossbedding. Random lenses of conglomeratic sandstones and chert conglomerates are similar to Morrison conglomerates, but only a few of the pebbles exceed 2 cm. in diameter.

The identification by Reeside and Cobban (1951, p. 1892) of Lower Cretaceous ammonites in the Mowry formation indicates that the Dakota sandstone of the Uinta Mountains is Lower Cretaceous in age.

Mowry Formation

The lower ten to forty feet of the Mowry are soft, black, paper-thin shale containing fish scales, thin quartzitic sandstones and selenite beds. The interbedding of quartzitic Dakota sandstone with Mowry shales attests to the gradational nature of the contact. The Mowry formation above the soft shale is characterized by siliceous shale, which weathers silvery gray. Unweathered Mowry is olive gray, thin-bedded and slate-like, with abundant fish bones and scales. A thin quartzitic sandstone occurs near the middle of the Mowry in some places.

Walton (1944, p. 101) applies the name Mancos formation to all the beds between the Dakota and the Mesaverde formations at the Colorado-Utah border, but this interval is now commonly divided into the Mowry, Frontier, and Mancos formations. The soft shale in the base of the Mowry occupies a stratigraphic position similar to that of the Thermopolis shale of Wyoming, though it is not definitely a correlative.

The Mowry formation in the eastern Uinta Mountains is the lithologic and stratigraphic equivalent of the Aspen formation, as well as of the type section of the Mowry in southwestern Wyoming. The age has generally been accepted as Lower Cretaceous since Cobban and Reeside (1951, p. 1892) identified ammonites of the genus *Gastroplites* in the Mowry shales.

Frontier Formation

A series of gray shales, sandy shales, and sandstones above the Mowry shale comprise the Frontier formation. The shales become more sandy upward until sandstones and shales are interbedded in equal amounts approximately 100 feet above the Mowry. The upper part of the formation is composed chiefly of ripple-marked, yellowish-gray, fine and medium-grained, micaceous, slightly calcareous, thin-bedded sandstones. The relative amount of sandstone and the average grain size decrease eastward.

A massive, medium- to coarse-grained sandstone forty feet from the top of the formation forms a prominent ridge and dip slope at West Draw. It thins to eight feet at Cocklebur Creek and pinches out a short distance to the east. Round, calcite "Cannon ball" concretions up to three feet in diameter are common in this bed. The carbonaceous shale above this sandstone has supplied some lignite from a mine approximately four miles west of West Draw. An aragonite bed with "cone in cone" structure lies 84 feet above the base of the formation at the K Ranch. The top of the Frontier formation intertongues with the overlying Mancos shales; therefore, the top is not at the same horizon in every section. Fossils from the top sandstone of the

Frontier formation are *Scaphites* sp. and *Inoceramus* sp. The lower sandstones bear *Paranomia* sp. and fragments of *Inoceramus* sp., *Inoceramus liabiatus* (Scholtheim), and unidentifiable teeth. The age of the Frontier formation is questionable. The Frontier of Wyoming, the Ferron sandstone of central Utah, and the Carlile formation of southeastern Colorado may be equivalent to the Frontier sandstones of the eastern Uinta Mountains.

INTERPRETATION OF SEDIMENTS

JURASSIC

Triassic redbed deposition in the eastern Uinta Mountains was followed by deposition of the eolian Navajo formation. In early Upper Jurassic time the Twin Creek sea invaded the Cordilleran geosyncline, for a time inundating the shelf area of central Montana, Wyoming, eastern Utah, northwestern Colorado, and eastern Idaho. The eastern Uinta Mountain area was part of the shelf zone, which was periodically submergent and emergent. Deposition of the Twin Creek limestone in the trough to the west was contemporaneous with deposition of the redbeds and gray marine shale and limestone of the Carmel in the eastern Uinta Mountains. Fluctuations of the sea resulted in deposition of eolian sandstone tongues in the base of the Carmel, while east of the shelf, deposition of the Entrada-Navajo formation continued uninterrupted.

After a brief interval the Twin Creek sea withdrew to the west, and while the red siltstones and shales of the Preuss formation were being deposited in the Cordilleran trough, the eolian sands of the Entrada formation were deposited in the Uinta Mountain area. The abrupt lateral change from eolian Entrada sandstone to Preuss redbeds takes place in approximately the same locality as the underlying change from the Carmel formation to Twin Creek limestone (Figure 2).

The Logan sea, which transgressed from the north over the Entrada desert in Upper Jurassic time, was part of the last major invasion by the Sundance sea. The sea transgressed slowly, reworking some Entrada sands to form the basal Curtis sandstone. The Logan sea advanced farther into Colorado and eastern Utah than did the Twin Creek sea, as evidenced by deposition of the Curtis formation in Eagle County, Colorado (Sheridan, 1950, p. 32).

Retreat of the Logan sea left eastern Utah and western Colorado a broad, flat flood plain, with many lakes and sluggish rivers in which the Morrison beds were deposited. Mountains to the west supplied chert for the conglomerates.

CRETACEOUS

The Transition beds above the Morrison are a series of interbedded terrestrial and marine sediments resulting from the oscillations of the transgressing Cretaceous sea. The Cretaceous sea, advancing westward, deposited the Dakota formation. The littoral deposition was interrupted locally by torrentially crossbedded sandstones and conglomerates deposited by eastward-flowing streams. As the Cretaceous sea advanced westward the Mowry formation and shales of the Frontier were deposited. During Carlile time, sand from the west ultimately forced the shoreline eastward, and part of the Uinta Mountains area was swamp for a period during Frontier deposition. The sea finally advanced westward over the marine and paludal Frontier formation and continued to deepen and spread, depositing the Mancos shale.

ACKNOWLEDGMENTS

Permission and assistance of The Ohio Oil Company to prepare and publish this paper are gratefully acknowledged. Appreciation is expressed to Mr. H. F. Murray for his supervision, to Dr. John Chronic for paleontological guidance, and to the University of Colorado geology staff. The field work was made possible through the cooperation of the National Park Service and Mr. Jess Lombard, Superintendent of Dinosaur National Monument.

REFERENCES

Abrassart, C. P. (1951), "Stratigraphy and Sedimentation of the Juniper Mountain Area, Colorado," M. S. Thesis, Univ. of Colo.

Baker, A. A., Dane, C. H., Reeside, J. B., Jr. (1936), "Correlations of the Jurassic Formations of Parts of Utah, Arizona, New Mexico and Colorado," U. S. Geol. Survey Prof. Paper 183.

――――― (1947), "Revised Correlation of Jurassic Formations of Parts of Utah, Arizona, New Mexico and Colorado," Amer. Assoc. Petrol. Geol. Bull., vol. 31, pp. 1664-1677.

Bartram, J. W. (1937), "Upper Cretaceous of the Rocky Mountain Area," Amer. Assoc. Petrol. Geol. Bull., vol. 21, pp. 899-913.

Cobban, W. A., and Reeside, J. B., Jr. (1951), "Lower Cretaceous Ammonites in Colorado, Wyoming, and Montana," Amer. Assoc. Petrol. Geol. Bull., vol. 35, pp. 1892-1893.

Eardley, A. J. (1944), "Geology of the North-Central Wasatch Mountains, Utah," Geol. Soc. Amer. Bull., vol. 55, pp. 819-894.

Goddard, E. N., Trask, P. D., DeFord, R. K., Rove, O. N., Singewald, J. T., Jr., and Overbeck, R. M. (1948), "Rock Color Chart," National Research Council, Washington, D. C.

Heaton, R. L. (1933), "Ancestral Rockies and Stratigraphy of the Rocky Mountain Region," Amer. Assoc. Petrol. Geol. Bull., vol. 17, pp. 109-168.

――――― (1939), "Contribution to Jurassic Stratigraphy of the Rocky Mountains," Amer. Assoc. Petrol. Geol. Bull., vol. 23, pp. 1153-1177.

――――― (1950), "Late Paleozoic and Mesozoic History of Colorado and Adjacent Areas," Amer. Assoc. Petrol. Geol. Bull., vol. 34, pp. 1659-1698.

Huddle, J. W., and McCann, F. T. (1947), "Pre-Tertiary Geology of the Duchesne River Area, Duchesne and Wasatch Counties, Utah," U. S. Geol. Survey, Oil and Gas Investigations, Preliminary Map 75.

Imlay, R. W. (1947), "Occurrence of Middle Jurassic Rocks in Western Interior of the United States," Amer. Assoc. Petrol. Geol. Bull., vol. 29, pp. 1019-1027.

――――― (1948), "Characteristic Marine Jurassic Fossils From the Western Interior of the United States," U. S. Geol. Survey Prof. Paper 214-B, pp. 13-23.

Kay, LeRoy, Carnegie Institute, Pittsburg, Penn., personal communication.

Kiersch, G. A. (1950), "Small Scale Structures and Other Features of Navajo Sandstone, Northern Part of San Rafael Swell, Utah," Amer. Assoc. Petrol. Geol. Bull., vol. 34, pp. 923-942.

Kinney, D. M. (1951), "Geology of the Uinta River and Brush Creek-Diamond Mountain Area, Duchesne and Uintah Counties, Utah," U. S. Geol. Survey, Oil and Gas Investigations, Preliminary Map OM 123.

Krumbein, W. C., and Pettijohn, F. J. (1938), Manual of Sedimentary Petrography, Appleton-Century-Crofts, Inc., New York, N. Y., pp. 80, 412-457.

Masters, J. A. (1951), "Frontier Formation," M. S. Thesis, Univ. of Colorado.

Sears, J. D. (1925), "Geology and Oil and Gas Prospects of Part of Moffat County," U. S. Geol. Survey Bull., 751.

Sheridan, D. S. (1950), "Permian (?), Triassic and Jurassic Stratigraphy of the McCoy Area, West Central Colorado," M. S. Thesis, Univ. of Colorado.

Stokes, W. L. (1944), "Morrison and Related Deposits in and Adjacent to the Colorado Plateau," Geol. Soc. Amer. Bull., 55, pp. 951-992.

Thomas, C. R., McCann, F. T., and Raman, N. D. (1945), "Mesozoic and Paleozoic Stratigraphy in Northwestern Colorado and Northeastern Utah," U. S. Geol. Survey, Oil and Gas Investigations, Preliminary Chart 16.

Thomas, H. D., and Krueger, M. L. (1946), "Late Paleozoic and Early Mesozoic Stratigraphy of the Uinta Mountains, Utah," Amer. Assoc. Petrol. Geol. Bull., vol. 30, pp. 1255-1293.

Untermann, G. E., and Untermann, B. R. (1954), "Geology of Dinosaur National Monument and Vicinity, Utah-Colorado," Utah Geol. Min. Survey Bull. 42.

Walton, P. T. (1944), "Geology of the Cretaceous of the Uinta Basin, Utah," Geol. Soc. Amer. Bull., vol. 55, pp. 91-130.

U. S. Department of Interior, Grazing Service

Paleozoic - pre-Cambrian unconformity, Cold Springs Mountain, southwest portion, T. 11 N., R. 101 W., Moffat County, Colorado.

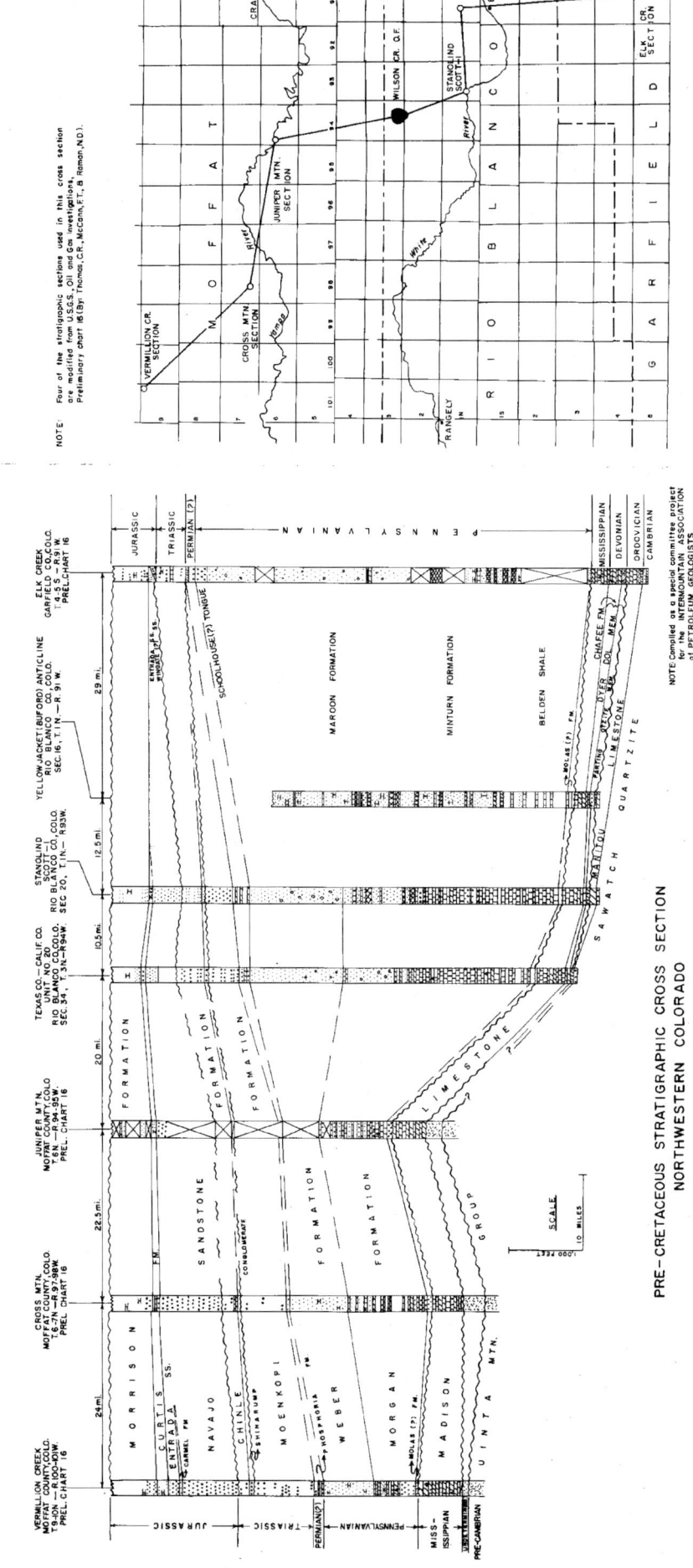

MICROFOSSILS OF THE CURTIS FORMATION, EASTERN UINTA MOUNTAINS, UTAH-COLORADO

By DON L. EICHER
University of Colorado, Boulder, Colorado

INTRODUCTION

The Curtis formation was deposited by the latest and most widespread Jurassic sea to invade the western interior of the United States. It is correlated with the Stump formation of eastern Idaho and western Wyoming and with the Swift formation of Montana, central and eastern Wyoming and western South Dakota. On the basis of ammonites, Imlay (1948) has demonstrated that the Curtis and its correlatives are equivalent in time with the Oxfordian stage of the Upper Jurassic of Europe.

Glauconitic, light-colored, calcareous sandstone and greenish-gray, fissile shale typify the Curtis formation in the eastern Uinta Mountain area. Associated with these rocks are lenticular, clastic limestones, commonly formed of oolites or shell fragments, which are always in some degree glauconitic. In this area the Curtis is overlain gradationally by the Morrison formation and underlain by the Entrada sandstone. Isopachs of the Curtis formtaion are shown on the index map of the cross-section.

Only within recent years have Oxfordian microfossils from the western interior received detailed study. Swain and Peterson (1951, 1952) and Peterson (1954) have studied the occurrence and distribution of the Swift ostracods in eastern Wyoming and adjacent areas, and Loeblich and Tappan (1950) have studied the Foraminifera from the Swift (Redwater shale) in the Black Hills area of South Dakota. These studies have provided an interesting basis for comparison of the Curtis microfaunas of the eastern Uinta area. For this report microfossil samples were collected at regular stratigraphic intervals from seven localities shown on the index map.

FORAMINIFERA

Twenty-one species of Foraminifera have been recognized from the Curtis formation. They occur in the shales and many of the fine-grained sandstones throughout the vertical extent of the rocks. Ten representative species are illustrated in plate I. In general the specimens are well-preserved, but seldom are they very abundant. Some species are represented by very few individuals. Because of their relative rarity, Foraminifera are of no apparent value in delineating faunizones within the Curtis. The following is a list of the species:

Ammobaculite cobbani, *Massilina* sp., *Lenticulina audax*, *Astacolus agalmatus*, *A. aphrastus*, *Vaginulinopsis epicharis*, *V. enodis*, *Dentalina leptosyna*, *D. liota*, *Nodosaria daedala*, *N. lirulata*, *N. mecista*, *Lingulina ampliata*, *L. camerata*, *L. tumida*, *Tristix reesidei*, *Vaginulina lechriosa*, *Citharina entypomatus*, *Citharinella rhomboidea*, *Paleopolymorphina elacatoides*, and *Spirillina sp*. These forms were described by Loeblich and Tappan (1950) from the Black Hills of South Dakota.

Lagenids make up by far the dominant portion of the Curtis foraminiferal population. The long-ranging genus *Lenticulina* is the most common, and *Astacolus* is next in abundance. *Ammobaculites*, *Massilina*, and *Spirillina* are the only non-lagenid forms in the fauna. Though considerably diversified, the Curtis foraminiferal fauna does not show the variety of species reported by Loeblich and Tappan from the Black Hills area. This comparative lack of variety is largely a result of the near absence of arenaceous forms in the eastern Uinta area. Eight arenaceous species are included in the fauna from the Black Hills. According to Loeblich and Tappan, ". . . arenaceous Foraminifera comprise a fairly large percentage of the total fauna, as specimens of these are quite abundant." Studies of living Foraminifera show that dominantly arenaceous assemblages indicate cold-water environments. This single notable difference between the faunas of the Uintas and the Black Hills may indicate that sedimentation occurred in warmer waters in the eastern Uinta Mountain area.

OSTRACODA

Ostracods are more common than Foraminifera in the Curtis throughout the area. Following is a list of the species arranged in order of abundance: *Aparchitocythere typica*, *Progonocythere hieroglyphica*, *Camptocythere elliptica*, *Cytherella paramuensteri*, *Leptocythere imlayi*, *Cythereis zygoventralis*, *Cytherura lanceolata*, *Monoceratina sundancensis*, *Cytherella ventropleura*, and *Protocythere quadricarinata*. Nine of these species are illustrated in plate I. These forms have all been described by Swain and Peterson (1951, 1952) from the Swift formation of South Dakota, Wyoming, and Montana. The relative abundance of the individual species in the Curtis compares favorably with that given by Peterson (1954, p. 497) for the Swift formation. However, one species, *Progonocythere crowcreekensis*, reported to be relatively abundant in the Swift, is con-

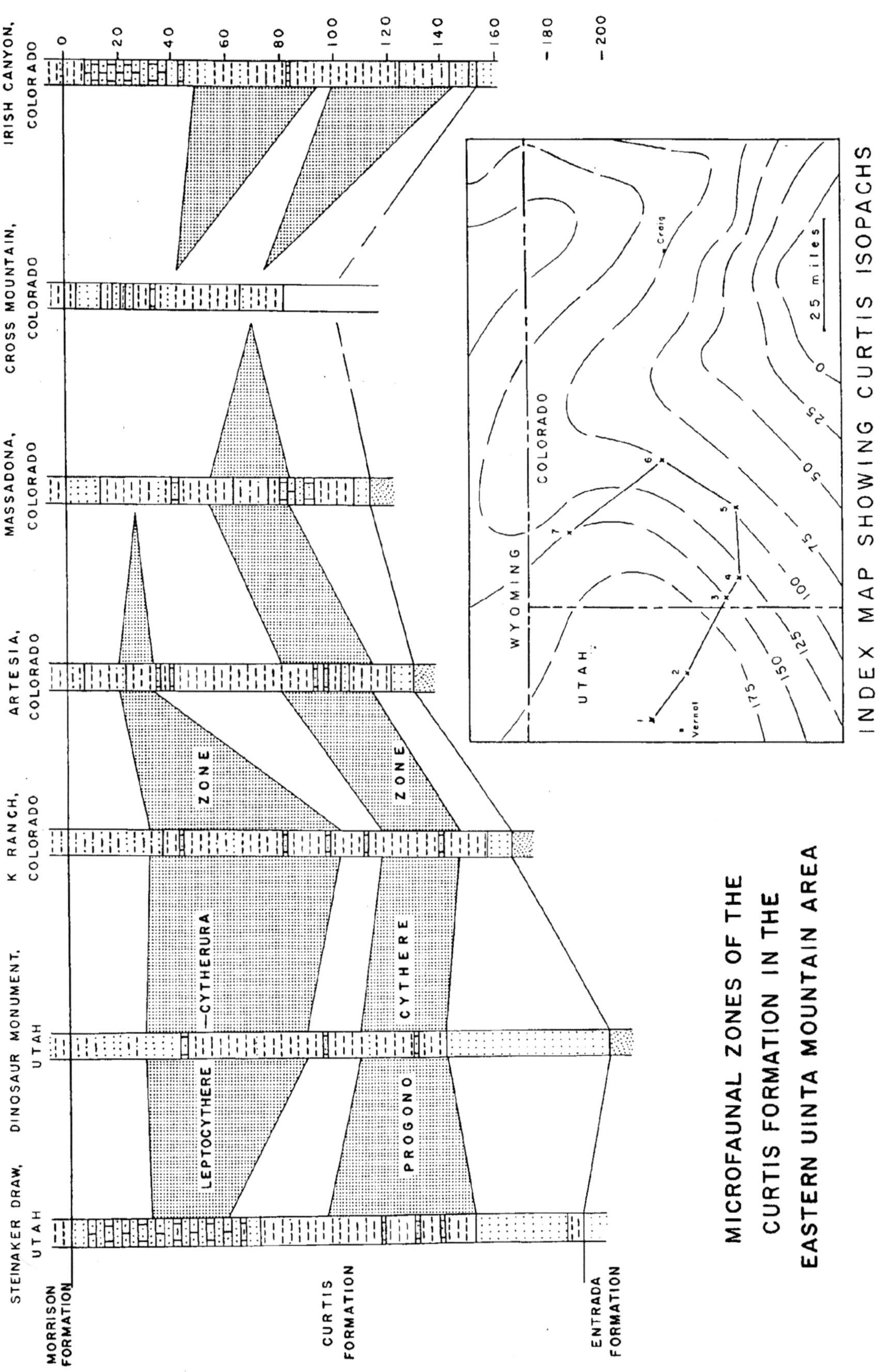
MICROFAUNAL ZONES OF THE CURTIS FORMATION IN THE EASTERN UINTA MOUNTAIN AREA
INDEX MAP SHOWING CURTIS ISOPACHS
1 STEINAKER DRAW, UTAH
2 DINOSAUR MONUMENT, UTAH
3 K RANCH, COLORADO
4 ARTESIA, COLORADO
5 MASSADONA, COLORADO
6 CROSS MOUNTAIN, COLORADO
7 IRISH CANYON, COLORADO
MORRISON FORMATION
CURTIS FORMATION
ENTRADA FORMATION
LEPTOCYTHERE—CYTHERURA ZONE
PROGONOCYTHERE ZONE
0
20
40
60
80
100
120
140
160
180
200
WYOMING
UTAH
COLORADO
Vernal
Craig
25 miles
0
25
50
75
100
125
150
175
Figure 1.

spicuously absent from the Curtis formation. Though it would hardly be expected that ostracod species would occur in precisely the same relative quantity in so widely separated areas, this single species, very common to the Swift formation, might well be expected to occur in the Curtis formation, particularly in places where populations are prolifically developed. Slight environmental differences between the two areas may be an explanation for this small discrepancy. Peterson (1954, p. 500) postulated that Swift sedimentation occurred in cool water. If, as was suggested in the case of the Foraminifera, the water became somewhat warmer toward the eastern Uinta area, this change may have been sufficient to eliminate *Progonocythere crowcreekensis* from the Curtis biocoenose.

STRATIGRAPHIC SIGNIFICANCE OF OSTRACODS

Aparchitocythere typica is the most common and widespread microfossil in the Curtis formation. It occurs throughout the vertical and lateral extent of the rocks, with few exceptions, wherever they are fossiliferous, and may be considered as typical of the Curtis formation as a whole. *Cytherella paramuensteri* and *Camptocythere elliptica*, though less common, have a stratigraphic distribution similar to *A. typica* and occur widely throughout the Curtis sequence. Some ostracods, however, occur only in certain stratigraphic intervals within the Curtis formation, and their abundance within these intervals delimits two distinct ostracod zones.

Progonocythere hieroglyphica, the second most common ostracod, occurs only in the lower portion of the Curtis formation, and gives rise to what is here designated as the *Progonocythere* zone. This zone is also characterized by *Protocythere quadricarinata* which is recorded from the Irish Canyon locality, section 7.

Overlying the *Progonocythere* zone is an assemblage of ostracods which includes *Leptocythere imlayi*, *Cytherura lanceolata*, *Cythereis zygoventralis*, *Cytherella ventropleura*, and *Monoceratina sundancensis*. The stratigraphic interval in which these forms occur is here called the *Leptocythere-Cytherura* zone. This zone occurs above the *Progonocythere* zone in the Curtis formation in the eastern Uinta Mountain area. The distribution and occurrence of the two zones through seven sections around the eastern end of the Uinta Mountains is shown in the cross-section.

James Peterson (1954) reported an almost identical zonal occurrence of ostracods in the Swift formation of Montana, northern and eastern Wyoming, and western South Dakota. However, *Progonocythere crowcreekensis* occurs in the lower zone of this area whereas it was not found in the Curtis of the eastern Uinta Mountain area. Also, *Monoceratina sundancensis* occurs in the Swift through a fairly broad stratigraphic range, not necessarily restricted to either of the zones. This ostracod is very rare in the Curtis formation, and the fact that the two samples in which it was found were in the *Leptocythere-Cytherura* zone may be coincidental.

In his study of the marine Upper Jurassic of eastern Wyoming and adjacent areas, Peterson (1954, p. 497) recognized in the Swift the "*Aparchitocythere typica* zone", and within it, two provisional ostracod subzones, the "*Progonocythere* subzone" and the "*Leptocythere imlayi - Cytherura lanceolata* subzone". Concerning the latter Peterson (1954, p. 499) stated, "These two ostracod subzones are provisionally designated, with the belief that further study will substantiate their validity". The zones which are, in this paper, called the "*Progonocythere* zone" and the "*Leptocythere-Cytherura* zone" of the Curtis formation are believed to be exact correlatives of the subzones reported from the Swift formation in and adjacent to eastern Wyoming. Thus, the validity of Peterson's provisionally designated subzones is substantiated and the geographic distribution of the Curtis-Swift ostracod zones in the western interior is greatly expanded. This discovery establishes a microfaunal tool by which detailed Upper Jurassic correlations may be carried over considerable areas in the western interior.

In the present study the two ostracod zones were recognized in sections 1, 2, 3, 4, and 7. Only the *Progonocythere* zone was found in section 5, and samples from section 6 contained none of the zonal forms. The apparent failure of the faunal zones to carry eastward from their excellent development to the west and north may be a function of the proximity of the Curtis shoreline and its concomitant shallower water or abnormal salinity. Sections 5 and 6 are only a few miles from this ancient feature as shown on the index map whereas sections 1 and 7 are about twice as far away.

A problem in defining the basal contact of the Curtis centers about a flat-bedded, glauconitic sandstone which overlies the highly cross-bedded sandstone of the Entrada formation. This flat-bedded sandstone, highly variable in thickness, was included in the upper part of the Entrada by Thomas, McCann, and Raman (1944). However, on the basis of bedding, glauconite content, and Curtis mollusks in its top, Bradley (1952, p. 16) considered it a basal sandstone of the Curtis. With the exception of the parallel bedding and glauconite grains, field exposures of this sand unit closely resemble the Entrada in color, texture, and resistance to erosion. That the entire flat-bedded sand unit was deposited under Curtis marine conditions is proven by microfossils found in it at localities 2 and 5. At locality 2, near the Dinosaur Monument headquarters buildings, the maximum development of 63 feet of this sand was

measured. Curtis ostracods and Foraminifera were found throughout its thickness. It is suggested that this basal Curtis sandstone was eroded from an uneven Entrada topography by the advancing Curtis sea, transported by currents, mixed with glauconite and microfaunas, and redeposited in the marine environment.

ACKNOWLEDGMENTS

This report is a portion of a study undertaken at the University of Colorado as an M.S. thesis under the supervision of Dr. John Chronic whose advice and assistance is gratefully acknowledged. Thanks are due to the Gulf Oil Company for financial assistance during the course of this study, and to The Carter Oil Company for use of their electric log files.

REFERENCES

Bradley, W. A. (1952), "Jurassic and Pre-Mancos Cretaceous Stratigraphy of the Eastern Uinta Mountains, Colorado-Utah," Unpublished Master's Thesis, University of Colorado.

Imlay, R. W. (1948), "Characteristic Marine Jurassic Fossils From the Western Interior of the United States," U. S. Geol. Survey Prof. Paper 214-B, pp. 13-33, Pls. 5-9.

Loeblich, A. R., Jr., and Tappan, Helen (1950), "North American Jurassic Foraminifera: I. The Type Redwater Shale (Oxfordian) of South Dakota," Jour. Paleon., vol. 24, pp. 39-60, pls. 11-16.

Peterson, J. A. (1954), "Marine Upper Jurassic, Eastern Wyoming," Bull. Amer. Assoc. Petrol. Geol., vol. 38, pp. 463-507.

Swain, F. M. and Peterson, J. A. (1951), "Ostracoda From the Upper Jurassic Redwater Shale Member of the Sundance Formation at the Type Locality in South Dakota," Jour. Paleon., vol. 25, pp. 796-807, pls. 113-114; 1 fig.

———— (1952), "Ostracoda From the 'Upper Sundance' Formation of South Dakota, Wyoming, and Southern Montana," U. S. Geol. Survey Prof. Paper 243-A, pp. 1-17, pls. 1, 2.

Thomas, C. R., McCann, F. T., and Raman, N. D. (1946), "Mesozoic and Paleozoic Stratigraphy in Northwestern Colorado and Northeastern Utah," U. S. Geol. Survey, Oil and Gas Investigations, Preliminary Chart 16.

EXPLANATION OF PLATE I

Fig. 1—*Monoceratina sundancensis* Swain and Peterson, 1951; right side (X52). 2—*Progonocythere hieroglyphica* Swain and Peterson, 1951; right side (X52). 3—*Camptoythere elliptica* Swain and Peterson, 1952; interior of a right valve (X52). 4—*Cytherura lanceolata* Swain and Peterson, 1951; left side (X58). 5—*Protocythere quadricarinata* Swain and Peterson, 1952; left side (X48). 6, 7—*Aparchitocythere typica* Swain and Peterson, 1952. 6, interior of a left valve (X55); 7, interior of a right valve (X48). 8—*Cythereis zygoventralis* Swain and Peterson, 1952; right side (X50). 9—*Cytherella paramuensteri* Swain and Peterson, 1952; left side (X48). 10—*Leptocythere imlayi* Swain and Peterson, 1951; interior of a left valve filled with matrix (X48). 11—*Vaginulina lechriosa* Loeblich and Tappan, 1950; side view showing oblique sutures (X52). 12—*Massilina sp.*; side view (X52). 13—*Astacolus aphrastus* Loeblich and Tappan, 1950; side view showing long, narrow chambers (X44). 14—*Nodosaria lirulata* Loeblich and Tappan, 1950; view of a broken specimen (X58). 15—*Lingulina tumida* Loeblich and Tappan, 1950; side view (X52). 16—*Dentalina leptosyna* Loeblich and Tappan, 1950; side view showing one nearly straight side (X48). 17—*Astacolus agalmatus* Loeblich and Tappan, 1950; side view of a large specimen (42). 18—*Lenticulina audax* Loeblich and Tappan, 1950; side view of a large, transparent specimen (X42). 19—*Vaginulinopsis enodis* Loeblich and Tappan, 1950; side view showing rounded last chamber (X52). 20—*Citharina entrypomatus* Loeblich and Tappan, 1950; side view showing oblique chambers (X58).

Eicher: Plate I.

THE CRETACEOUS ROCKS OF NORTHWEST COLORADO

By CHARLES C. O'BOYLE

Brainerd and Van Tuyl, Consulting Geologists, Denver, Colorado

GENERAL FEATURES

The Cretaceous rocks of northwest Colorado, while varying in detail, are similar in gross lithologies and stratigraphic sequence to Cretaceous rocks throughout the greater portion of the Rocky Mountain province. A basal conglomeratic sandstone above an unconformity is overlain by a siliceous and arenaceous sequence, which grades upward into calcareous neritic shales and limestones. These beds are overlain in turn by a silty marine shale, the sandiness of which increases upward until the section is predominantly sand and the marine facies change to near-shore and terrestrial deposits. The youngest beds are primarily terrestrial and are marked at the top by another unconformity. This sequence of formations was deposited on the western margin of a north-south trending Cretaceous sea. The section consists essentially of arenaceous sediments derived from the west, which interfinger eastward with the main body of marine shale. The strand line oscillated in an east-west direction throughout Cretaceous time.

The subdivision of this Cretaceous sequence into various formations and members of formations represents the efforts of geologists for many years. The names applied to separate lithologies by these workers were, in many instances, carried in from adjacent areas, while others applied purely local names to the various units.

The great thickness of marine shale within this system has been referred to for many years as the Mancos throughout western Colorado. However, recent work has shown that two formations, the Mowry and Frontier of adjacent areas, are present within this sequence. The hiatus between lower and upper Cretaceous time separates these formations. The term Mancos will be restricted to the marine shale which lies above the Frontier and below the Mesaverde sandstones.

The accompanying figure is included to facilitate correlation of these Cretaceous formations of northwest Colorado with equivalent formations to the east and west.

LOWER CRETACEOUS SYSTEM

Dakota (?) Sandstone

This thin, but remarkably continuous, formation exhibits a three-fold lithologic division throughout northwestern Colorado. The basal and uppermost beds consist of massive, carbonaceous sandstones with discontinuous bedding. Both sandstones are conglomeratic in places. On the outcrop these sands are white, buff to yellowish-brown, with limonite staining. Individual beds are composed of medium to coarse, subangular to rounded quartz grains, with relatively abundant feldspars. The cementing material is generally silica, but clay and limonite are not uncommon. Conglomeratic lenses within the sandstones are composed of gray and black, well-rounded chert as its principal constituent, but rounded to angular pieces of limestone, quartz, feldspar and pre-Cambrian rocks are locally abundant. Dividing these sandstones is a dark greenish-gray, and in places variegated, argillaceous sandy shale. These shales are quite similar to those of the underlying Morrison formation. Coal and plant fragments are found irregularly throughout this interval.

While individual members range in thickness, the formation as a whole has a remarkably uniform thickness of between 90 and 150 feet, and on the outcrop generally forms a pronounced hogback. Subsurface thicknesses, as a rule, are less than surface measured sections. An erosional unconformity at the base of the lower sandstone marks the division between Jurassic and Early Cretaceous time. Because both the Dakota (?) and underlying Jurassic Morrison formations contain intraformational unconformities, and as the shales and sandstones of both formations are quite similar, the recognition of this major unconformity at the surface and in the subsurface is frequently difficult. This circumstance may account for the large variance in thickness shown in measured sections, even within local areas.

While the coals found within the Dakota (?) do not compare with those from the Mesaverde, they have been mined for local consumption in various places because of convenience. The formation produces gas in seven fields in northwestern Colorado, but has a striking paucity of oil, though shows have been found in several structures within the region. At present, only non-commercial uranium mineralization has been found in these sandstones.

Mowry Shale

Overlying the Dakota (?) sandstone and in gradational contact with it are beds of black, carbonaceous marine shale in the lower part, which grade upward into the typical siliceous shales of the Mowry formation. These siliceous shales weather light silver-gray to

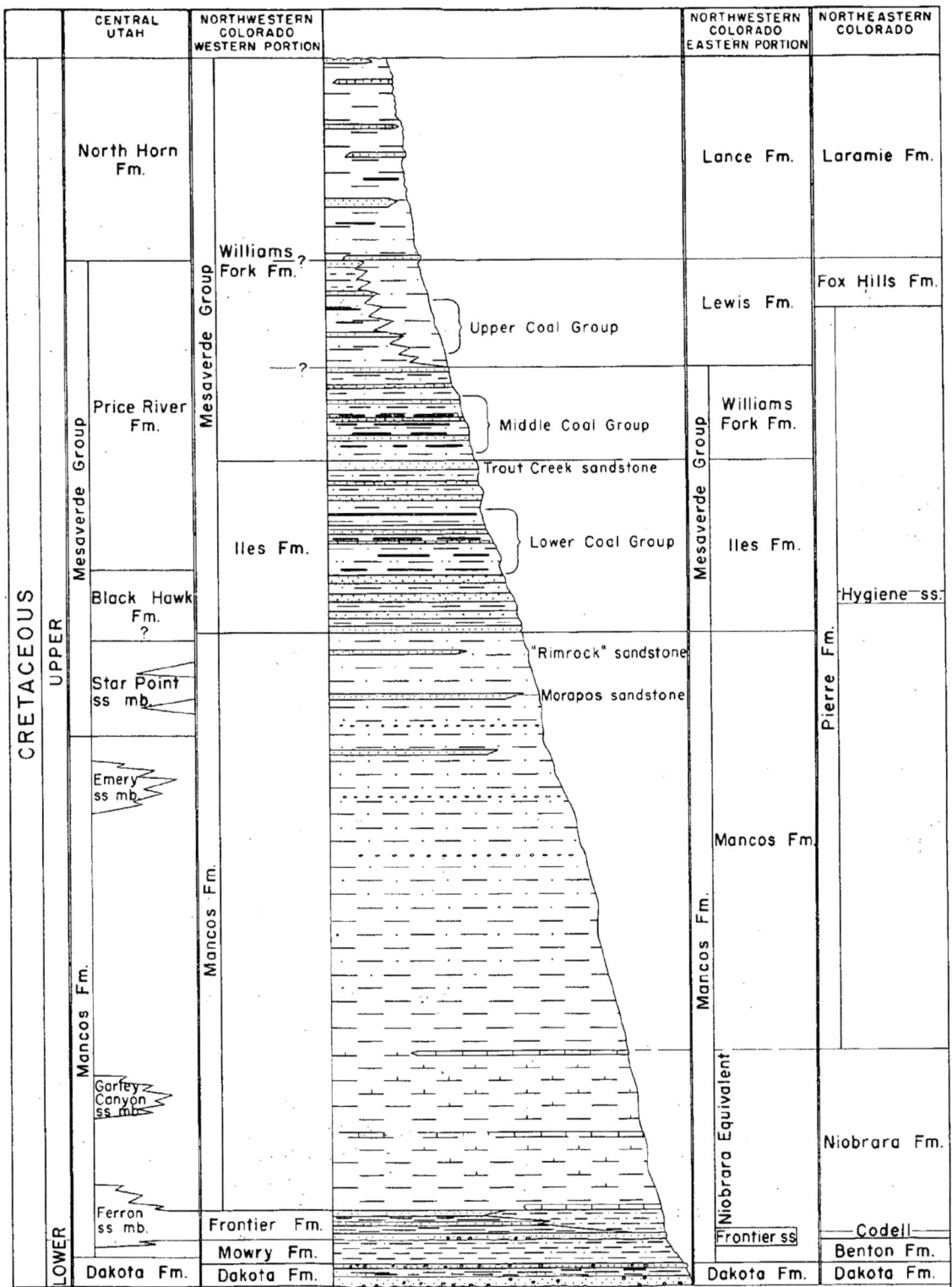

GENERALIZED COLUMUNAR SECTION OF THE CRETACEOUS ROCKS OF NORTHWESTERN COLORADO

bluish-gray on the outcrop and generally support a stand of evergreens. Thin beds of dark gray limestone and bentonite occur within the formation. Selenite is found along joints and bedding planes and fish scales are relatively abundant. The basal black, less siliceous shale may be equivalent to the Thermopolis shale of Wyoming. The top of these beds marks the division between the lower and upper Cretaceous. The thickness of this formation ranges between 100 and 200 feet.

In the Moffat field, the Mowry shale has produced oil from fractures. The bentonite beds within this formation in western Colorado are not of commercial importance at present.

UPPER CRETACEOUS SYSTEM

Frontier Formation

Early geologists recognized the presence of a somewhat sandy facies lying above the siliceous Mowry shale. While this facies was stratigraphically equivalent to the Frontier formation of Wyoming, its lack of massive sandstones in parts of the area, and poorly marked contacts, precluded their separating this sandy facies as a formation from the enclosing Mancos shale. Recent work, both on the surface and in the subsurface, has shown that this sandy sequence is actually equivalent to the Frontier formation of Wyoming and in part at least to the Ferron sandstone of central Utah. This series of beds consists of alternate sands and sandy shales. The sandstones are massive to thin-bedded, calcareous and fine-grained. Ripple marks are common. In places the sandstones contain rounded chert fragments. Thin coals and brown weathering limestone concretions ranging in size from a few inches to six or eight feet in diameter are sometimes found within these beds. In most of northwestern Colorado these sandstones are essentially thin-bedded and shaly, and because of their calcareous content have very low porosity and permeability and do not form reservoir beds of consequence. The sandstone beds, where sufficiently well-developed, form low rounded hills behind the more pronounced Dakota (?) hogback. The formation is approximately 250 feet thick in the northwestern part of the region and thins southward and eastward.

While shows of gas have been found in these beds, there has been no commercial production of oil or gas from this formation in northwestern Colorado.

Mancos Shale

Overlying the Frontier formation and in gradational contact with it is the great mass of sandy marine shales of the Mancos. In northwestern Colorado the thickness of these beds ranges from 4,000 to more than 5,000 feet. It is calcareous in its lower part and this portion of the formation is equivalent to the Niobrara of eastern Colorado, though over much of the area these beds are so poorly developed that they are not recognized as a distinct formation. Above this calcareous zone the shale becomes increasingly sandy until, near the top, well-developed massive sandstones occur. These sandstones represent the basal major tongues of the overlying Mesaverde formation and form scarps which break the topographic monotony of the Mancos shale. These scarps or rims have been given local names, such as the "rim rock" at Rangely and the Morapos sandstone of the Meeker area. The Mancos consists essentially of a dark gray, calcareous clay shale that weathers light-gray to yellow-brown. Gypsum in the form of selenite and calcite geodes are common throughout the formation. Calcite seams are present where the shale has been fractured by uplift. Locally, persistent beds of limonitic concretions are generally the only mappable horizons within the main portion of this formation.

Fractures in the Niobrara equivalent at the base of this formation, have produced over two million barrels of oil at Tow Creek. At Rangely, calcite seams or dikes within this shale have produced approximately five million barrels of oil from shallow depths.

Mesaverde Group

During Mesaverde time, sandstones from the west progressively pushed the strand line eastward, with minor retreats, until the Cretaceous sea disappeared from the region. This intertonguing of marine shales and marine and non-marine sandstones is best exhibited in the Book Cliffs along the southern border of the Piceance basin. This relatively resistant formation is expressed topographically as a series of ledges and steep slopes which form an escarpment of magnitude above the valleys of the Mancos shale. In the subsurface, this group is difficult to subdivide and consists of a monotonous sequence of sands, shales and coals of similar lithologies. The group can be divided into the Iles and Williams Work formations at the surface.

Within this group are the great coal reserves of Colorado, measured in billions of tons. Although at present these coals are not mined as extensively as in the past, they will undoubtedly be utilized in the future as a major source of energy.

Iles Formation

Overlying the Mancos shale and gradational with it are the predominantly marine sandstones and shales of the Iles formation. These sandstones are fine-grained, buff to yellow-brown, and are generally more massive and continuous than those of the overlying Williams Fork. Interbedded with these sandstones are sandy

shales, carbonaceous shales and beds of coal. The coal content increases eastward and these coals are known as the "lower coal group" throughout much of the area. The Iles formation maintains a uniform thickness throughout of 1300 to 1500 feet. The upper contact is placed at the top of a persistent, massive, white sandstone, called the Trout Creek in many reports. Over much of the area, this sandstone forms a striking escarpment.

Williams Fork Formation

The beds conformably overlying the Iles formation represent a transition from predominantly marine to littoral and lagunal deposits. These beds on the outcrop are, in most places, pale- to bright-red, due to the burning of interbedded coals, in contrast to the buffs and yellow-browns of the underlying Iles. The great coal deposits of northwestern Colorado lie within this formation and are known as the "middle and upper coal groups". Lithologically, these beds consist of thin to massive, discontinuous, medium- to fine-grained sandstones, with clay forming a great portion of the cementing material. Gray, sandy shales lie between these sandstones and include thin to massive coal beds. The thickness of this formation, as now mapped, ranges considerably from less than 1500 feet to more than 5000 feet. This great range in thickness is due to facies changes within the formation and to the erosional unconformity at its top. In the northern and eastern portions of the region, the Lewis marine shale, described below, separates the Williams Fork from the overlying Lance. In the southwestern part of the area south of Axial basin, this shale is not present, and the interval between the Iles and the Paleocene is lithologically similar throughout, and all beds within this section are included in the Williams Fork formation.

Lewis Shale

In the northern and eastern parts of this area, the gray marine Lewis shale lies above the Mesaverde group. This formation represents a wedge of marine shale in the predominantly arenaceous upper portion of the Cretaceous of northwestern Colorado. It is composed of soft, sandy, clay shale, calcareous in places, containing small lenticular masses of limestone and ferruginous concretions. To the east and north it is between 1600 and 2000 feet thick, thinning rapidly to the southwest, and is not recognized south of Axial basin. The contact between the underlying Mesaverde and overlying Lance formations is gradational throughout the area where this shale is present.

Lance Formation

The Lance formation, the uppermost of the Cretaceous formations in this area, is essentially a terrestrial deposit. It consists of a succession of gray to yellow shales, thin coal beds, and massive, discontinuous crossbedded sandstones. These beds closely resemble those of the underlying Mesaverde, but as a rule are less resistant and form a similar, but less rugged, type of topography.

Coals within the Lance are not as thick, extensive, or as good in quality as those of the Mesaverde, but they do add extensively to the coal reserves of the region. Shows of gas have been found in this formation in the Hiawatha field.

The top of the Lance formation is marked by a profound erosional unconformity which separates the Cretaceous from overlying Paleocene and Eocene beds. This unconformity is generally marked by a massive conglomerate throughout the area.

REFERENCES

Cobban, W. A., and Reeside, J. B., Jr. (1951), "Lower Cretaceous Ammonites in Colorado, Wyoming, and Montana," Bull. Amer. Assoc. Petrol. Geol., vol. 35, no. 8, (August 1951), pp. 1892-1893.

.... (1952), "Correlation of the Cretaceous Formations of the Western Interior of the United States," Geol. Soc. of America Bull., vol. 63, no. 10, (October 1952), pp. 1011-1044.

.... (1952), "Frontier Formation, Wyoming and Adjacent Areas," Bull. Amer. Assoc. Petrol. Geol., vol. 36, no. 10, (October 1952), pp. 1913-1961.

Coffin, R. C., Perini, V. C., Jr., and Collins, M. J. (1920), "Some Anticlines of Western Colorado," Colorado Geol. Survey Bull. 24.

Erdmann, C. E. (1934), "The Book Cliffs Coal Field in Garfield and Mesa Counties, Colorado," U. S. Geol. Survey Bull. 851.

Gale, H. S. (1908), "Geology of the Rangely Oil District, Rio Blanco County, Colorado," U. S. Geol. Survey Bull. 350.

.... (1910), "Northwestern Colorado and Northeastern Utah," U. S. Geol. Survey Bull. 415.

Katich, P. J. (1953), "Source Direction of Ferron Sandstone in Utah," Bull. Amer. Assoc. Petrol. Geol., vol. 37, no. 4, (April 1953), pp. 858-861.

Sears, J. D. (1924), "Geology and Oil and Gas Prospects of Moffat County, Colorado, and Southern Sweetwater County, Wyoming," U. S. Geol. Survey Bull. 751-G.

Walton, P. T. (1944), "Geology of the Cretaceous of the Uinta Basin, Utah," Geol. Soc. of Amer. Bull., vol. 55, no. 1 (January 1944), pp. 91-130.

Young, R. G. (1955), "Sedimentary Facies and Intertonguing in the Upper Cretaceous of the Book Cliffs, Utah-Colorado," Geol. Soc. of Amer. Bull., vol. 66, no. 2 (February 1955), pp. 177-202.

Untermann, G. E., and Untermann, B. R. (1950), "Petroleum Geology of the Uinta Basin," Guidebook to the Geology of Utah No. 5, Intermountain Assoc. of Petrol. Geol.

.... (1953), "Field Conference in Northwestern Colorado," Guidebook, Rocky Mountain Assoc. of Geol., (May 1953).

.... (1954), "Geology of Dinosaur National Monument and Vicinity, Utah-Colorado," Utah Geol. and Min. Survey Bull. 43, (June 1954).

.... (1954), "The Oil and Gas Fields of Colorado," a Symposium publ. by the Rocky Mountain Assoc. of Geol.

EARLY CENOZOIC HISTORY OF THE SAND WASH BASIN, NORTHWEST COLORADO

By HOWARD R. RITZMA

General Petroleum Corporation, Salt Lake City, Utah

Paleocene and Eocene time saw marked changes in the structural framework and depositional pattern of the Sand Wash Basin area. These changes are outlined briefly.

THE GEOLOGIC SETTING

During much of Cretaceous time, the Sand Wash Basin was the site of marine deposition in the widespread Cretaceous sea. To the west lay an orogenic belt that had advanced steadily from west to east from Nevada to central Utah during the span of Mesozoic time.

From time to time low uplifts and shoaling interrupted the Cretaceous seaway, and prominent among these was a low north-south trending fold located along the present Colorado-Utah border and extending north for a short distance into Wyoming. This uplift is now reflected in the Douglas Creek Arch and the south portion of the Rock Springs Uplift. The whole Rock Springs Uplift as presently constituted does not coincide with the old high (Ritzma, 1955).

The south end of the Rock Springs Uplift and the Douglas Creek Arch are closely linked. The unconformities along their flanks are similar and the continuity of the two is fairly complete except where the later Uinta uplift has erased direct evidence. Similarity of the two uplifts was pointed out by Gow (1950, p. 139).

It is useful to think of the Douglas Creek Arch and the south end of the Rock Springs Uplift as a foreland fold in front of the Cretaceous orogenic belt of central and northern Utah. As expressed in the Mancos and lower Mesaverde exposures of the Book Cliffs and elsewhere in northwest Colorado, the Douglas Creek Arch began as a gentle warping which acted as a shoal in the Cretaceous sea. This caused thinning in the Mancos and minor deposition of silts and sands (Young, 1955, x-sect.). There is little to indicate similar conditions in the south end of the Rock Springs Uplift. As orogenic pulses moved eastward, the orogenic area overran the foreland fold; and the foreland itself became a positive area of low to moderate relief with its crest continually stripped of older deposits and accelerated deposition taking place off its eastern flank. Rather than one simple arch it was undoubtedly a complex of folds. Strong northeast-southwest trends within the larger north-south uplift still persist in the alignment of the axes of Canyon Creek and Hiawatha Anticlines. However, in a gross way it appears that the Douglas Creek Arch-Rock Springs Uplift structural element was a large anticlinal nose plunging northward. The south end, the Douglas Creek Arch portion, was most active first and continued to be so through its history. At least until the close of Paleocene time, the Uintas had not entered the structural picture.

PALEOCENE (FORT UNION)

The Cretaceous sea withdrew slowly to the east and the foreland continued to rise until it was an active positive area. Into the troughs to the east, in what are now the deeps of the Piceance and Sand Wash Basins, abnormal thicknesses of uppermost Cretaceous and Paleocene sediments were deposited, beds almost unrepresented immediately west of the foreland in the Uinta Basin.

At this same time, the Park Range, a typical north-south trending Laramide uplift had risen to the east. This is marked in the Fort Union formation by an extensive basal conglomerate which, in the Elkhead Mountains at its easternmost exposure, is 75 to 100 feet thick. Thirty five miles west, north of Maybell, the conglomerate has thinned to a feather edge. Without this distinctive unit, the Lance-Fort Union contact becomes difficult to define. This conglomerate contains abundant material from the pre-Cambrian core of the Park Range, many pebbles of fossiliferous Paleozoic chert and limestone, some brown slabby silicified wood and scattered cobbles of deeply weathered dacite and rhyolite. These volcanic flow rocks are widely distributed along the outcrop from Baggs to south of Craig, and their source poses an interesting problem. Paleogeography of the Paleocene is depicted in Figure 1A.

Above the basal conglomerate, the Fort Union consists of sandstone, shale, carbonaceous shale, lignite, and coal, all typical of beds deposited in a humid swampy low-lying basin. Some of the bedding is remarkably persistent, and the shales and sands are generally better indurated than those of younger formations.

EOCENE (HIAWATHA), THE RISE OF THE UINTAS

Uplands to the east and west of the Sand Wash Basin area continued to rise in lower Eocene (Hiawatha) time.

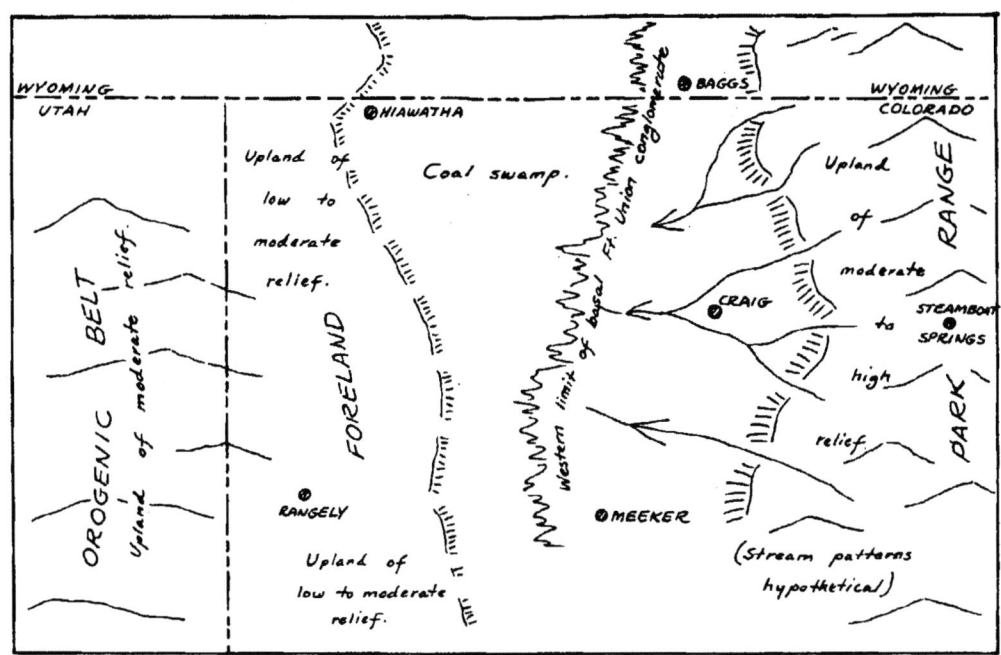

A. Paleogeography, Fort Union (Paleocene) time.

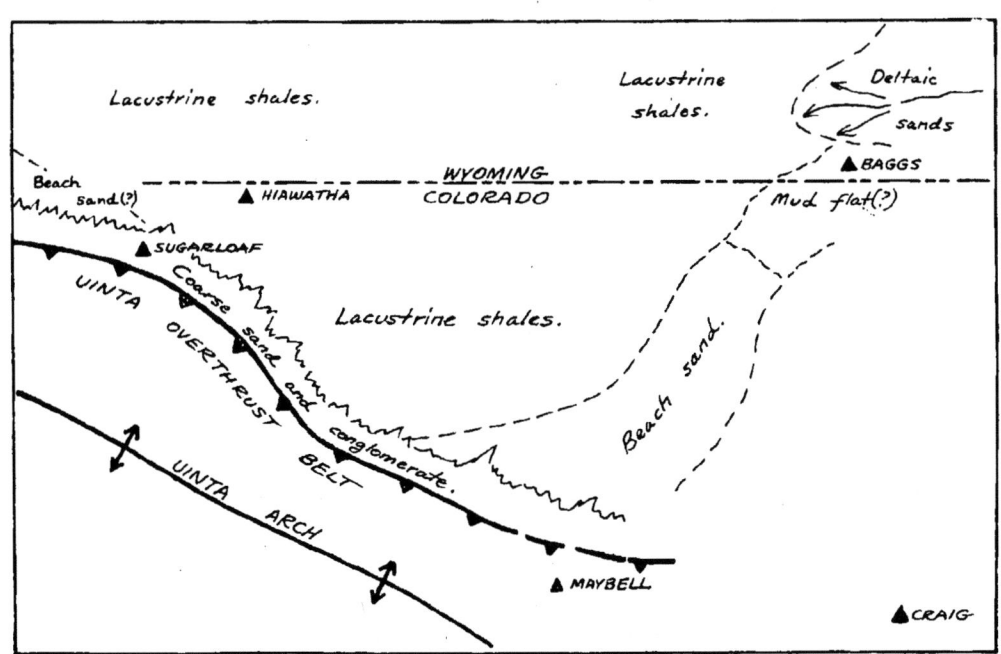

B. Paleogeography, Tipton (Lower Eocene) time.

Figure I – STRUCTURAL AND STRATIGRAPHIC PATTERNS, SAND WASH BASIN, NORTHWEST COLO.

The Uinta Arch began to grow and a strong east-west trend was rather abruptly superimposed across the older north-south trend. As orogenic events are viewed this was not too sudden an event. Although its entire history is confined to one vertebrate faunal zone of the Eocene, this was a span of five or so million years.

The sedimentary pattern of the Hiawatha records that folding proceeded through the deposition of the Hiawatha, culminating in major overthrusting during the Tipton stage of the Green River Lake. In T. 10 N., R. 100 W., along Vermilion Creek, these events are remarkably preserved. From east to west in a distance of two miles, the Cathedral Bluffs is seen to rest conformably on Tipton, then bevels folded and faulted Tipton and finally overlaps the Sparks-Uinta Overthrust and the overthrust sheet with little or no disturbance. Apparently the Uintas ceased to be an active tectonic element for the remainder of Eocene time.

UINTA TECTONICS

The Uinta Arch undoubtedly owes its somewhat unique east-west alignment to the resistant, homogenous mass of pre-Cambrian metasediments that makes up the core of the range from its western extremity to Juniper Mountain in Colorado, the farthest east outlier of the range. In this distance of some 150 miles, the dominant trend of the metasediments, particularly the massive quartzites, closely parallels the axis of the Uinta Arch, even to the pronounced southeast swing in the axis just east of the Utah-Colorado boundary. The ancient east-west trough in which the Uinta Mountain group was deposited continued to receive some additional thicknesses of sediments into Paleozoic time as well.

The development of troughs in which thick uppermost Cretaceous and early Cenozoic sediments accumulated (i.e.: the south portion of the Bridger Basin, the southwest part of the Sand Wash Basin and the north portion of the Uinta Basin) was coincident with the rise of the Uinta Arch. The sinking of these troughs and the adjacent rise of the arch seem definitely linked to the presence of this shallow-rooted pod of pre-Cambrian metasediments and Paleozoic sediments imbedded in the earth's crust.

The Uinta Arch was also subjected to other complex stresses. A definite westerly component of compressive stress is shown in the trends of flanking anticlines and complementing folds in the pre-Cambrian core of the range (Hansen, 1954, pp. 14-15). The Uintas were probably pushed eastward by the same west to east orogenic pulses that had caused the orogenic belts of Nevada, Idaho and central Utah to migrate steadily in this direction through Cretaceous and early Cenozoic time. This eastward push of the shallow rooted Uinta Arch is reflected also by major thrusting of the east end of the Uintas over the deep of the Sand Wash Basin in Colorado; separation of the Cross Mountain (Colorado) segment of the Uinta Arch, a block rotated clockwise by a westerly stress component; and by Juniper Mountain (Colorado), an uplift bounded by high angle reverse faults on the south and east indicating a push from the northwest.

The Uinta Arch is bounded on both its north and south flanks by high angle reverse faults or low angle overthrusts which tend to thrust the mountain mass over the adjacent or subjacent basin. The maximum displacement to the south is on the west end of the south flank; the maximum to the north is on the east end of the north flank. Hence, the Uintas were also subjected to counter-clockwise rotational motion with the older north-south Douglas Creek-Rock Springs element as pivotal area.

The Uinta Fault on the north flank of the Uinta Arch is a high angle reverse fault which becomes progressively a low angle overthrust as traced to the east. The fault is a stretch thrust in which the dip of the fault plane steepens rapidly toward the roots of the fault where the rupture involves more competent beds. Therefore, where pre-Cambrian quartzite is thrust against competent Paleozoics near Flaming Gorge, the fault is nearly vertical. To the east, where the fault involves pre-Cambrian over incompetent Mancos and Hiawatha shales and claystone, the inclination lessens to 15° and perhaps to nearly horizontal. In a thrust of this type, the angle of the fault at a particular place is a function of the degree to which erosion has stripped back the thrust sheet to its roots. The low angle nature of the Uinta Fault east of Clay Basin is strongly indicated by much indirect evidence in surface outcrops.

DEPOSITION CONTROLLED BY THE UINTAS

The rise of the Uintas is recorded in the Hiawatha and Tipton deposits of the adjacent basins. The Hiawatha and Tipton of the Vermilion Basin area contain increasingly coarse clastics composed mainly of fragments of Paleozoic and Mesozoic rocks. Above the Tipton is an unmistakable influx of coarse material derived from the pre-Cambrian Uinta Mountain group indicating a widespread breach of the Uinta Arch to its core. Hansen (1954, p. 11) notes this in the Clay Basin area 25 miles west.

The Tipton tongue of the Green River formation contains an unusually complete and well exposed record of the configuration of the Sand Wash Basin in that rather brief instant of geologic time (Fig. 1B). Areal extent of the Tipton has heretofore been limited

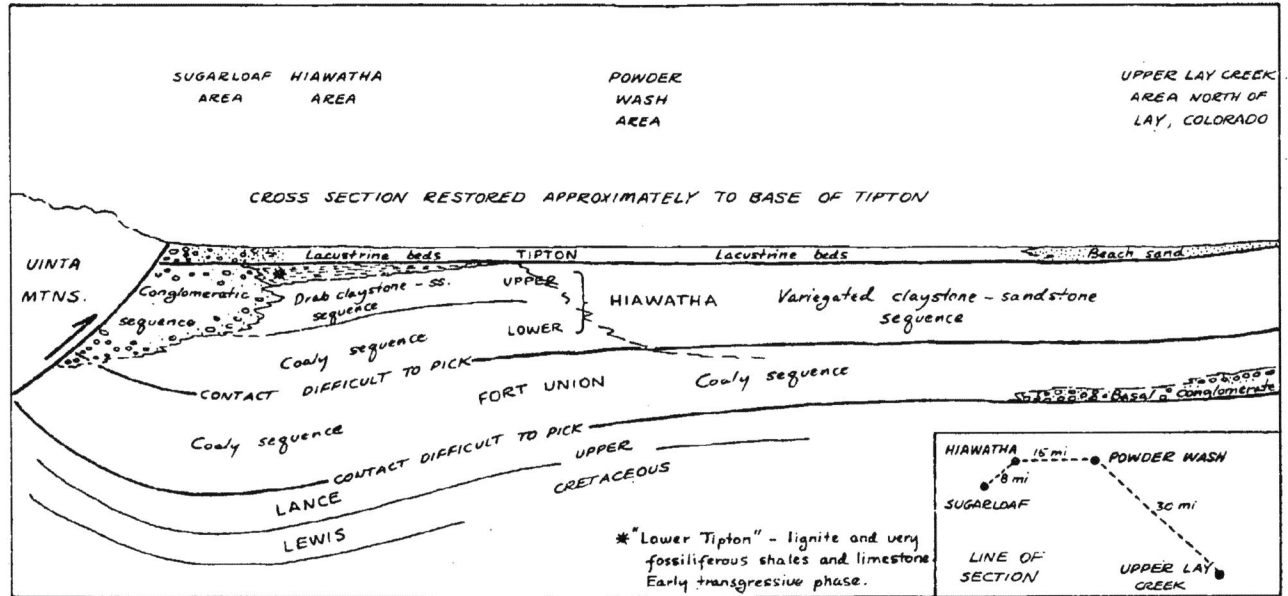

Figure 2 — IDEALIZED CROSS SECTION THROUGH SAND WASH BASIN, COLO. SHOWING FACIES CHANGES IN EARLY CENOZOIC FORMATIONS

to the Vermilion Basin area, but this writer recognizes the Tipton 25 miles south and east on the south flank of the Sand Wash Basin. The Tipton is extended to include the conspicuous sandstone unit mentioned by Sears (1924, p. 292) and shown on the map accompanying this bulletin by hachures through T. 8 N., Rs. 95 and 96 W. The stratigraphic position of this unit, its lithology and fossils found in the subjacent sandy shales make certain identification of the unit as a shore phase of the Tipton.

Change of the Tipton from lacustrine beds to very coarse bouldery conglomerate is well displayed north of Diamond Peak in T. 12 N., R. 102 W. (Colorado). This remarkable facies change over a distance of two miles was first noted by Schultz (1920, p. 31).

After Tipton time, the Uintas ceased to be active although a large volume of sediment continued to be swept off the mountains into the adjacent basins. The Cathedral Bluffs tongue of the Knight (Wasatch) is largely flood plain sediment. Succeeding it is the main body of the Green River formation which was deposited as the Green River Lake flooded to its maximum extent. This, in turn, is succeeded by interfingering lacustrine and flood plain deposits known as Bridger (or Sand Wash). Major orogenic activity in the western part of the Sand Wash Basin had subsided and gentle downwarping and basin filling was the main depositional pattern.

DEPOSITIONAL ENVIRONMENT

Van Houten's (1948) investigations and conclusions on depositional environments in Cenozoic basins in the Rocky Mountains fit very well the known pattern of the Sand Wash Basin. The idealized cross section (Figure 2) shows facies of pre-Tipton deposits of the basin as they correspond to varying conditions and climates of deposition.

The area in Paleocene time was a low-lying humid basin occupied mainly by coal swamps (Figure 1A). Coalescing fans encroached on the coal swamps from the rising Park Range to the east; and to a lesser extent, clastics filtered into the Fort Union from the low uplift to the west. As the basin was elevated and surrounding highlands rose, clastics continued to encroach on the coal swamp until it occupied only the deepest part of the basin (lower Hiawatha). On the margin, variegated sediments were deposited on the savannas and flood plains. Continued elevation, basin filling and a drier climate destroyed the coal swamp environment. Upper Hiawatha deposits in the basin deep are sparingly carbonaceous but drab-colored, reflecting the persistence of humid conditions with an environment in which ferric oxide was reduced by carbonaceous matter in the sediments. Variegated beds in the upper Hiawatha were widely and extensively developed over the rest of the basin. Flooding in of the Tipton stage of the Green

River Lake followed with a brief period of lacustrine depositional environment. The Cathedral Bluffs variegated beds appear to be a widespread general response to drier climate and the return of flood plain and savanna deposition. Succeeding lacustrine invasion and retreat is reflected in the Laney and the Bridger (Sand Wash).

REFERENCES

Gow, Kenneth (1950), "Douglas Creek Gas Field," Intermountain Assoc. Petrol. Geol., Guidebook to Geology of Utah, no. 5, pp. 139-46.

Hansen, W. R. (1954), "Laramide Faulting and Orogeny on the North Flank of Uinta Mountains in Eastern Daggett County, Utah," Colo. Scient. Soc. Proc., vol. 17, no. 1, pp. 1-29.

Ritzma, H. R. (1955), "Late Cretaceous and Early Cenozoic Structural Pattern, Southern Rock Springs Uplift, Wyoming," Wyo. Geol. Assn. Tenth Annual Field Conf. Guidebook, (in print).

Schultz, A. R. (1920), "Oil Possibilities In and Around Baxter Basin, in the Rock Springs Uplift, Sweetwater County, Wyoming," U. S. Geol. Survey Bull. 702.

Sears, J. D. (1924), "Geology and Oil and Gas Prospects of Part of Moffat County, Colorado and Southern Sweetwater County, Wyoming," U. S. Geol. Survey Bull. 751-G.

Young, Robert G. (1955), "Sedimentary Facies and Intertonguing in the Upper Cretaceous of the Book Cliffs, Utah-Colorado," Bull. Geol. Soc. America, vol. 66, no. 2, (cross section).

JOHN CHARLES FREMONT

John Charles Fremont, brilliant American Pathmarker, was born in Savannah, Georgia, January 21, 1813. He secured an appointment to West Point where, in 1838, he was graduated and commissioned second-lieutenant in the Topographical Corps. As an officer in that organization, his first important assignment was as chief associate in the expeditions headed by the distinguished French scientitst, Joseph Nicoles Nicollet. Their second expedition was planned for the scientific examination of the South Pass and Wind River Mountain area in south central Wyoming in order "to facilitate the rapidly increasing emigration to the Pacific Northwest." Fremont was again chief assistant; but Nicollet's health prompted a change of plans, and in December 1841, he replaced the French scientist as chief of command.

In 1841, Fremont married Jesse Benton, daughter of the influential senator, Thomas Hart Benton. This was a most fortunate marriage, for Jesse Barton proved to be one of the most talented and charming women in Washington. It was she who edited Fremont's copious notes and prepared them for publication. And by this marriage, the young explorer obtained the support of the American statesman who was most interested in the mapping and colonizing of the American West.

During the period 1841 to 1853, Fremont conducted five scientific expeditions, three of which, the second, third and fifth, penetrated the Great Basin and the region beyond to the Pacific. Two of Fremont's expeditions crossed northwest Colorado. Eastbound in June 1844, the Fremont party sojourned briefly in Browns Park near the ruins of Fort Davy Crockett. Leaving Browns Park by way of Bull and Irish Canyons, they made their first noon camp at Irish Lake. Crossing the Vermilion Basin, the expedition traversed the Powder Rim and ascended the Little Snake River and St. Vrains Fork (Battle Creek) to cross the Continental Divide into the Platte Valley. In 1845, westbound, Fremont descended the White River to the Green.

The United States Government published in full the scientific observations made by Fremont and his equally brilliant associates with the result that subsequent explorers, travellers and colonists became for the first time relatively well informed about the hitherto unknown West.

Besides his career as an explorer, Fremont played an important role in the military and political annals of our country. In 1846 he was directly responsible for the Bear Flag Revolt and later participated in the conquest of California. During the Civil War he played an important role in the Missouri Campaign. In 1847, he was military governor of California and, in 1850, was chosen first United States Senator from that state. In 1856, Fremont was first Republican candidate for the presidency of the United States. He died at New York City, July 13, 1890.

—L. H. Creer

EARLIEST EOCENE VERTEBRATES FROM THE SAND WASH BASIN, NORTHWEST COLORADO

By MALCOLM C. McKENNA

Museum of Paleontology, University of California, Berkeley, California

During the past four years it has been found that deposition of the Wasatch formation began much earlier than had heretofore been suspected. In the northeastern corner of the Sand Wash basin the Four Mile local fauna (McKenna, 1954) has been discovered in sediments of the Hiawatha member of the Wasatch formation, below the Tipton tongue of the Green River formation. The Four Mile assemblage is much older than the Dad local fauna (Wood, et al., 1941), which has been recovered from beneath the Tipton tongue some miles to the north of Baggs, Wyoming. The accompanying index map shows the more important Four Mile localities. Approximately nine thousand complete small mammal teeth have been recovered by use of an underwater screening process from about twenty localities in the following sections in the Craig Quadrangle, Colorado. Alheit Pocket, section 25, T. 12 N., R. 92 W. (along route of Field Conference in the first patch of badlands west of Colorado Highway no. 13, on the Snake River Bridge road); Timberlake Quarry, etc., section 28, T. 12 N., R. 91 W.; section 2, T. 11 N., R. 91 W.; Sand Quarry, etc., section 24, T. 12 N., R. 91 W.; section 36, T. 12 N., R. 91 W.; section 1, T. 11 N., R. 91 W.; section 11, T. 11 N., R. 91 W.; Despair Quarry, etc., section 12, T. 11 N., R. 91 W.; Anthill Quarry, section 13, T. 11 N., R. 91 W.; Kent Quarry, etc., section 18, T. 11 N., R. 90 W.; and section 19, T. 11 N., R. 90 W. All these localities lie on essentially one plane dipping southwest from 1° to 3°, resulting from the post-Wasatchian development of the Baggs anticline. The Baggs anticline is responsible for bringing basinward sediment of the Wasatch to the surface in a region where usually the exposed earlier Wasatch sediments are too coarse to bear abundant vertebrate remains. The University of California project has not yet been extended to the area north of the Baggs anticline except for some brief sampling of the Dad local fauna.

Generally speaking, the late Wasatchian sediments of the southeast Washakie basin and northeast Sand Wash basin are red and white banded, with few prominent yellow or lavender outcrops, while the early Wasatchian contains noticeably more outcrops of these colors. No unconformity has been found between the early and late Wasatchian sediments, but repeated attempts to find fossils of middle Wasatchian age have failed miserably so far.

The Dad local fauna occurs in approximately the top hundred feet of the Hiawatha member of the Wasatch for a distance of about ten miles southward from Dad, Wyoming, along the prominent cliffs west of the road to Baggs. The Four Mile local fauna, on the other hand, occurs at least five hundred feet below the top of the Hiawatha southeast of Baggs. The exact stratigraphic relations have not been worked out in detail, as mapping is not yet complete. Judging from the two faunas, however, a time lapse between them on the order of several million years is indicated.

Taxonomic work on the Four Mile local fauna is in progress, and is now completed for the insectivores, marsupials, and multituberculates. The following forms have been identified:

Order MULTITUBERCULATA
    Ptilodontidae
        Ectypodus tardus (Jepsen, 1930)
        Neoliotomus ultimus (Granger and Simpson, 1928)
        Uncertain form A.
        Uncertain form B.

Order MARSUPIALIA
    Didelphidae
        cf. *Peratherium chesteri* Gazin, 1953

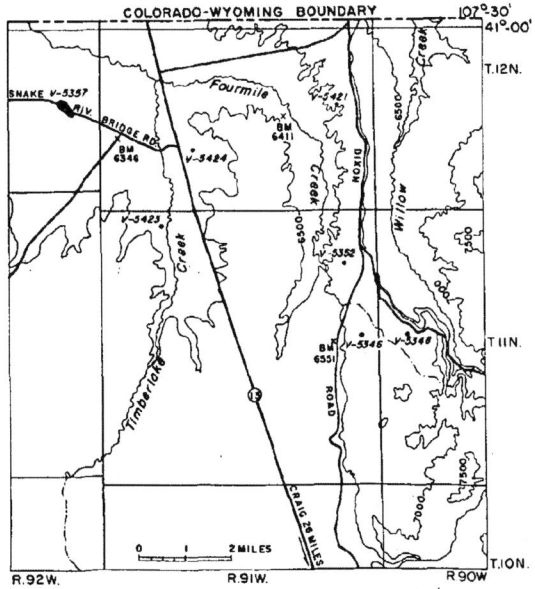

Figure 1. Index map.

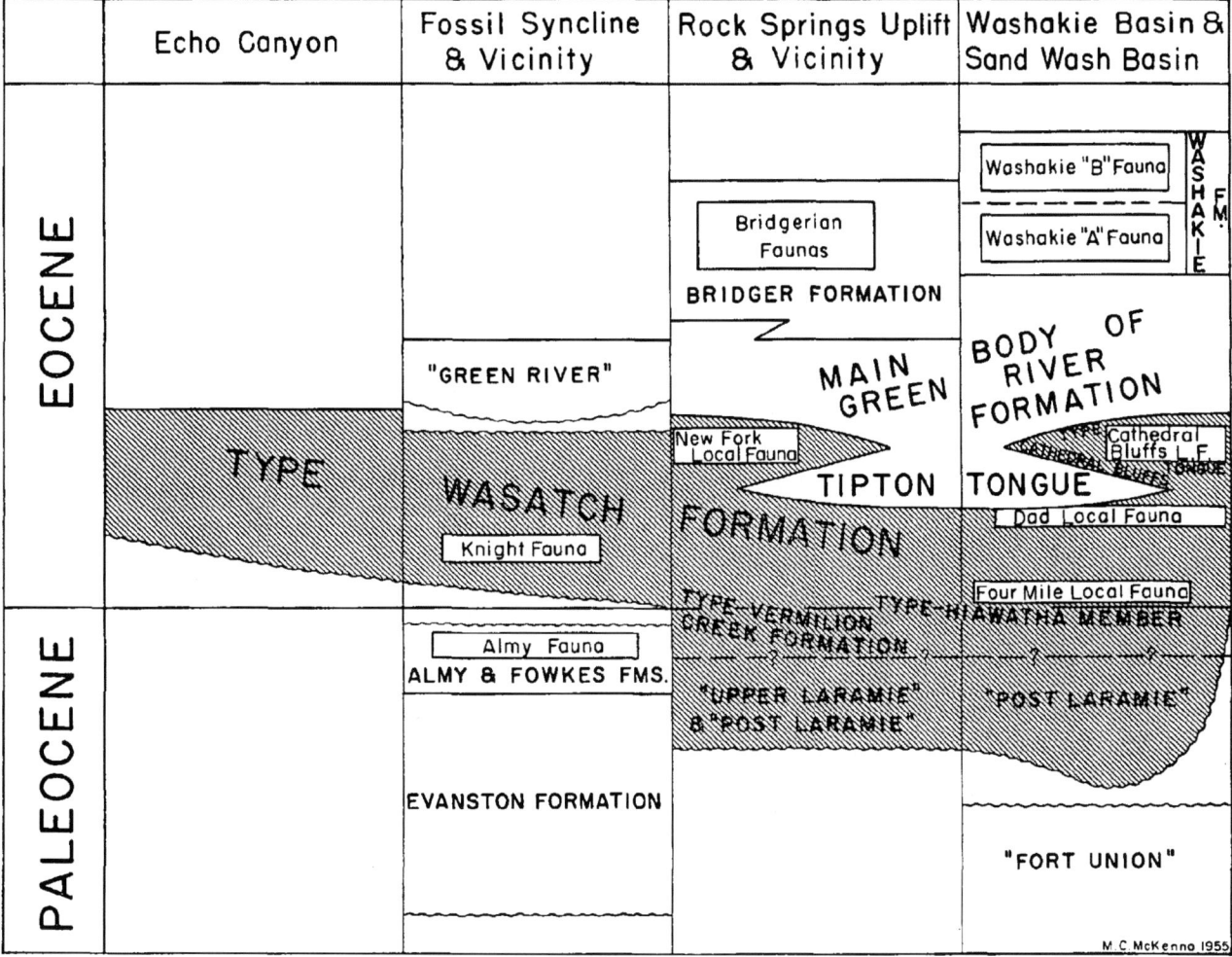

Figure 2. Correlation of Four Mile local fauna.

Order INSECTIVORA
    Deltatheridiidae
        Didelphodus absarokae (Cope, 1881)
        ?*Didelphodus* sp.
    Pantolestidae
        cf. *Palaeosinopa didelphoides* (Cope, 1881)
        undescribed pantolestid or leptictid
    Mixodectidae
        Cynodontomys, new species
    ?Soricidae
        New genus and species
    Leptictidae
        cf. *Diacodon tauri-cineri* Jepsen, 1930
        Diacodon, new species
        cf. *Entomolestes nitens* Matthew, 1918
        ?*Nyctitherium* sp.

Except for the aberrant Condylarth *Apheliscus*, the remainder of the fauna has not yet been studied in detail. The following genera are known to be present, however:

Order PRIMATES
    Plesiadapis
    Tetonius
    Minute genus of anaptomorphs
    Apatemyid near *Apatemys*

Order TILLODONTIA
    Esthonyx

Order RODENTIA
    At least three species of *Paramys*-like ischyromyids

Order CARNIVORA
    Chriacus
    Thryptacodon
    Dissacus or *Pachyaena*
    Large Oxyaenid
    Sinopa
    Didymictis
    Viverravus
    Several additional forms

Order CONDYLARTHRA
    Apheliscus nitidus
    Haplomylus

Hyopsodus (Very primitive forms)
Phenacodus sp. A.
Phenacodus sp. B.
Ectocion
?Proto-perissodactyl

Order PERRISSODACTYLA
Hyracotherium type A.
Hyracotherium type B.

Order ARTIODACTYLA
Wasatchia
Diacodexis

In addition, abundant fish, lizard, crocodile, chelonian, and other vertebrate remains have been collected but not yet studied.

The closest relationships of the Four Mile local fauna appear to lie with the assemblage that has been collected from Granger's (1914) Sand Coulee beds, a series of about two hundred feet of red banded shales lying between two gray shales near the base of the Willwood formation in the Bighorn basin. The actual age of the Four Mile local fauna could be somewhat older, however, judging from the abundance of multituberculates, *Plesiadapis,* and very primitive species of the condylarth and insectivore genera. Comparison of the fauna with the poorly known assemblage described by Gazin (1942) from beds west of La Barge, Wyoming, believed equivalent to Veatch's (1907) Almy formation has not yet been made, but a preliminary estimate would be that Gazin's material is older. The fauna of the Indian Meadows formation of Love (1939) is more nearly comparable to the Four Mile assemblage, but the relationships are not yet known in detail. The accompanying correlation chart attempts to show the stratigraphic and chronologic relationships of the Four Mile local fauna; the details of stratigraphic nomenclature will be discussed fully elsewhere.

All major quarries in the Four Mile area were discovered on the basis of very scanty surface showings. In general, the ratio of teeth eventually recovered by washing to those originally seen on the surface runs in the neighborhood of one hundred to one, with the average discovery site yielding around ten teeth from surface exposures. Generally, if washing is attempted at any site yielding more than one tooth on the surface, enough data to determine the locality's age can be amassed in a day or two. The process shows much promise for the future and will be employed in the 1955 season to work out the faunal sequence in the Godiva Ridge region in the Wasatch formation. Mammalian microfossils bear essentially the same relation to continental stratigraphy as foraminifera bear to marine stratigraphy, and it is hoped that they will become as useful.

REFERENCES

Gazin, C. L. (1942), "Fossil Mammalia From the Almy Formation in Western Wyoming," Jour. Wash. Acad. Sci., vol. 32, no. 7, July 15, 1942, pp. 217-220.

Granger, Walter (1914), "On the Names of Lower Eocene Faunal Horizons of Wyoming and New Mexico," Bull. Amer. Mus. Nat. Hist., vol. 33, art. 15, pp. 201-207.

Love, J. D. (1939), "Geology Along the Southern Margin of the Absaroka Range, Wyoming," G.S.A. Special Papers No. 20.

McKenna, M. C. (1954), "Earliest Wasatchian Vertebrates From the Hiawatha Member of the Knight Formation, Moffat County, Colorado," Bull. G.S.A., vol. 65, no. 12, part 2, Dec. 1954, p. 1283.

Veatch, A. C. (1907), "Geography and Geology of a Portion of Southwestern Wyoming, with Special Reference to Coal and Oil," U. S. Geol. Survey Prof. Paper No. 56, pp. i-vi, 1-178.

Wood, H. E., 2nd, et al. (1941), "Nomenclature and Correlation of the North American Continental Tertiary," Bull. G.S.A., vol. 52, pp. 1-48.

THE ELKHEAD MOUNTAINS VOLCANIC FIELD, NORTHWESTERN COLORADO

By BYRL D. CAREY, JR.
The California Company, Casper, Wyoming

INTRODUCTION

The Elkhead Mountains are the erosional remnants of an isolated Tertiary volcanic field located in Moffat and Routt Counties of northwestern Colorado. The mountains are situated a few miles south of the Wyoming-Colorado state line and east of Colorado State Highway 13 along the drainage divide between the Yampa and the Little Snake Rivers. The boundaries of the Elkhead Mountains volcanic field extend considerably beyond the mountains, volcanic flows and/or dikes being found as far north as Battle Mountain, Wyoming and as far south as Cedar Mountain near Craig, Colorado. The westernmost volcanics occur as Fortification dike which extends several miles west of the highway between Baggs, Wyoming, and Craig, Colorado. The south and east boundaries of the volcanic field are rather ill-defined in that the volcanics from the Elkhead Mountains merge with those that extend north from the White River Plateau and those that extend north and west from the Rabbit Ears and Yampa regions. Hahns Peak is generally considered as being somewhere near the eastern end of the volcanic field.

CHRONOLOGY OF INVESTIGATIONS

With the exception of the areas in the immediate vicinities of Hahns Peak and Fortification dike, the Elkhead Mountains have received little geologic investigation since the time of the early regional surveys. During the 1870's the range of mountains was visited by members of the Powell and Hayden surveys, but very little is written concerning the geology in their subsequent reports. In contrast to these two surveys, the geologists participating in the exploration of the Fortieth Parallel (King Survey) studied the Elkhead Mountains in considerable detail. The reports of Emmons (1877), King (1878), and Zirkel (1876) still contain most of the published information treating with the geology of the Elkhead Mountains and the volcanic field found therein.

White (1899, plate XXXIV) included in his discussion of the geology of northwestern Colorado, a geologic map which depicts the Elkhead Mountains and the general distribution of the eruptive rocks. It is probable, however, that the information was obtained from the publications of the King Survey.

The discovery of placer gold at the base of Hahns Peak in 1864 led to considerable interest in the immediate area by the mining industry. At a result of this interest, a number of brief geologic accounts dealing primarily with the Hahns Peak gold field were published. Among these are the reports by Draper (1897), Gale (1905), Lakes (1909), and George and Crawford (1909). The latter describes the geology of the gold field in considerable detail. Production statistics and/or estimates of the mineral resources of the Hahns Peak district are included in reports by Lee (1901), Parsons and Liddell (1903), White (1906), Worcester (1919), Henderson (1926), and Vanderwilt (1947). The most comprehensive information concerning the geology of the Hahns Peak area is contained in theses by Barnwell (1955) and Hunter (1955).

The publications of the King Survey and those concerned with the geology of the Hahns Peak area constitute the major portion of the information on the geology of the Elkhead Mountains. Some minor occurrences of high grade coal, however, are briefly discussed in a report by Chisolm (1887). Fortification dike has been studied in detail, perhaps, because of the unusual rock types, the impressive erosional form, and the proximity to an interstate highway. Ross (1926) discusses the petrography of several samples collected from the dike along with several flow rocks found near the town of Craig.

In addition to the published material, considerable information concerning the geology of the Elkhead Mountains has been obtained in the past few years by geologists working within the oil industry. Unfortunately, much of this information is contained in company files and is not available for publication.

DESCRIPTIVE GEOLOGY

The Elkhead Mountains are contained in an area of approximately 18 townships between the exterior lines of Townships 9 and 11 North, Ranges 85 and 90 West. The outer boundaries of the Elkhead Mountains volcanic field extend beyond the mountain range and are contained between the exterior lines of Townships 6 and 13 North, Ranges 85 and 91 West. There is no adequate published geologic map of the Elkhead Mountains and the volcanic field. The map which best depicts the geography of the area is the U. S. Depart-

ment of Agriculture Forest Service map of Routt National Forest.

The volcanic field is located regionally on the axis and the south flank of a large east-west trending structural arch which plunges westward into the basin from the west flank of the Sierra Madre Mountains. The axis of the arch coincides fairly closely with the Colorado-Wyoming state line.

The Elkhead Mountains are composed of the erosional remnants of the volcanics and the associated Tertiary sediments. The large flat-topped mountains among which are Mt. Welba and Mt. Oliphant consist of thick flows of lava overlying the Browns Park and North Park (?) formations. The steep-sided conical-shaped mountains similar to Hahns Peak and Bears Ears are the remnants of large volcanic stocks or plugs.

The Browns Park formation overlies with considerable angular unconformity all formations from the Mancos shale on the east end of the mountains in the vicinity of Hahns Peak to the Wasatch formation on the west end in the vicinity of Mt. Welba. The basal Browns Park conglomerate is well exposed at a number of localities throughout the volcanic field. The thickness of the Browns Park formation in the vicinity of Hahns Peak is estimated by Barnwell (1955, p. 32) as being approximately 1,200 feet.

On Sand Point, northwest of Mt. Welba, are excellent exposures of the transition zone between the Browns Park formation and the overlying volcanics. Initial volcanism can be detected by the appearance of small fragments of volcanic debris imbedded in a typical Browns Park sandy matrix. The fragments become larger higher in the section, and at the top of Sand Point the sediments consist of a volcanic breccia with boulders a foot or more in diameter. It is probable that the several hundred feet of volcanically derived sediments immediately beneath the high lava flows on Mt. Welba are the local equivalent of the North Park formation

The masking effect of the volcanics and the Tertiary sediments makes it impossible to resolve the structure of the Paleozoic and Mesozoic rocks beneath the Elkhead Mountains except by geophysical methods. Structural control by some of the intrusives, however, is very much in evidence. A large number of dikes throughout the volcanic field trend approximately N. 60° W. A swarm of dikes trending in this direction is particularly apparent throughout Township 10 North, Ranges 86 and 87 West. Since a considerable number of local post-volcanism normal faults also trend in this direction, it is probable that a well developed fracture system exists in the basement rocks.

PETROGRAPHY

The petrographic descriptions of the Elkhead Mountains volcanics are confined principally to five reports. The most comprehensive is that by Zirkel (1876) in which a number of rock samples collected from localities throughout the volcanic field are described. A few of the descriptions include chemical analyses. Ross (1926), in addition to his descriptions of the rock samples collected from the Fortification dike, included in his report descriptions of an analcite basalt collected from Breeze Mountain four and one-half miles southeast of Craig and an olivine basalt collected from Buck Peak six miles southeast of Craig. The reports by George and Crawford (1909), Barnwell (1955), and Hunter (1955) describe several samples of rhyolite porphyry and olivine basalt from the Hahns Peak area. A chemical analysis of a nepheline tephrite from the Elkhead Mountains is reported by Cross (1904, p. 187).

Emmons (1877, pp. 167-168) pointed out the unusual nature of the volcanic rock suite in the Elkhead Mountains. He wrote:

> From a mineralogical point of view, the eruptive rocks of the region form a remarkably interesting and peculiar group, being characteristically different from any of the wide-spread groups of volcanic rocks, which cover so large an area in the western portion of the region embraced within our explorations. They consist mainly of quartziferous trachytes and nepheline basalts . . . These trachytes, besides the normal constituents, sanidine, hornblende, and mica, contain also a relatively large proportion of augite, and in some cases a considerable amount of olivine . . . The basalts of the region are no less remarkable than the trachytes, being the only representatives of the group of nepheline-basalts found within the limits of our exploration.

To the student of petrography who believes in an orderly, simple classification of volcanic rocks, the rock suite in the Elkhead Mountains constitutes something very close to a petrographic nightmare. Zirkel (1876, p. 160) describes a rock from Whitehead Peak which contains both quartz and olivine. It was the first observation of this mineral combination. He describes a nepheline-bearing trachyte from the mouth of Slaters Fork which consists of phenocrysts of yellow-brown mica in a groundmass of sanidine and nepheline. The rock contains no hornblende or quartz, but exhibits well developed crystals of augite and olivine. The trachyte from Steves Ridge is said to resemble closely the Drachenfels trachyte from the Rhine River area.

The basalts from the Elkhead Mountains belong, with few exceptions, to the group of nepheline basalts containing little or no feldspar, but in general considerable olivine with lesser amounts of augite and magnetite. Occasionally biotite and some triclinic feldspar is found (Emmons, p. 176). The writer has examined a number of thin sections of the volcanics and what

has been described as nepheline would now be termed analcite.

The rock types found in the Fortification dike are no less unusual than the others among the suite. Ross (1926) describes the main rock type as a finely-crystalline soda-verite, and compares the rock with a verite described by de Yorza from Fortuna, Murcia, Spain, and one described by Osann from Cabo de Gato, Almeria, Spain. The dike rock consists essentially of phenocrysts of biotite, augite, and olivine in a soda-sanidine groundmass. Other rocks found within the dike are described as analcite syenites and soda-syenites.

Emmons (1877, p. 173) believes that the acidic intrusive forming Hantz Peak (Hahns Peak) is the only occurrence of rhyolite in the region. If this is true the stock forming Hahns Peak may be more closely associated with the volcanics in the vicinity of Steamboat Springs than with those of the Elkhead Mountains volcanic field.

AGE OF THE VOLCANISM

The volcanism in the Elkhead Mountains is either late Browns Park and/or North Park in age. A provincial age determination is impossible until Tertiary stratigraphers can agree on the provincial ages of the two formations. For many years the Browns Park formation has been considered as being Upper Miocene and the North Park as definitely Pliocene. A number of vertebrate paleontologists, however, have questioned these age determinations. Additional faunal collections from areas adjacent to the type areas have prompted a re-evaluation of the faunal collections upon which the formational age determinations are based.

A suggestion is now before the Committee on the Continental Cenozoic of the Society of Vertebrate Paleontologists to consider the Browns Park formation as Middle Miocene (Hemingfordian), and, perhaps, partly Lower Miocene (Arikareean) in age. The North Park formation is now known definitely to be at least partly Upper Miocene (Barstovian) in age (McGrew, 1955). If these modifications are accepted the volcanism in the Elkhead Mountains must be considered as occurring between Middle Miocene and Lower Pliocene time.

ECONOMIC GEOLOGY

The Hahns Peak mining district has been an intermittent producer of gold, silver, copper, and lead since 1866. The principal product until approximately 1895 was placer gold. Since that time attempts have been made to produce lode deposits of gold, silver, lead, copper, zinc, and molybdenum which are found within the acidic intrusives and the adjacent sedimentary and pre-Cambrian rocks. Most ventures failed to recover their investments. Mining activity is still in progress but on a very limited scale. The total value of all mineral production from the district through 1954 has been estimated as between $400,000 and $500,000 (Barnwell, 1955, pp. 74-84).

In addition to the Hahns Peak area, placer gold has been observed at a number of localities throughout the Elkhead Mountains. Several gold-bearing placer deposits are indicated on the original U. S. General Land Office survey plats.

Recently, the Elkhead Mountains have received considerable attention by the oil industry. Only future drilling can evaluate the oil and gas potential of the Paleozoic and Mesozoic sediments buried beneath the volcanic field.

REFERENCES

Barnwell, W. W. (1955), "The Geology of the South Hahns Peak District," Unpublished masters thesis, Univ. of Wyoming, Laramie, Wyo.

Chisolm, F. F. (1887), "The Elk Head Anthracite Coal Field of Routt Co., Colorado," Colo. Sc. Soc. Proc., no. 2, pp. 147-149.

Cross, Whitman (1904), "Analyses of Rocks From the Laboratory of the United States Geological Survey, Tabulated by F. W. Clarke," U. S. Geol. Survey Bull., no. 228, p. 187.

Draper, Marshall (1897), "Hahns Peak Mining Region, Colorado," Colliery Engineer and Metal Miner, vol. 17, no. 10, May, pp. 437-438.

Emmons, S. F. (1877), "Descriptive Geology," Rept. U. S. Geol. Expl. 40th Par. (King Survey), vol. 2, pp. 167-180.

Gale, H. S. (1905), "The Hahns Peak Gold Field, Colorado," U. S. Geol. Survey Bull., no. 285, pp. 28-34.

George, R. D., and Crawford, R. D. (1909), "The Hahns Peak Region, Routt County, Colorado," Colo. Geol. Survey, 1st Rept., 1908, pp. 189-229.

Henderson, C. W. (1926), "Mining in Colorado," U. S. Geol. Survey Prof. Paper no. 183, 263 pp.

Hunter, J. M. (1955), "The Geology of the North Hahns Peak District," Unpublished masters thesis, Univ. of Wyoming, Laramie, Wyo.

King, Clarence (1878), "Systematic Geology," Rept. U. S. Geol. Expl. 40th Par., vol. 1, pp. 545-726.

Lakes, Arthur (1909), "The Hahns Peak Mining Region, Routt County, Colorado," Mining Science, vol. 60, Sept. 30, pp. 292-296.

Lee, H. E. (1901), Second Report of the Bureau of Mines, Colorado, pp. 173-174.

McGrew, P. O. (1955), Personal communication, Univ. of Wyoming, Laramie, Wyo.

Parsons, H. F., and Liddell, C. A. (1903), "The Coal and Mineral Resources of Routt County, Colorado," Colo. Sch. Mines Bull., vol. 1, no. 4, pp. 47-59.

Ross, C. S. (1926), "A Colorado Lamprophyre of the Verite Type," Am. Jour. Sci., 5th ser., vol. 12, Sept., pp. 217-229.

Vanderwilt, J. W. (1947), "Metals, Nonmetals, and Fuels, Pt. 1 of Mineral Resources of Colorado," Colo. Mineral Resources Board Bull., pp. 1-290.

White, C. A. (1889), "On the Geology and Physiography of a Portion of Northwestern Colorado and Adjacent Parts of Utah and Wyoming," U. S. Geol. Survey An. Rept., no. 9, pp. 677-712.

White, E. L. (1906), Report of the Bureau of Mines, Colorado, 1905-1906, 127 pp.

Worcester, P. G. (1919), "Molybdenum Deposits of Colorado," Colo. Geol. Survey Bull., no. 14, pp. 78-80.

Zirkel, Ferdinand (1876), "Microscopial Petrography," Rept. U. S. Geol. Expl. 40th Par. (King Survey), vol. 6, pp. 159-162, 256-258.

A REVIEW OF THE BROWNS PARK FORMATION

By BYRL D. CAREY, JR.
The California Company, Casper, Wyoming

INTRODUCTION

Brown's Park, located in the extreme northwest corner of Colorado and the adjacent northeast corner of Utah, is the type area for a mid-Tertiary formation which crops out extensively throughout the region. It was from this mountain park that Powell (1876, p. 44) described the Browns Park group, now known as the Browns Park formation.

The Browns Park is of considerable interest to the practicing geologist in that much of the middle and late Tertiary history of the region must be deciphered from evidence obtained from beds of this formation. Two events in the Tertiary geologic history of northwest Colorado which are intimately connected with several aspects of the Browns Park formation are (1) the beginning of the extensive volcanism, the effects of which modify much of the present topography of the region, and (2) the post-Browns Park collapse of the east end of the Uinta Mountain arch and the west end of the Axial Basin anticline, which formed the Uinta Mountain graben.

CHRONOLOGY OF INVESTIGATIONS

Sedimentary rocks which are now included in the Browns Park formation were examined by members of all three of the early regional surveys (Hayden, 1871; King, 1877; Powell, 1876; and White, 1878, 1889). Following the early surveys a number of geologists became concerned with the Browns Park formation while engaged in the detailed mapping of portions of northwest Colorado (Gale, 1910; Sears, 1924; Hancock, 1925; and Bradley, 1936, 1945).

The reports of the early surveys generated considerable controversy as to the relationship that the course of the Green River exhibited with the Uinta Mountain Uplift. After more details of the geology of the region became known, it was found that the partial burial of the Uinta Mountains by the Browns Park formation was influential in determining the present course of the Green River (Hancock, 1915, and Sears, 1924a).

For many years the Browns Park formation as described from the eastern end of the Uinta Mountains was thought to be the same as the Uinta formation of the Uinta Basin and the two names were used interchangeably. Irving (1896) was the first geologist to present fossil evidence that the Browns Park formation was not Eocene, but probably was either Miocene or Pliocene in age. It was not until over thirty years later that sufficient faunal collections were found to substantiate this early age determination (Peterson, 1924, 1928). Although most vertebrate paleontologists agree that the age of the formation is Miocene and/or Pliocene, refinements to this age assignment are still being made (McGrew, 1951, 1953, and 1955).

AREAL DISTRIBUTION OF THE FORMATION

The principal exposures of the Browns Park formation extend from Brown's Park east along the Uinta Mountain graben and the Axial Basin anticline to a locality in T. 5 N., R. 94 W., east of Juniper Mountain. These outcrops continue eastward along the south flank of the Sand Wash Basin to Cedar Mountain, northwest of Craig, Colorado. Outcrops of considerable areal extent have been described and, in some cases, mapped throughout the Sand Wash Basin between Cold Springs Mountain and Cherokee Ridge, along the Cherokee Ridge between Baggs, Wyoming, and Powder Wash, Colorado, along the north flank of the White River Uplift in the vicinity of Pagoda, and throughout the Elkhead Mountains. Among the detailed geologic maps which depict part of these outcrops are those by Gale (1910, plates XVI, XVIII), Hancock (1925, plate XIX), Sears (1924, plate XXXV), and Bradley (1936, plate 34; 1945).

The recently published geologic map of Wyoming indicates outcrops of Browns Park formation over a sizeable area along the west and north flanks of the Sierra Madre Mountains and along the north end of the Saratoga Valley (Love, et al., 1955).

The thickest deposits of the Browns Park formation are generally found either in down-dropped areas associated with extensive normal faulting as in the Uinta Mountain graben (Bradley, 1936, p. 184), or preserved beneath high-level lava flows as at Mount Welba in the Elkhead Mountains.

LITHOLOGY

The Browns Park formation has been separated in the literature into two units. The lower unit is a conglomerate of variable thickness which consists primarily of pre-Cambrian cobbles and is termed the basal Browns Park conglomerate. The upper unit is composed of beds of chalky-white and grayish-white sandstone, tuffaceous sandstone, thin beds of chert, and occasional beds of vitric tuff and fresh water limestone.

Most of the sandstones in the upper unit are soft, friable, and calcareous.

Along the south flank of the Sand Wash Basin and east of the Little Snake River, the upper part of the formation is highly cross-bedded and suggests a windblown deposit (Bradley, 1936, pp. 182-183; Sears, 1924, p. 295). At these localities, the cross laminations are characterized by extreme variability in both direction and amount of dip. Dips up to 32° are quite common. North of Powder Wash, the sand grains found in the upper part of the formation contain frosted and pitted surfaces and are generally well rounded and sorted. Bradley (1936, p. 183) believes that the sand grains from this area were well rounded in the formation from which the Browns Park was derived, and that the grains had been secondarily enlarged before they were reworked in the Browns Park formation.

The basal conglomerate of the Browns Park formation generally consists of cobbles and pebbles of rock-types comprising the pre-Cambrian core of the nearest mountain range. Near the east end of the Uinta Mountains, Bradley (1936, p. 182) describes the conglomerate as consisting essentially of cobbles of Uinta Mountain quartzite. In the Axial quadrangle, Hancock (1925, p. 24) describes the conglomerate as consisting of pebbles of schist, gneiss, coarse and fine-grained granite, white and reddish quartzite, and white and reddish vein quartz.

The thickness of the basal conglomerate has been reported as varying from 0 to 300 feet. The upper unit was reported by Sears (1924, p. 286) as having an aggregate thickness of 1200 feet, however, a thickness in excess of this estimate has been penetrated in drilling within the Uinta Mountain graben. The estimate by Powell (1876, p. 40) of 1800 feet for the total thickness of the formation appears to be fairly representative for northwestern Colorado.

VOLCANICS CONTAINED IN THE BROWNS PARK FORMATION

According to Bradley (1936, p. 183) the lower part of the Browns Park formation along the south side of the Washakie Basin contains considerable amounts of glass tuffs. The glass shards in these tuffs have a refractive index ranging from a little below to a little above 1.50. With the shards are found a moderate quantity of mineral fragments, among which the heavy minerals consist chiefly of biotite and magnetite with lesser amounts of apatite and zircon. Feldspars are rare, but grains of sanidine are found together with plagioclase crystals that range in composition from calcic oligoclase to calcic andesine. The mineral suite represented by the fragments points to a source area in which the volcanic rocks ranged in composition primarily from andesites to rhyolites.

The volcanics observed in the upper part of the Browns Park formation have a different composition. Hancock (1925, p. 24) describes a Browns Park type sandstone capped by a volcanic conglomerate, which is exposed about 6 miles southeast of Pagoda and about 100 feet below the top of the basalt flows that form the Flat Tops. The conglomerate is composed entirely of well-rounded pebbles and boulders of basalt, some of which are as much as 18 inches in diameter.

In the Elkhead Mountains a transition zone several hundred feet in thickness is present in which typical Browns Park sandstones grade upward into volcanic breccias consisting of basalt boulders imbedded essentially within a Browns Park sandstone matrix.

Several pure, bluish-gray beds of volcanic ash are exposed near the top of the Browns Park section in the canyon where Moffat County 318 crosses Vermilion Creek. These beds are considerably thicker and appear to be of a different chemical composition than those found in the lower part of the section. (Field conference will pass these beds.)

Hancock (1915, p. 187) reported finding a portion of the Browns Park sandstone resting upon the basalt flow which caps Cedar Mountain northwest of Craig. The sandstone contains numerous rounded masses of the basalt and suggests that the basalt on the mountain is upper Browns Park in age.

AGE OF THE BROWNS PARK FORMATION

The Browns Park sediments have been found to overlie unconformably all formations from the pre-Cambrian through the Middle Eocene Bridger formation. Considerable differences existed among the early geologists, however, as to a more exact age determination, and the latitude of opinion ranged from Upper Eocene (King, 1877, p. 222) to Pleistocene (Sinclair, 1906, p. 278). The scarcity of fossils required that most age arguments be based upon physical criteria.

The first collection of diagnostic vertebrate fossils was found near Sunbeam, Colorado. A study of the fauna resulted in dating the formation between uppermost Oligocene and middle Miocene (Peterson, 1924, p. 300). Additional fossils found near the Greystone post office prompted Peterson (1928, p. 88) to change the age to upper Miocene or lower Pliocene. The younger age was based essentially upon the discovery of a primitive new species of probescidean from a stratigraphic horizon fairly high in the section. The U. S. Geological Survey has tentatively classified the formation according to Peterson's age determination (Wilmarth, 1938, p. 275).

Additional faunal collections from the Browns Park formation in south-central Wyoming (McGrew, 1953, pp. 62-64; 1955) have caused some vertebrate paleontologists to favor a lower and middle Miocene age for at least part of the formation.

The relationship of the volcanics to the Browns Park formation, particularly at Cedar Mountain, along with the available fossil evidence, suggest that the upper part of the Browns Park formation near the type area is equivalent, at least in part, to the North Park formation exposed at North Park in north-central Colorado and Saratoga Valley in south-central Wyoming.

A suggestion is now before the Committee on the Continental Genozoic selected by the Society of Vertebrate Paleontologists to consider the Browns Park formation as Middle Miocene (Hemingfordian) and, perhaps, partly Lower Miocene (Arikareean) in age. The North Park formation is now known definitely to be at least partly Upper Miocene (Barstovian) in age (McGrew, 1955).

At one time, it was believed (Sears, 1924, p. 296) that the Bishop conglomerate was the basal conglomerate of the Browns Park formation. Bradley (1936, pp. 181-182) has demonstrated, however, that the Gilbert Peak erosion surface, upon which the Bishop conglomerate was deposited, is distinctly older than the Bear Mountain erosion surface upon which the basal Browns Park conglomerate was deposited.

ENVIRONMENT OF DEPOSITION

Bradley (1936, p. 184) believes that the basal conglomerate of the Browns Park formation is part of a shifting gravel mantle which was deposited as a result of a change in the climate to greater aridity. As evidence supporting the arid conditions under which much of the sandstones were deposited, he cites the occurrence of halite molds in beds exposed near the west end of Browns Park and the presence of wind-faceted cobbles, wind-blown sand, and possible dune deposits, in part of the formation exposed along the south side of the Washakie Basin.

The thick sections of Browns Park found in the down-faulted areas may be partially attributed to subsidence of the fault troughs contemporaneous with deposition. Bradley reports (1936, p. 185) that during deposition of the Browns Park formation, the east end of the Uinta Mountain arch began to fail. A local angular unconformity within the Browns Park formation is exposed a few miles west of Ladore post office. The major collapse of the Uinta Mountain Arch, however, occurred in post-Browns Park time. Sears (1924, pp. 287-298) has discussed in detail the evidence for this collapse and described the extent of the resulting graben.

REFERENCES

Bradley, W. H. (1936), "Geomorphology of the North Flank of the Uinta Mountains," U. S. Geol. Survey Prof. Paper 185-I, pp. 182-185, plate 34.

——— (1945), "Geology of the Washakie Basin, Sweetwater and Carbon Counties, Wyoming and Moffat County, Colorado," U. S. Geol. Survey Oil and Gas Inv., Prel. map no. 32.

Gale, H. S. (1910), "Coal Fields of Northwestern Colorado and Northwestern Utah," U. S. Geol. Survey Bull. no. 415, pp. 86-89, plates xvi, xviii.

Hancock, E. T. (1915), "The History of a Portion of Yampa River, Colo., and Its Possible Bearing on That of Green River," U. S. Geol. Survey Prof. Paper 90-K, pp. 183-189.

——— (1925), "Geology and Coal Resources of the Axial and Monument Butte Quadrangles, Moffat County, Colorado," U. S. Geol. Survey Bull. no. 757, pp. 23-25, plate xix.

Hayden, F. V. (1871), "Preliminary Report of the United States Geological Survey of Wyoming and Portions of Contiguous Territories," Second Ann. Rept. U. S. Geol. Survey, p. 64.

Irving, J. D. (1896), "The Stratigraphic Relations of the Brown's Park Beds of Utah," N. Y. Ac. Sci., Trans. 15, pp. 252-259.

King, Clarence (1877), "Systematic Geology," Rept. U. S. Geol. Expl. 40th Par., vol. 2, p. 222.

Love, J. D., Weitz, J. L., and Hose, R. K. (1955), Geologic Map of Wyoming, U. S. Geol. Survey.

McGrew, P. O. (1951), "Tertiary Stratigraphy and Paleontology of South-central Wyoming," Wyo. Geol. Assoc. Guidebook, 6th ann. field conf., pp. 54-57.

——— (1953), "Tertiary Deposits of Southeastern Wyoming," Wyo. Geol. Assoc. Guidebook, 8th ann. field conf., pp. 61-64.

——— (1955), Personal communication, Univ. of Wyoming, Laramie, Wyo.

Peterson, O. A. (1924), "Discovery of Fossil Mammals in the Brown's Park Formation of Moffat County, Colo.," Carnie Mus. Annals, vol. 15, p. 300.

——— (1928), "The Brown's Park Formation," Carnegie Mus. Mem., vol. 11, no. 2, pp. 87-130.

Powell, J. W. (1876), "Report on the Geology of the Eastern Portion of the Uinta Mountains and a Region of Country Adjacent thereto," U. S. Geol. and Geog. Survey Terr., pp. 40, 44, 168-169.

Sears, J. D. (1925), "Geology and Oil and Gas Prospects of Part of Moffat County, Colorado, and Southern Sweetwater County, Wyoming," U. S. Geol. Survey Bull. no. 751-G, pp. 295-296, plate xxxv.

——— (1924a), "Relations of the Browns Park Formation and Bishop Conglomerate, and Their Role in the Origin of Green and Yampa Rivers," Geol. Soc. America Bull., vol. 35, pp. 279-304.

Sinclair, W. J. (1906), "Volcanic Ash in the Bridger Beds of Wyoming," Bull. Amer. Mus. Nat. Hist., vol. XXII, p. 278.

White, C. A. (1878), "Report on the Geology of a Portion of Northwestern Colorado," U. S. Geol. and Geog. Survey Terr., Tenth Ann. Rept., p. 38.

——— (1889), "On the Geology and Physiography of a Portion of Northwestern Colorado and Adjacent Parts of Utah and Wyoming," Ninth Ann. Rept. U. S. Geol. Survey, p. 691, plate LXXXVIII.

Wilmarth, M. G. (1938), "Lexicon of Geologic Names of the United States," U. S. Geol. Survey Bull. no. 896, pp. 194, 275.

CORRELATION OF CENOZOIC DEPOSITS OF NORTHWESTERN COLORADO

By M. DANE PICARD
Salt Lake City, Utah
and
PAUL O. McGREW
University of Wyoming, Laramie, Wyoming

This paper is a cooperative project between the Intermountain Association of Petroleum Geologists and the Wyoming Geological Association, in an attempt to correlate the Cenozoic deposits of northwestern Colorado with surrounding areas. This information is presented as a correlation chart (Fig. 1).

In most studies of Tertiary rock units, the number of dashed lines, question marks, and other noted anomalies of correlation runs high. They often exceed the number of definitely known age determinations. The present effort is no exception.

MECHANICAL FEATURES OF CHART

The columns of the chart are arranged so that they encircle the major area of interest (northwestern Colorado). Similar stratigraphic sequences are shown in the same or contiguous columns. Where part of the section has been removed by erosion, or was never deposited, a vertical-line pattern is used.

STRATIGRAPHIC NOTES

General

In many cases the time correlation of specific units between basins is highly tentative because of the lack of known vertebrate faunas. Absolute time correlation across the chart is not possible at this time and may never be. Constant revisions will be made in the future.

Fort Union Formation

The Fort Union formation is approximately 700 feet thick at the Hiawatha field in northwestern Colorado.

"Wasatch Group"

The Wasatch Group (Paleocene-Eocene) and correlative units are in a state of flux. Essentially, the term "Wasatch" indicates an environment of deposition (predominantly fluvial) and what is believed to be, by stratigraphic position, older Tertiary strata. The stratigraphy of "Wasatch type rocks" must be studied in more detail in limited areas before a consistent regional picture can be worked out.

The writers have departed from previous correlation charts (Jones, et al., 1954) in extending the base of the "Wasatch Group" in the Uinta basin downward to the top of the Upper Cretaceous. This was done primarily because of comparable stratigraphic thicknesses in the central part of the Uinta basin and the Wasatch Plateau area. In the Wasatch Plateau Spieker (1946) has demonstrated that the Cretaceous-Tertiary boundary occurs within the North Horn formation. It also seems possible that this system boundary occurs within the "Wasatch Group" of the central Uinta basin area. Stratigraphic relationships on the eastern edge of the Uinta basin are not clear.

A maximum thickness of 1,750 feet has been given (Nightingale, 1930) for the Cathedral Bluffs tongue along Kinney Rim in northwestern Colorado. The main body of the Wasatch formation is approximately 4500 feet thick at the Hiawatha field.

Green River Formation

The Green River formation of the Rocky Mountains is a complex stratigraphic unit. Fundamentally the formation is the product of fluctuating lacustrine and fluvial environmental conditions (Bradley, 1931, Dane, 1954, and Picard, 1955).

Bradley (1945) measured 1200 feet of Laney shale member in the vicinity of Lookout Mountain, Colorado. In the same area the Tipton Tongue member was approximately 300 feet thick.

Bishop Conglomerate

Scattered remnants of the Bishop conglomerate (Miocene) are present in northwestern Colorado. The exact age of the formation is very much in doubt.

Browns Park Formation

The Browns Park formation (Miocene (?)) is discussed elsewhere in this guidebook.

CORRELATION TABLE OF CENOZOIC FORMATIONS OF NORTHWESTERN COLORADO AND ADJACENT AREAS
COMPILED BY: M. DANE PICARD AND PAUL O. McGREW
APRIL, 1955
FIGURE —1—

SELECTED REFERENCES

Bradley, W. H. (1931), "Origin and Microfossils of the Oil Shale of the Green River Formation of Colorado and Utah," U. S. Geol. Survey Prof. Paper 168, 58 pp.

——— (1945), "Geology of the Washakie Basin, Sweetwater and Carbon Counties, Wyoming, and Moffat County, Colorado," ibid, Oil and Gas Investigations Prel. Map 32.

Dane, Carle H. (1954), "Stratigraphic and Facies Relationships of Upper Part of Green River Formation and Lower Part of Uinta Formation in Duchesne, Uintah, and Wasatch Counties, Utah," Bull. Amer. Assoc. Petrol. Geol., vol. 38, no. 3, pp. 405-25.

Jones, D. J., Picard, M. D., and Wyeth, J. E. (1954), "Correlation of Non-Marine Cenozoic of Utah," ibid, vol. 38, no. 40, pp. 2219-22.

Merriam, D. F. (1953), "Tertiary Geology of the Piceance Basin," The Compass, vol. 31, no. 3, pp. 155-71.

McGrew, P. O. (1950), "Tertiary Vertebrate Fossils of the Green River Basin," Guidebook, Fifth Annual Conference, Wyoming Geol. Assoc., pp. 68-74.

——— (1953), "Tertiary Deposits of Southeastern Wyoming," Guidebook, Eighth Annual Conference, Wyoming Geol. Assoc., pp. 61-65.

Nightingale, W. T. (1930), "Geology of Vermillion Creek Gas Area in Southwest Wyoming and Northwest Colorado," Bull. Amer. Assoc. Petrol. Geol., vol. 14, no. 8.

Picard, M. Dane (1955), "Subsurface Stratigraphy and Lithology of Green River Formation in Uinta Basin, Utah," Bull. Amer. Assoc. Petrol. Geol., vol. 39, no. 1, pp. 75-102.

Spieker, Edmund M. (1946), "Late Mesozoic and Early Cenozoic History of Central Utah," U. S. Geol. Survey Prof. Paper 205-D, pp. 117-162.

Wiggins Studio, Craig

Amphitheatre Mountain and Trappers Lake in the Flat Tops.

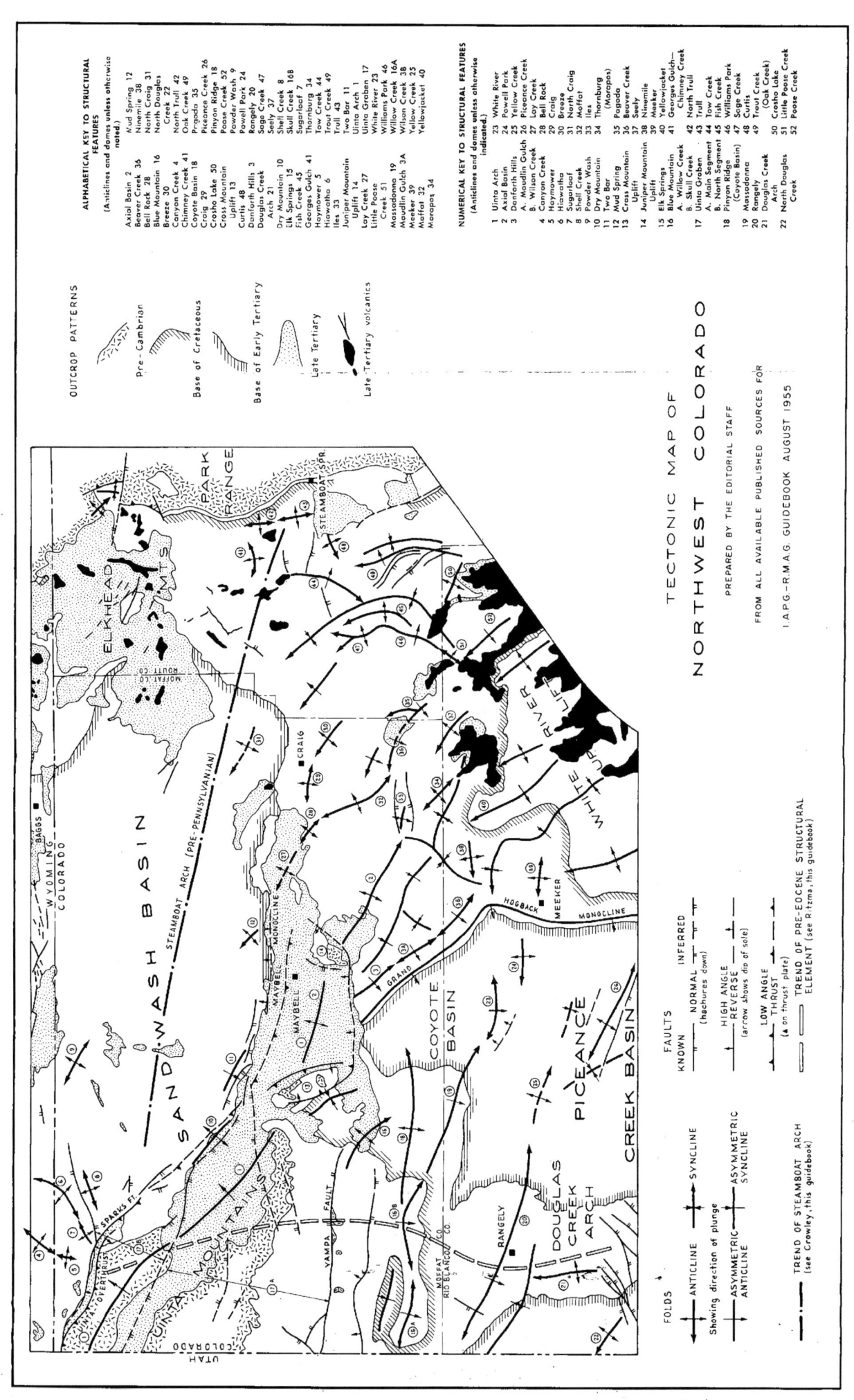

A STRUCTURAL HISTORY OF NORTHWESTERN COLORADO AND PARTS OF NORTHEASTERN UTAH

By A. J. CROWLEY

Continental Oil Company, Denver, Colorado

FOREWORD

A resume of structural history such as follows, is more fitting for a lengthy dissertation of the professional paper type. In a discussion of this sort, necessarily limited in length, it is virtually impossible to cite all evidence at hand; consequently the discussion, without doubt, appears dogmatic. The conclusions drawn and presented herein are well supported by field evidence, and in many cases represent the results of more than ten years of continuing investigation of the region by the writer.

Briefly, the discussion is confined to regional movements and generalities. To specifically describe and discuss each fold and fault system would require enough space to completely fill this publication. Therefore, the writer, while apologizing for the rather high handed treatment of the subject matter, feels that the best possible method of attack has been chosen in view of the limitations. He regrets that this does not include a more specific discussion of local features such as Wilson Creek, Rangely, or the Douglas Creek Arch, but believes the major tectonic features are more important to the structural history of the region than are the subsidiary folds.

HISTORY

Paleozoic

The Paleozoic structural history of Northwestern Colorado is but little understood because of the depth of burial of this system of rocks in most places. This has precluded widespread drilling in most of the area and what we may infer is largely from outcrops of these beds which are at best only peripheral to the region and are probably not truly representative of those buried beneath thick Mesozoic and Tertiary sediments.

From outcroppings in the general region of Steamboat Springs we may infer that pre-Pennsylvanian uplift took place in this region. By studying the east side of the Rocky Mountains, in the general vicinity of Boulder, Colorado, and northward and southward from both Boulder and Steamboat Springs, we may further infer that the direction of the axis of this uplift was roughly east-west and that it was probably a western extension of ancient Siouxia. From studying this uplift in well borings in Eastern Colorado and Nebraska, we may still further infer that the folding was probably pre-Mississippian in age; for various members of the Mississippian on-lap this ancient uplift. Westward of the general area under discussion, we find our first real evidence of pre-Mississippian movement. In the Uinta Mountains, in the canyon of the Whiterocks River, Mississippian rocks of Burlington-Keokuk to Kinderhook age rest unconformably upon Cambrian shales. The unconformity is angular and the angle is approximately eleven degrees. Thus, from regional inference we can say that movement took place in the general area in post-Cambrian and pre-Mississippian time. Further than this we cannot go at present. Thus, we begin our structural history of Northwestern Colorado with an east-west trending arch through the vicinity of Steamboat Springs which had its probable beginnings in post-Cambrian - pre-Mississippian times, and persisted well into the Pennsylvanian, probably through rejuvenation. The arch was a broad feature and its effects were felt at least as far south as Township 1 North, while to the north it probably affected areas north of the Wyoming-Colorado line. For purposes of convenience, and for lack of a better name, it will be referred to as the "Steamboat Arch".

We have no evidence whatever for tectonic movements in Northwestern Colorado during Mississippian time. This period seems to have been a quiet one and it resulted in the deposition of marine limestones, except over previously uplifted areas.

The Pennsylvanian opened as a time of little movement and, during the Morrowan and part of the Atokan, black shales were widely deposited. Following the deposition of these shales widespread uplifting took place throughout the general Colorado Rocky Mountain chain. The Pennsylvanian movement, which seemingly began in early Des Moines time, had little effect upon the western portion of the region and, except for those parts of Northwestern Colorado adjacent to the present Rocky Mountain chain and the Uncompahgre Uplift, it might better be termed a period of subsidence and marine deposition. A large evaporite basin was developed in the southeastern part of the area under discussion, and seems to have centered geographically somewhere near the northwest edge of the present White River Plateau. The salt basin possibly formed in an embayment between the ancestral Uncompahgre Uplift to the south, and the Steamboat Arch to the north.

During Pennsylvanian time the western part of the area was receiving more or less continuous marine deposition, but the southern and eastern parts of Northwest Colorado were covered by conglomerates and arkoses derived from the newly formed uplifts.

These conditions seem to have persisted with but minor interruptions throughout nearly all of the early and middle Pennsylvanian.

Toward the close of the Pennsylvanian the seas began to recede and the typically marine limestone, shale, and evaporite deposits gave way to clastics. These clastics, predominantly arkoses adjacent to the uplifts, graded laterally into fine grained, cross-bedded sandstones to the west and northwest. These conditions gave origin to the typical Weber sandstone and seem to have persisted well into Wolfcamp time at the close of Weber deposition and the beginning of the marine encroachment responsible for the deposition of the Phosphoria formation in the northwestern part of the area.

Permian and Triassic

Save for some minor movements, and possible rejuvenation of parts of the Uncompahgre Uplift, the Permian seems to have represented a period of quiescence and stable, uniform conditions. The Triassic likewise appears to have been a period of little fluctuation although the seas appear to have advanced and receded several times during Moenkopi time without leaving much trace except for a few marine fossils and some evaporites. One period, and only one during the Triassic appears to record any sharp movement or uplift. This was during the deposition of the Shinarump conglomerate. This formation is so widespread and so relatively uniform in thickness that it must represent the widespread, spasmodic and contemporary rejuvenation of many uplifts that supplied the sediments. It appears to be a piedmont type of deposit formed by the coalescence and the reworking of alluvial fans. The angular character of its constituent grains belies the idea of any considerable amount of transportation.

Following the deposition of the Shinarump, conditions settled down to what must have been a rather constant rain of volcanic ash upon a low lying platform. This is evidenced by the highly bentonitic Chinle formation. No other evidence of local Triassic vulcanism is known within Northwest Colorado.

Jurassic and Cretaceous

The Jurassic came and went, with the area fluctuating between marine, and low lying non-marine conditions. Only during the deposition of the Curtis formation and, doubtfully, during Carmel time was the area submergent to any degree. For the most part the area received continental sediments, and doubtless, during part of the Jurassic, it contributed sediment to other areas as is attested by the absence of some members present elsewhere, and the relative thinness of those that are present. This is particularly true of the San Rafael Group.

The Cretaceous was by and large a period of submergence with non-marine conditions being fluctuatingly persistent through Dakota (?) and into Frontier time as evidenced by the coal in the latter formation in the Uinta Mountains. Following Frontier time the territory submerged to receive the thick Mancos marine shale deposits.

After receiving up to 5000 feet of Mancos shale in places, the area became again emergent to receive the non-marine Mesa Verde deposits, and about the middle of Mesa Verde time the northern portion of the area again submerged to receive the Lewis shale. After Lewis time the entire area emerged and has remained so to the present.

Tertiary

There is no evidence in Northwest Colorado of mountain making movements between the end of the Pennsylvanian and the close of the Cretaceous. However, sometime shortly after the close of the Cretaceous, the area began to feel the effects of the Laramide revolution and a tremendous epoch of mountain building began. Sometime during the Eocene the Uinta Mountains were first formed and the present Rocky Mountain chain was rejuvenated and modified.

Much faulting and fracturing took place. The southern fault system associated with the Uinta Mountains was then formed. Although it cannot be as accurately dated, the northern system of Uinta faults probably also had its origin at this time. There followed a period of quiescence during which the Uintas, and the Rockies, were peneplaned. The mountains were eroded to old topography and monadnocks were developed. These monadnocks form the topmost peaks of the Uintas at this time. It is probable that the Green River lakes were formed during this period of peneplanation.

This was followed, also in the Eocene, by a second period of elevation, and it is believed that the Uinta and Duchesne River beds were laid down subsequent to this deformation. Faulting does not seem to be a predominant feature of this period of uplifting although rejuvenation of older faults undoubtedly took place. Peneplanation again followed, and such beds as the Duchesne River lapped well up into what are now the mountains. This is well demonstrated at Little Mountain northwest of Vernal, Utah.

This second peneplanation was extremely extensive. Its traces can be found throughout the area under discussion. It extended over the entire Uinta Basin at least as far south as the present face of the Book Cliffs. It is undoubtedly responsible for the high level flatlands in the vicinity of Rabbit Ears Pass, the top of the Gore range, and in many other areas in the mountains of Colorado. It probably extended north to coincide and merge with the peneplain on the south end of the Wind River Mountains in Wyoming. Everywhere it truncates Eocene and older rocks. The Bishop conglomerate was laid down locally on this surface. Local vulcanism apparently took place about the time peneplanation was complete, for the lower part of the Browns Park formation in the Washakie Basin consists mainly of glass tuffs. This vulcanism was possibly the precursor of the next period of movement.

Following this peneplanation there occurred another great period of differential uplift accompanied by collapse. The Rocky Mountain system and the Uinta Mountains reached their present development at that time; subsequent erosion has only partially modified them from what they were at the close of this period of orogeny.

Collapse took place, both in the Uinta Basin and at the east end of the Uinta Mountains. The present day Uinta Basin appears to owe its form to collapse rather than actual differential uplifting of the Uinta Mountains. The visible portions of the second period peneplain on the Roan Plateau are remarkably coincident on sea level datum with the same peneplain on the fore-front of the Uinta Mountains. Moreover the peneplain can be traced very well from the edge of the Book Cliffs down into the Uinta Basin. It has a uniform approximately three degree tilt from south to north until the axis of the basin is reached. In places the peneplain can be observed to be bent to correspond to the degree of this last age of folding, and is a good guide to its intensity. This is most notable in the Vermilion Creek area in Colorado in section 36, T. 10 N., R. 101 W. There the peneplain can be seen tilted 13° S.W. and truncating northeast dipping Mesozoics and Paleozoics.

Among the best evidences of collapse in the Uinta Basin are the many tension cracks, some of which are now filled with gilsonite. The collapse began in the eastern Uintas during the deposition of the Browns Park formation and subsequent to the deposition of the Bishop conglomerate which, in places, is the equivalent of the basal Browns Park conglomerate. A local unconformity showing the beginnings of the collapse can be seen a few miles west of Ladore, Colorado. The collapse continued and much of it took place after the end of deposition of the Browns Park. This collapse involved the whole east end of the Uinta Range and considerable area in Colorado and Wyoming. A reasonable assumption is that the Uinta Basin collapse took place at the same time as did that on the east end of the Uintas. However, only the east Uinta collapse can be dated accurately.

There is no good evidence of collapse adjacent to the present Colorado Rocky Mountain uplift. It is thought that structural features here are due largely to uplift.

Following this last period of deformation there occurred another volcanic period which seems to have persisted well into the Quaternary. North and west of Steamboat Springs it evidenced itself in numerous intrusives, while to the south, flows are much more common. The lava caps of the White River Plateau are typical of the latter.

Abundant balanced rocks, and the lack of recorded earth tremors in general, now indicate that the area is at least temporarily quiescent, and has been for some time.

REFERENCES

Blackwelder, E. (1915), "Post-Cretaceous History of the Mountains of Central Wyoming," Jour. Geol., vol. 18, pp. 193-207.

Bradley, W. H. (1936), "Geomorphology of the North Flank of the Uinta Mountains," U. S. Geol. Survey Prof. Paper 185-i, pp. 163-201.

Hancock, E. T. (1915), "The History of a Portion of the Yampa River, Colorado, Etc.," U. S. Geol. Survey Prof. Paper 90-k, pp. 183-189.

Powell, J. W. (1876), "Geology of the Eastern Portion of the Uinta Mountains," U. S. Geol. and Geogr. Survey of the Territories, 218 pp.

Sears, J. D. (1924), "Relations of the Browns Park Formation, Etc.," Bull. G.S.A., vol. 35, pp. 279-304.

GEOMORPHOLOGY OF NORTHWESTERN COLORADO

By WM. LEE STOKES
University of Utah, Salt Lake City, Utah

INTRODUCTION

First detailed investigations of the topographic features of northwestern Colorado were made by members of the Hayden surveys in 1876 and 1877 (Hayden, 1878). At this time a number of parties traversed the region and accumulated information which was published in 1878. Hayden wrote regarding this work: ". . . the area under investigation was located in the interior of the country, far remote from settlements, and among hostile bands of Ute Indians that attacked three of the parties the previous year." (Hayden, 1878, p. xiii). The information gathered and recorded by White, Endlich, Peale, Wilson, Gannett, Crittenden, and Bechler of the Hayden parties is very complete and still forms a sound basis for understanding the geology and topography of the region. Not to be overlooked is the work of Powell although it applied more specifically to the Uinta Range (Powell, 1876). King's survey of the 40th parallel also included the area but not much was contributed by it. Later investigators include Hancock, (1915); Sears, (1924); and Bradley (1936).

PHYSIOGRAPHIC DIVISIONS

The northwestern corner of Colorado contains portions of three great physiographic provinces; the Colorado Plateau, the Rocky Mountains, and the Great Plains. The area south of the Uinta Range and Yampa Plateau partakes of the characteristics of the Colorado Plateau with generally flat lying sediments and mild structural deformation. The Uinta Range and its depressed eastward continuations are included in the Northern Rockies province while the White River Plateau and Elk Head Mountains are part of the Southern Rockies. The area north of the Uinta Arch is a part of the Great Plains province and the structure and stratigraphy is related to the Green River Basin of Wyoming. The physiographic subdivision of the region reflects the merging of major tectonic elements which have been in existence through much of geologic time.

MINOR PHYSIOGRAPHIC FEATURES

The Hayden surveys of the region were conducted partly to determine the possible utilization of the land by future colonists; and therefore, much attention was paid to details of drainage, water supplies and routes of travel. The geologists found it convenient to discuss the topography in terms of such fundamental land forms as mountains, hills, plateaus, valleys, badlands, parks and basins. They realized that only the Uinta and Park uplifts could properly be classed as mountain ranges; but they recognized smaller mountains such as the Williams River, Elk Head, and even the relatively small Cross and Juniper mountains. For the higher and more impressive flat-topped eminences such as the Yampa and White River areas they reserved the designation of plateaus. Less conspicuous and more subdued are the Danforth Hills and the Gray Hills.

The terms "parks" and "basins" were used in somewhat specialized sense by the Hayden topographers. Parks, according to White, include only those "expansions of the river valleys that contain more or less broad spaces of comparatively level land, a part of which is susceptible of cultivation by irrigation . . . The term basin I shall use for similarly excavated or hollowed spaces in the general surface of the country through which no river or perennial stream flows by which its lands may be irrigated." (White, p. 9).

Along the White River are located Agency Park, Powell Park, and Raven Park; while along the Yampa are Canon Park and Lily Park. Island Park, Echo Park and Browns Park are located on the Green River. The Midland Basin, Axial Basin and Coyote Basin are relatively large tracts of comparatively level land with no through-flowing streams. Most of the names applied by the Hayden Surveys are still in use; but Junction Mountain has become Cross Mountain, Yampa Mountain has become Juniper Mountain, Raven Park is better known as the Rangely area, Yampa Plateau is now known as Blue Mountain, and the Escalante Hills have become Douglas Mountain.

DRAINAGE

Green River, the major tributary of the Colorado, passes through the western part of the area. It is joined by the Yampa which flows westerly through numerous canyons and gorges to the point of confluence at Pats Hole. Further southwest in Utah the Green is joined by the White River which flows roughly parallel with the Yampa and about 30 miles south of it.

The Green River enters Colorado flowing through Browns Park to the point where it turns abruptly southward into Lodore Canyon to cross the Uinta Range. Near the junction of the Green and Yampa is the controversial Echo Park dam site and complex meander patterns.

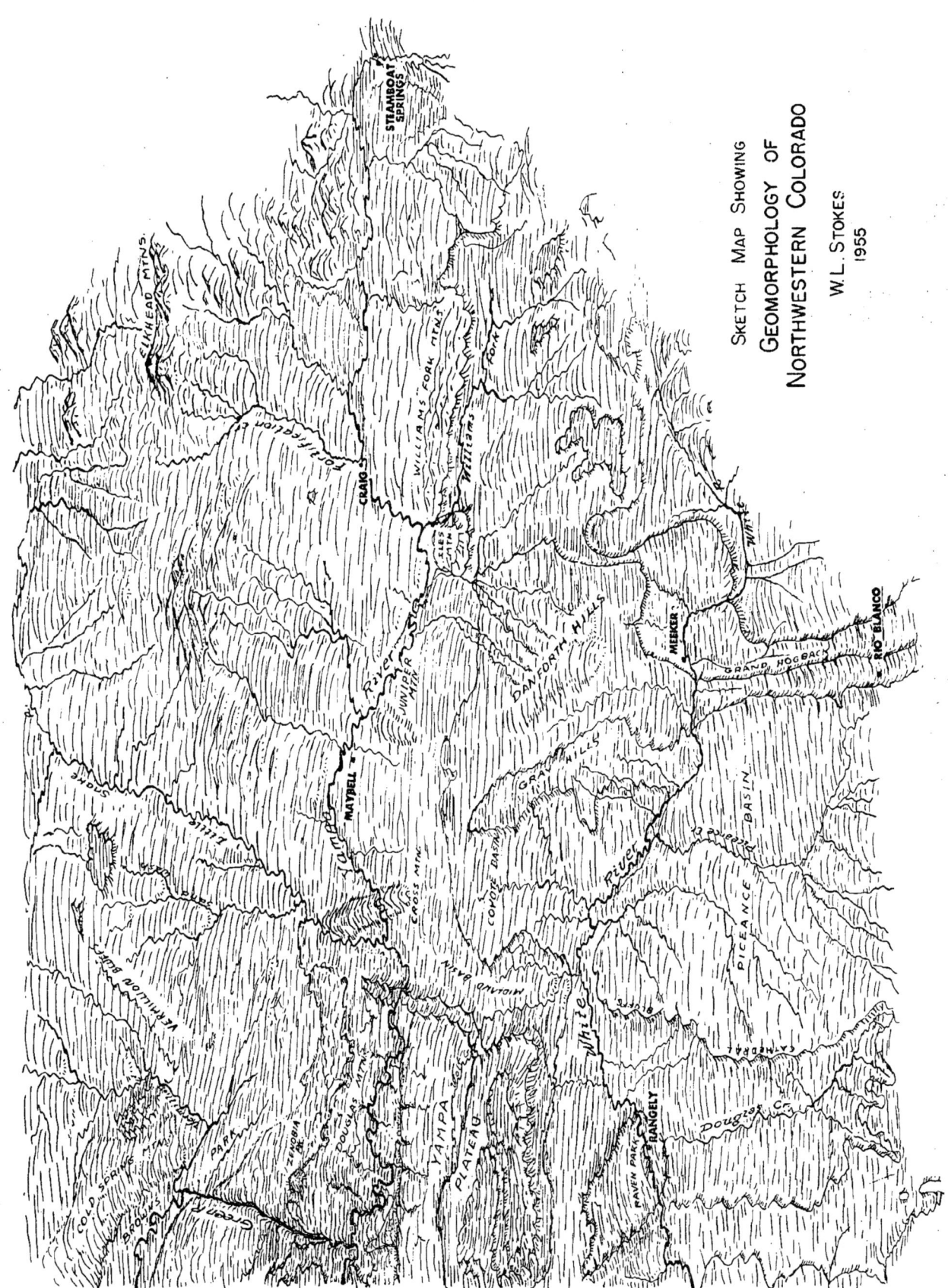

The Yampa River drains about 5000 square miles and has its headwaters in the high Park and Elkhead mountains east of the area. Its chief tributaries are Elkhead Creek, Williams Fork, and Little Snake River. Its course across Yampa Plateau is marked by incised meanders and rugged side canyons.

White River drains an area of about 3,060 square miles. It rises in the White River plateau. Most of the tributaries enter from the south, the most important being South Fork, Piceance Creek, and Douglas Creek. Although it is not so deeply entrenched as the Yampa, there are many long stretches along the course of the White River that are walled by steep cliffs.

CLIMATIC INFLUENCES

Most of northwest Colorado is semiarid. Soil cover is thin, and bare rock surfaces including cliffs, dip slopes, and badlands are common. Differential erosion has left the harder rocks standing in relief almost in direct proportion to their resistance to erosion. Debris is rapidly removed and vegetation is scanty. The major structural features are plainly depicted in the topography. A. C. White observed long ago: "The geologist may go to the top of any of the mountains or higher hills and see the geological structure of the neighborhood spread out below him like a well-drawn picture; and even the geological structure of the more distant parts of the landscape may often be accurately determined from these elevated points of observation". (White, p. 6).

Pedimentation has left many gravel surfaces and there are river terraces along most of the streams. Mature soils have formed along restricted parts of the stream courses and are marked by agricultural towns and isolated ranches.

SPECIAL PROBLEMS

The striking spectacle of rivers plunging directly into and across uplifts, fault scarps and folds regardless of topography or easier routes of travel presented a challenge to the imagination of the early investigators. Hayden apparently was the first to conceive the idea of antecedent streams but the concept was elaborated by Powell and is usually ascribed to him. According to this theory the master streams maintained their courses during and in spite of uplift and were able to cut downward at least as fast as folds or faults came into existence across their paths. This would assume that the Yampa and Green Rivers were once flowing on a nearly level and structurally simple plain and that the folds and faults which they now transect came into existence so slowly that the streams were not diverted.

The antecedent theory failed to account for many of the observed facts and was replaced by the theory of superposition. In 1915 Hancock published his observations on the Yampa River, and concluded that the river was superposed onto the older preexistent topography through a covering of the Browns Park formation. Thus the deep, narrow canyons across Jupiter and Cross mountains were not cut by the river as the uplifts arose across its course but were carved as the river was let down by erosion from a more ancient position on the surface of the Browns Park formation.

Sears (1924) presented additional evidence for the idea of superposition and also gathered proof of a great structural collapse of the eastern end of the Uinta arch which depressed a broad belt lying between the Yampa fault on the south and a system of faults and flexures on the north. This collapse commenced during the deposition of the Browns Park formation and continued for a lengthy time thereafter. On the westward-sloping surface thus created, the Yampa took up a meandering course which was gradually impressed on the older rocks.

Still another interesting physiographic problem is that involving the course of the Green River along Browns Park and through Lodore Canyon. Sears and Bradley have both analyzed this problem with essentially similar results. Ancient Browns Park valley was cut before the collapse of the Uinta arch and during the period that the upper portions of the Green were flowing out of the Green River basin by some unknown eastward route. Later the Browns Park river valley was filled with the Browns Park formation and the Uinta arch collapsed allowing the Yampa to run westward.

After the collapse phase, a tributary of the Browns Park stream working headward into the Green River Basin captured and diverted the upper portion of the ancestral Green River so that it came to flow southward into and along the same course as the ancient Browns Park stream. When this stream first occupied Browns Park it probably continued eastward at least around the end of the present pre-Cambrian exposures or perhaps even out of the region entirely. Several ideas have been proposed to account for its diversion into Lodore Canyon. The first appeals to simple superposition from a higher Browns Park surface. This would require that the Browns Park formation be at least 2000 feet thicker than at present and that the coincidence of the present Green River and the upper part of Browns Park be accidental since the ancient valley would have had no surface expression when the Green flowed across it. The second idea appeals to stream piracy on a level considerably higher than now but not overtopping the shoulders of Lodore Canyon. A northward extension of one of the tributaries of south-flowing Pot Creek

may have cut through a narrow divide and tapped the Green River to turn it into the Lodore Canyon. Bradley points out that the Lodore section is different from the portion above and below in its steeper gradient and lack of meanders.

REFERENCES

Bradley, Wilmot H. (1936), "Geomorphology of the North Flank of the Uinta Mountains," U. S. Geol. Survey Prof. Paper 185-i, pp. 163-201.

Hancock, E. T. (1915), "The History of a Portion of the Yampa River, Colorado, and Its Possible Bearing on That of the Green River," U. S. Geol. Survey Prof. Paper 90-k, pp. 183-189.

Hayden, F. V. (1878), "Colorado and Parts of Adjacent Territories," Tenth Annual Report of the U. S. Geol. and Geogr. Survey of the Territories, 546 pp.

Powell, J. W. (1876), "Geology of the Eastern Portion of the Uinta Mountains," U. S. Geol. and Geogr. Survey of the Territories, Washington, D. C., 218 pp.

Sears, J. D. (1924), "Relations of Browns Park Formation and the Bishop Conglomerate in the Origin of Green and Yampa Rivers," Geol. Soc. Amer. Bull. 35, pp. 279-304.

White, A. C. (1878), "Report on the Geology of a portion of Northwestern Colorado," in Tenth Annual Report of the Geol. and Geogr. Survey of the Territories, pp. 3-59.

ROUTT COUNTY WAR

As the White River Utes made their exodus to Utah, settlers moved close on their heels to take over the former reservation on the Western slope. By the late 1870's and early 1880's cattlemen had located large herds in Routt, Moffat, Rio Blanco and Garfield counties. No sheepmen made homes in the region, but Wyoming sheep were gradually moving along the mountain slopes for summer range.

The clash of the two groups was inevitable. "Deadlines", boundaries of sheep and cattle country, were drawn and violated. Over-grazing and fencing led to violence as the two groups contended for range and water. On one occasion, near Meeker, 2000 sheep were killed when large bands were driven off the range. Sheep camps were shot up, herders ambushed. Reprisal was swift: cattle were shot, ranches raided, barns and haystacks fired. Heavy loss was incurred in human life and property on both sides. No one could be neutral in the conflict; citizens of the area were either "fur or agin" the woolly sheep.

The Federal Government stepped in to regulate grazing rights in the National Forests and end the range war. In 1915 sheep were permitted on the forest reserve where the grazing supervisor saw fit at a ratio of five sheep to one cow; and soon afterward, new areas were opened to sheepmen. Creation in 1912 of Moffat County, largely sheep country, out of Routt County, corrected some inequities and quieted local animosities.

The fight switched to the courts, and the carefully phrased legal term supplanted the six-gun.

State Historical Society of Colorado

GEOLOGY OF CROSS MOUNTAIN, MOFFAT COUNTY, COLORADO

By S. P. KANIZAY

Colorado School of Mines, Golden, Colorado

Cross Mountain, located at the easternmost end of the Uinta Mountains is, according to Forrester (1936), the most eastern structural unit of this range. Two major disturbances are marked in the sedimentary sequence exposed at Cross Mountain: one at the end of the pre-Cambrian, the second during the Laramide orogeny before the Browns Park formation was deposited.

STRATIGRAPHY

Sediments ranging from pre-Cambrian through Tertiary are exposed at Cross Mountain. The stratigraphic section is depicted in Figure 1. An areal geologic map of Cross Mountain is included in the pocket at the back of this guidebook.

Lying unconformably on the pre-Cambrian Uinta Mountain Group are beds tentatively identified as the Cambrian Sawatch formation. These are mainly red and gray quartzites and pebble conglomerates resembling the Sawatch formation of Colorado more than the Lodore shale of the Uinta Mountains. The Devonian Chaffee formation, a thin-bedded sequence of pink, dense, slope-forming dolomitic limestone overlies the Sawatch. The Chaffee grades into the overlying Madison and would be difficult to delimit were it not for an edgewise conglomerate at the base of the Madison.

Ascending the Madison section, the limestone becomes more coarse crystalline and vuggy. The Chaffee and Madison formations both have brown jasperoid chert lenses. The Morgan (Pennsylvanian) consists of three lithologic units: basal cross-bedded sandstone and silty shale; a middle sandstone, limestone and chert sequence and an upper limestone unit.

The Weber formation overlying the Morgan is a tan, cross-bedded sandstone. The overlying Phosphoria is little more than reworked Weber sandstone, horizontally bedded, approximately 20 feet thick. It probably represents the initial stages of the advance of the Moenkopi sea.

The Moenkopi overlying the Phosphoria consists of a basal unit of gray shale and grades upward into tan, very fine-grained sandstones and yellow shales. The dark reddish-black Shinarump conglomerate unconformably overlies the Moenkopi and grades upward into the red sandstones, siltstones and variegated shales of the Chinle formation. A notable change in lithology takes place between the Chinle and the overlying Navajo-Entrada deposits. The Navajo-Entrada here is a massive cross-bedded sandstone. Except for two thin red calcareous shale lenses, there is no evidence of the Carmel formation which, to the west, occurs between the Navajo and Entrada. The two shale units occur about 100 feet above the base which leaves a much thicker sequence above to be credited to the Entrada.

The marine Curtis formation overlies the Navajo-Entrada. It consists of dark greenish gray, sandy shales and limestones, glauconitic and fossiliferous. The basal unit of the Morrison is a clean, white, coarse, cross-bedded massive sandstone. This is overlain by variegated shales and mudstones with some nodular limestone units.

The Dakota lies unconformably on the Morrison and consists of two tan conglomeratic sandstones separated by a shale interval which is largely covered. Mowry shale and Frontier sandstone succeed the Dakota. The Mowry is a black siliceous shale, and the Frontier is a limy sandstone with much intercalated shale. The overlying Mancos shale is almost entirely alternating tan and gray shale with a little sandstone.

The Browns Park formation, the uppermost unit of the section, lies with profound angular unconformity on the older beds. At Cross Mountain two units of the Browns Park are noted: on the west side of the mountain, only the basal conglomerate is present; and on the east, only the upper sand and ash beds are evident. The conglomerate is locally derived and consists mostly of limestone boulders and cobbles with considerable red and brown chert derived from the Morgan and Madison formations. The Browns Park sandstone is ashy, white, extremely cross-bedded and, in general, very loosely cemented.

STRUCTURE

Cross Mountain is a doubly plunging anticline bordered on the east and west by longitudinal rotational faults. On the west side, in the Little Snake River valley, a complimentary syncline is found. Both longitudinal faults are nearly vertical. The western longitudinal fault is normal, dipping to the west, the stratigraphic throw varying from approximately 100 feet on the north to about 5000 feet on the south. On the eastern side, faulting is again rotational, varying from less than 100 feet on the south to approximately 300 feet displacement at the northern end. Thus there is down-dropping of the flanks diagonally opposed. The eastern fault appears to be a reverse fault with dip to the west.

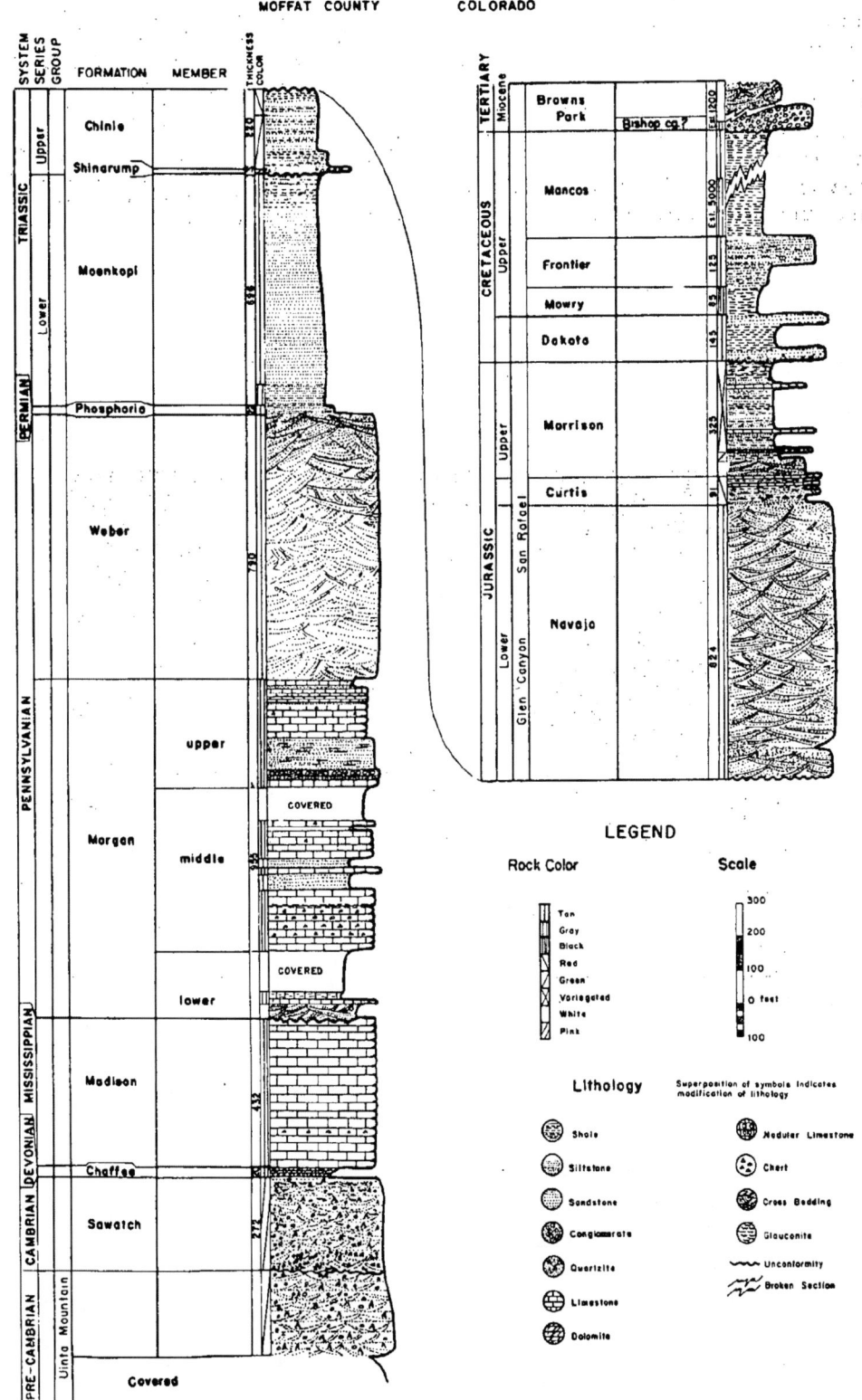

FIGURE 1.

The northern end of the mountain is cut by transverse faults which result in a series of fault blocks. Displacement is small, seldom exceeding 100 feet.

Folding of Cross Mountain largely preceded faulting, and in the later stages was the principal means of stress release. In the center of section 13, T. 6 N., R. 98 W., a fault of relatively small displacement, less than 100 feet, offsets the southern end of the eastern longitudinal fault. This fault appears to be a release fault with slumping of the western block.

Vertical forces acting in couple probably were responsible for the structure of Cross Mountain.

SELECTED BIBLIOGRAPHY

Forrester, J. D. (1948), "Structure of the Uinta Mountains," Geol. Soc. Am. Bull., vol. 48, p. 644.

Sears, J. D. (1924), "Geology and Oil and Gas Prospects of Part of Moffat County, Colorado and Southern Sweetwater County, Wyoming," U. S. Geol. Survey Bull. 751-g, p. 300.

Thomas, C. R., McCann, F. T., and Raman, N. D. (1945), "Mesozoic and Paleozoic Stratigraphy in Northwestern Colorado and Northeastern Utah," U. S. Geol. Survey, Oil and Gas Investigations, Preliminary Chart 16, Sheet No. 2.

Untermann, G. E., and Untermann, B. R. (1949), "Geology of Green and Yampa River Canyons and Vicinity, Dinosaur National Monument, Utah and Colorado," Bull. Amer. Assoc. Petrol. Geol., vol. 33, no. 5, pp. 683-694.

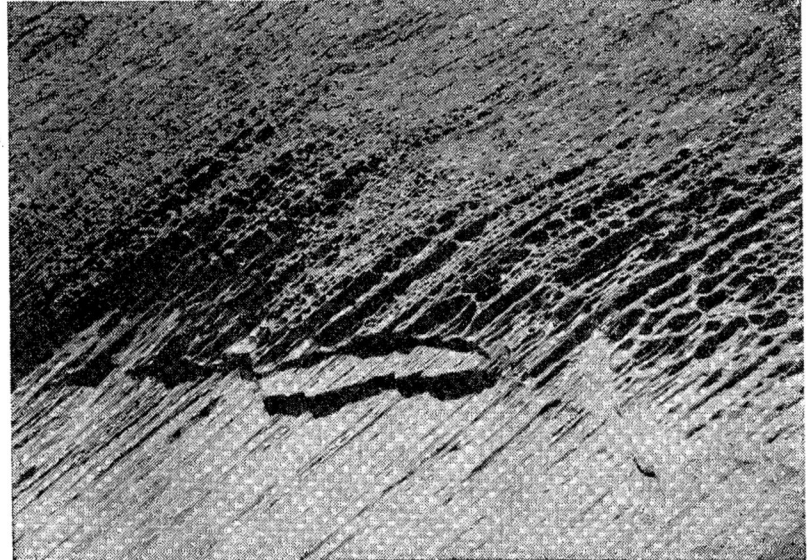

H. R. Ritzma

Wind Fretwork in the Entrada Sandstone, Cross Mountain.

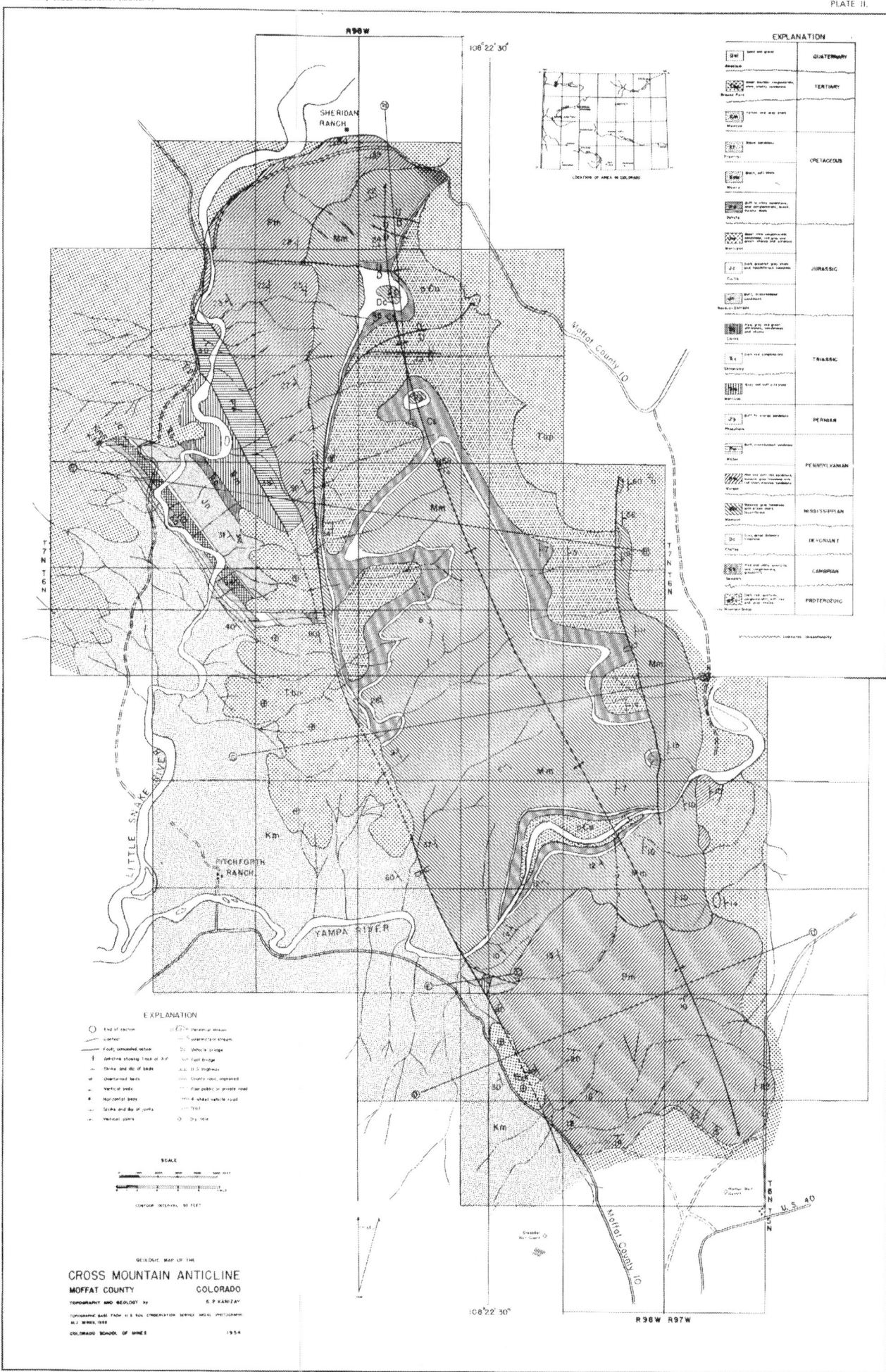

JUNIPER MOUNTAIN AREA, COLORADO*

By C. P. ABRASSART
Sinclair Oil and Gas Company, Casper, Wyoming
and
G. A. CLOUGH
Atlantic Refining Company, Casper, Wyoming

INTRODUCTION

The Juniper Mountain area is in T. 5 and 6 N., R. 94 and 95 W., Moffat County, Colorado. The accompanying areal geologic map includes 51 square miles, and extends about eight miles south of U.S. Highway 40.

Geologically, the Juniper Mountain area is involved in the folding of the Axial Basin anticline, and one axis of Juniper Mountain dome coincides with the axis of the Axial Basin anticline. The area lies between the Sand Wash Basin to the north and the Piceance Creek Basin to the south, 14 miles east of Cross Mountain. Topographic relief is approximately 2000 feet. Maximum elevation is 8000 feet; minimum elevation is slightly below 6000 feet. Juniper Mountain is an isolated mountain mass in an area of low relief. The deep canyon of the Yampa River crosses the northeast corner of the mountain, and canyons of intermittant streams dissect the mountain into ridges that radiate from the top. As a result of the deep canyons, many of the areas of the mountain are not easily accessible. A geologic map is included in the pocket of this guidebook.

PREVIOUS WORK

M. L. Thompson (1945) measured the Pennsylvanian section at the east end of the canyon of the Yampa River, but the measured beds are involved in at least three zones of faulting. R. L. Heaton and I. A. Keyte measured the same section, but generalized their descriptions (Univ. of Colo., Dept. of Geol. files). Thomas, McCann, and Raman (1945) measured a Pennsylvanian section in sections 16 and 17, T. 6 N., R. 94 W.; and the Madison formation and older rocks in section 24, T. 6 N., R. 95 W. H. Aurand (Univ. of Colo., Dept. of Geol. files) measured a post-Pennsylvanian section in T. 6 N., R. 94 W. Sears (1924) mapped the area, but did not differentiate the stratigraphy. Gale (1910) gave a general description of the stratigraphy of the area, and Girty (Gale, 1910) described a fauna from the Madison formation. Hancock (1925) measured a generalized section of Pennsylvanian rocks and identified a fauna.

---

*Partial rewrite of theses presented at Univ. of Colo. towards M.A. degree.

STRUCTURE

Juniper Mountain is a faulted, asymmetric domal fold, approximately 25 miles long. It is the major fold in the immediate area, and represents the culmination of the Axial Basin anticline, which trends northwest. The Axial Basin anticline is on trend with the Uinta Mountain anticline, the eastern nose of which is located about twenty miles to the northwest of Juniper Mountain.

Folds

Juniper Mountain dome is a sharp, large fold. It is asymmetrical, and the beds on the southwest flank dip from 20 to 30 degrees, where not influenced by faulting. The beds of the west flank also dip more steeply than the beds of the east flank. Both Axial Basin anticline and Juniper Mountain dome have similar asymmetry, inasmuch as the dips on the south flank of each structural feature are greater than the dips on the north flank. East of Juniper Mountain dome, the axis of Axial Basin anticline plunges gently to the southeastward.

The uplift and erosion at Juniper Mountain has been great enough to expose pre-Cambrian Uinta quartzite. The tilted strata of the dome crop out over an area seven miles long and three miles wide. The Browns Park formation (early Pliocene) surrounds the dome and is the only formation exposed for fourteen miles west of the dome. It is reasonable to expect that this feature has at least 3500 feet of closure on the Madison formation, and might have more than 5500 feet of closure, inasmuch as the amount of closure is limited only by Collom syncline to the south, and unknown structure to the west. The outcrop pattern of the Iles formation to the northeast and north of Juniper Mountain suggests that the north nose of this dome continues for some distance.

Faults

The dome is transected by two normal faults which occur on the east and west flanks. These faults were probably caused by the cessation of the forces which probably caused this localized, sharp uplift.

The west flank fault (V-V') is about one mile and a half long, strikes north 13 degrees east, has a throw

of about 108 feet, and is well exposed. (See cross-sections.)

Fault (W-W'), which occurs on the east flank of the dome is of unknown length, strikes north 45 degrees east, has at least 900 feet of throw, and is postulated on the basis of field evidence at two outcrops. Displacement of this fault is based on an outcrop of the Morgan formation, which dips 77 degrees east, and is thought to be in the downfaulted block. However, this Morgan outcrop may represent fault drag in the fault zone, and the displacement of the fault may be greater than postulated. The Moore and Gilmore, R. D. Seely No. 1 well, section 2, T. 5 N., R. 95 W., in which the Madison is estimated at a datum of plus 2488 feet, and in which the sedimentary section does not indicate steep dip also substantiates the existence of a large fault on the east flank of the structure.

Thrust Faulting

In section 18, T. 6 N., R. 94 W., the Yampa River has cut a canyon almost perpendicular to the strike of three zones of thrust faulting. The faults trend north and are extremely well exposed in beds of the Morgan formation on the north wall of the canyon. Beds on the east side of all these faults were thrust westward. These faults suggest oblique-directed forces that acted in an east-west direction.

Fault zone (Z-Z') which occurs farthest to the west, exposes two thrust faults in complex relationship. These faults have a total displacement of about 470 feet, and the Morgan formation is in juxtaposition with the Madison formation. This fault zone, which is exposed for about one and a half miles, increases in displacement to the north.

Two other zones of thrust faulting (Y-Y') and (X-X') are present, and these faults are essentially bedding-plane thrust faults. Changes in dip of the thrust fault planes, fault drag, flowage of incompetent beds, and other evidence related to faulting may be observed.

Regional Tectonics

Forrester (1937) wrote the following theory regarding the stresses of the eastern portion of the Uinta Mountains which may have affected the Axial Basin region:

"Deep-seated, localized stresses of the Uinta type, acting from probably a southerly direction, and controlled by the well defined Uinta trough, became progressively smaller as the trough shallowed eastward. These stresses gradually lost their potency to the east-west stresses prevailing throughout the Rocky Mountain region. The result of the combination of these stress relations, together with the southeastward swing of the trough, has caused all the folds in this area to trend southeast."

Forrester (1937) suggested that the Uinta Mountain vertical uplift in late Eocene time caused the warping which formed the Axial Basin anticline. The Axial Basin region adjoins the Uinta Mountain Range on the east, but was not folded during the Laramide Revolution as was the Uinta Mountain Range. Forrester (1937) further believes that the uplift by fault displacements in the Axial Basin region were relatively insignificant. He postulates that release of pressure on the uplifted portion of the Axial Basin anticline followed the warping, because strike-faulting in the northern part of the region caused south sides of the faults, which were the elevated sides during folding, to become the downthrown sides.

STRATIGRAPHY

Precambrian
Uinta Mountain Group

The Uinta Mountain group consists of conglomeratic sandstone, quartzite, and red, sandy shale. The beds are resistant, cliff-forming, and cross-bedded in places; color varies from light to dark red. These Uinta beds, which are unfossiliferous, represent only the uppermost part of a thick series of rocks, estimated by Sears (1924) to be not less than 12,000 feet thick. On the south side of Juniper Mountain, a pink quartzitic sandstone lies immediately below the pronounced disconformable surface that separates the Uinta group from the overlying Cambrian (?) beds. Leith and Van Hise (Forrester, 1937) consider the rocks of the Uinta group to be of Algonkian age, and believe that they resemble somewhat, in stratigraphic position, lithology, and thickness, the Belt Series of south-central Montana.

Cambrian (?)
Cambrian (?) Beds

A sequence of rocks, which ranges from 47 to 203 feet in thickness, overlies the Uinta quartzite and is separated from the Uinta group, disconformably. These rocks consist of variegated shale and sandy shale at the base, conglomeratic sandstone and quartzite above the shales. The shales are purple, red, gray, green and white mottled; they are thin-bedded, lenticular, and vary in thickness, but have a maximum thickness of eight feet. Both the conglomeratic sandstone and the quartzite are cross-bedded, unfossiliferous, resistant, and cliff-forming. On the basis of lithology, stratigraphic position, and apparent lack of fossils, by comparison these basal shale beds closely resemble the Lodore shales of Powell's "Lodore group" as suggested by Sears (1924).

Mississippian
Madison Formation

The Madison formation includes beds of limestone, dolomitic limestone, dolomite, and limy dolomite. Dolomitization appears to have occurred throughout the Madison formation. The limestones and dolomites are crystalline, and fine-grained crystalline predominate over medium-grained crystalline strata. The Madison formation is porous and cavernous, and contains many calcite-filled cavities. The beds are cliff-forming, and color varies from light or dark gray, to pink and buff. Thin chert lenses are found throughout.

The Mississippian-Pennsylvanian contact is placed at the base of the lowest sandstone above a thick series of limestones and dolomites of the Madison formation.

Lower Pennsylvanian
Morgan Formation

The Morgan formation of lower Pennsylvanian age is divisible into three mappable members:

The *lower Morgan member* is predominantly sandstone, but contains some thin limestones, dolomitic limestones, and limestone conglomerates. The sandstones consist of quartz grains in a cement of silica or iron oxide. They are white, buff, pink, and yellow; fine- to coarse-grained; cross-bedded; limy, and in places, shaly. The limestones are buff and black, finely crystalline, and sandy in places. The black color of some of the limestone beds is suggestive of "toxic" conditions.

The *middle Morgan member* includes thick limestones; nodular limestones and intercalated shales; red, purple, green, gray, and yellow shales; and a few thin sandstones and sandy shales. The limestones are white, gray, pink, buff, and bluish-gray, very fossiliferous, dense, and cliff-forming. Some zones of the limestone are composed almost entirely of fragments of fossils, of which fusulinid remains are most abundant. Red, gray, and black chert layers and nodules are interstratified throughout. The nodular limestones, which constitute about 30 percent of the middle Morgan beds, are gray, purple, buff, and black, and contain interstratified red, black, and purple shales.

The *upper Morgan member* is composed of thick limestones, sandstones, and some shale and limestone conglomerate beds. The limestones are gray to bluish-gray, very fossiliferous, and contain red, gray, and black chert lenses and nodules. Some zones of the limestone are composed almost entirely of fusulinids. The sandstones are composed of quartz grains in a cement of silica and iron oxide. They are fine- to medium-grained, cross-bedded, and buff, white, gray, and pink.

The Morgan formation is thought to be of lower Pennsylvanian age. The upper two members are assigned a lower Pennsylvanian age on the basis of microfossils identified by J. S. Williams, and fusulinids identified by L. G. Henbest (Thomas, 1945). Williams tentatively assigned a lower Pennsylvanian age to the lower Morgan member, because of lithologic composition, general sequence of beds, and the presence of gastropods in the lowest member at Vermillion Creek.

Miscrofossils present in the beds in the canyon of the Yampa River, sections 12 and 13, T. 6 N., R. 95 W., and in section 10, T. 6 N., R. 95 W., include: *Spirifer rockymontanus* (Marcou), *Neospirifer cameratus* (Morton), *Composita subtilita* (Hall), *Dictyoclostus portlockianus* (Norwood and Pratten), *Linoproductus* sp., *Composita ovata* (Mather), and *Dictyoclostus* sp.

Permian and Triassic

Although no Permian or Triassic beds are exposed on Juniper Mountain, their presence southeast of Juniper Mountain at Moore and Gilmore No. 1 Seely indicate that these beds are present under surface cover.

Permian
Weber Formation (Permian (?) and Pennsylvanian)

The Weber sandstone, as present at Cross Mountain and Vermilion Creek, and as observed in well samples from the Maybell Blue No. 1 well, consists of buff to light gray, fine-grained, well-sorted, calcareous sandstone.

Park City Formation

This formation is composed of beds of limestone, chert, shale, sandstone, and phosphate. It is equivalent, at least in part, to the Phosphoria formation.

Triassic
Moenkopi Formation

The Moenkopi includes beds of sandstone; siltstone; red, well-bedded shale; sandy shale; and some thin stringers of gypsum in the lower part. At Cross Mountain, the Moenkopi is about 630 feet thick.

Shinarump Formation

The Shinarump consists of cross-bedded sandstone and conglomerate. The conglomerate is composed of quartz grains and pebbles in a matrix of medium-grained sand. It is ridge-forming, and varies in thickness and character. It is possible that the Shinarump is quite thin, or perhaps even absent, at Juniper Mountain.

Chinle Formation

The Chinle formation overlies the Shinarump conglomerate, and includes beds of red mudstone, calcareous shale, and a few thin, fine-grained sandstone beds. At Cross Mountain, about 200 feet of predominantly red beds make up the Chinle. In subsurface, the Chinle ranges in thickness from 160 feet at Rangely oil field, to 360 feet in the Wilson Creek oil field.

Jurassic
Navajo Formation

The Navajo formation consists of thick beds of buff to light gray, fine- to medium-grained sandstone, which is characterized by eolian type cross-bedding. This sandstone forms abrupt scarps, and some outcrops weather to rounded surfaces. The Navajo is characterized by its well-rounded, frosted, and well-sorted quartz grains in a silica cement. It weathers to yellowish-brown, and is not readily differentiated from the overlying Entrada formation. At Split Mountain, which is about 70 miles west of Juniper Mountain, the Carmel formation lies between the Entrada and Navajo formations; at Juniper Mountain, the Carmel is absent.

Entrada Formation

The Entrada formation is composed of beds of buff to light gray, fine- to medium-grained sandstone. The quartz grains are rounded, frosted, and well-sorted; the cement is siliceous. The upper part is parallel-bedded, glauconitic, and hard; the lower part is cross-bedded. The fact that the Entrada and Navajo formations have similar attributes, implies similar origins. The rounded and frosted quartz grains, uniform sorting, and eolian-type cross-bedding suggest eolian origin for these formations.

Curtis Formation

The Curtis formation includes beds of light to dark green shale, sandy shale, thin-bedded limestone, and sandy limestone. At Juniper Springs, the top four feet consist of green, black, and red mottled paper shale. The Curtis formation is believed to be of upper Jurassic age.

Morrison Formation

The Morrison formation, of Jurassic age, is divided into three general units, on the basis of a measured section at Juniper Springs:

(1) The upper one-third consists of green, red, and varicolored shale, containing thin chert-conglomerate lenses. The shale, which fractures into prisms, is interstratified with thin, resistant limestone beds, containing some red and white calcite seams.

(2) The middle one-third consists of variegated shale and platy sandstone.

(3) The lower one-third is composed of massive, buff to white, parallel and cross-bedded sandstone, which contains quartz grains in a siliceous cement. Features of sedimentation and attributes of the sediments suggest a floodplain, swamp, and lake origin.

Cretaceous
Beds of Undetermined Age (Lakota (?))

Beds of undetermined age overlie the Morrison formation. These beds are 163 feet thick at Juniper Springs, where they crop out. They consist of medium- to coarse-grained, buff-colored sandstone; dark green and gray, conglomeratic quartzite; green and gray shale, and a few thin, fresh-water limestone beds. The conglomeratic quartzite consists of varicolored chert pebbles, up to two inches in diameter, in a matrix of coarse-grained sand. The cement is silica and iron oxide. The sandstones resemble those of the Dakota formation; whereas, the green and gray shales are typical of the Morrison. Stratigraphic position and lithology suggest that these unusual beds are the equivalent of the Lakota member of the Cloverly formation of Wyoming, and they may appropriately be considered transition beds, representing sedimentation during conditions which changed from fresh to brackish to marine water.

Dakota Formation

The Dakota formation is composed of beds of gray to buff, fine- to medium-grained sandstone; dark gray shale; and thin conglomerate lenses. At Juniper Springs, the Dakota is 35 feet thick. A marine beach or brackish-water origin is postulated for the Dakota formation.

Tertiary
Browns Park Formation

The Browns Park formation consists of well-bedded, fine to coarse-grained, white to pale greenish-yellow sandstone, interstratified with a few thin limestone beds and chert layers. At some exposures, it has a "salt and pepper" appearance. The quartz grains are well-rounded and frosted, and the cement is calcareous. The Browns Park formation rests with angular unconformity upon all older rocks of the Juniper Mountain area.

REFERENCES

Berkey, C. P. (1905), "Stratigraphy of the Uinta Mountains," Geol. Soc. Am. Bull., vol. 16, pp. 517-530.

Branson, C. C. (1939), "Pennsylvanian Formations of Central Wyoming," Geol. Soc. Am. Bull., vol. 50, pp. 1199-1266.

Brill, K. G., Jr. (1944), "Late Paleozoic Stratigraphy, West-Central and Northwestern Colorado," Geol. Soc. Am. Bull., col. 55, pp. 621-656.

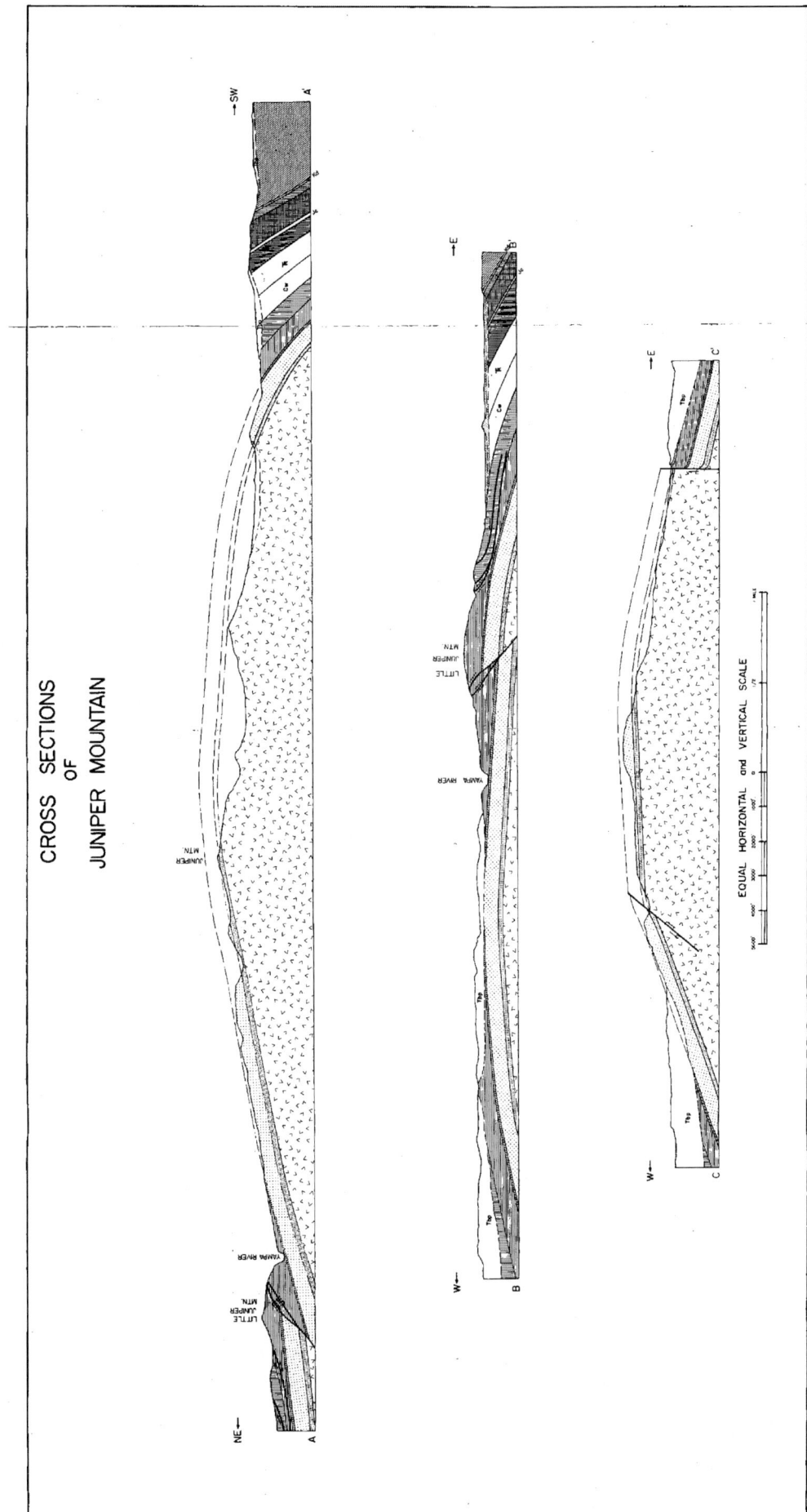

ABRASSART AND CLOUGH: FIGURE 1. CROSS SECTIONS OF JUNIPER MOUNTAIN.

Emmons, S. F. (1907), "Uinta Mountains," Geol. Soc. Am. Bull., vol. 18, pp. 287-302.
Forrester, J. D. (1937), "Structures of the Uinta Mountains," Geol. Soc. Am. Bull., vol. 48, pp. 631-666.
Gale, H. S. (1910), "Coal Fields of Northwestern Colorado and Northeastern Utah," U. S. Geol. Survey Bull. 415.
Girty, G. H. (1903), "The Carboniferous Formations and Faunas of Colorado," U. S. Geol. Survey Prof. Paper 16, pp. 1-546.
Hancock, E. T. (1925), "Geology and Coal Resources of the Axial and Monument Butte Quadrangles, Moffat County, Colorado," U. S. Geol. Survey Bull. 757.
Heaton, R. L. (1933), "Ancestral Rockies and Mesozoic and Late Paleozoic Stratigraphy of the Rocky Mountain Region," A. A. P. G. Bull., vol. 17, pp. 109-168.
Powell, J. W. (1876), "Report on the Geology of the Eastern Portion of the Uinta Mountains," U. S. Geol. and Geogr. Sur. Terr., 2nd Div., pp. 1-218.
Reeside, J. B., Jr. (1923), "Notes on the Geology of Green River Valley Between Green River, Wyoming and Green River, Utah," U. S. Geol. Survey Prof. Paper 132, pp. 35-50.
Sears, J. D. (1924), "Geology and Oil and Gas Prospects of Part of Moffat County, Colorado and Southern Sweetwater County, Wyoming," U. S. Geol. Survey Bull. 751-g, pp. 269-319.
Thomas, C. R., McCann, F. T., and Raman, N. D. (1945), "Mesozoic and Paleozoic Stratigraphy in Northwestern Colorado and Northeastern Utah," U. S. Geol. Survey, Oil and Gas Investigation Chart.
Thompson, M. L. (1945), "Pennsylvanian Rocks and Fusulinids of East Utah and Northwestern Colorado Correlated with Kansas Section," State Geol. Sur. Kan. Bull. 60, pl. 2.
Walton, P. T. (1944), "Geology of the Cretaceous of the Uinta Basin, Utah," Geol. Soc. Am. Bull., vol. 55.
Weeks, F. B. (1907), "Stratigraphy and Structure of the Uinta Range," Geol. Soc. Am. Bull., vol. 18, pp. 427-448.
Williams, J. S. (1943), "Carboniferous Formations of the Uinta and Northern Wasatch Mountains, Utah," Geol. Soc. Am. Bull. 54, pp. 591-624.

Composite Stratigraphic Section of the Juniper Mountain Area

CRETACEOUS

THICKNESS (in feet)	DESCRIPTION
	(Cretaceous and Jurassic measured in secs. 16 and 17, T. 6 N., R. 94 W.)
	Dakota formation (34.8 feet)
7.4	Sandstone, gray-buff, fine-coarse-grained; top 1.6' sugary; weathers brown-buff; degree of cementation varies; ridge-forming.
5.0	Sandstone, white-buff, sugary, coarse-grained; weathers buff; sorting poor.
6.6	Shale, dark gray, soft.
1.3	Shale, light-dark green, sandy.
2.3	Shale, light gray.
1.8	Shale, dark gray-black.
4.0	Sandstone, gray, fine-medium-grained, carbonaceous.
3.4	Sandstone, conglomeratic; black chert and quartz pebbles; carbonaceous material in fracture planes.
3.0	Sandstone, buff-gray; fine-coarse-grained; thin conglomerate lenses.
	Beds of undetermined age (163.2 feet) (Lakota ?)
98.0	Covered.
23.0	Sandstone, buff; coarse-grained, well-bedded, sugary; weathers brown; grades into conglomeratic sandstone at base, with pebbles of chert and quartz up to ¼" diameter; soft.
5.6	Quartzite, conglomeratic, dark green, hard.
1.3	Limestone, white, crystalline; green shale between thin limestone layers; conglomeratic.
11.9	Shale, green; conglomeratic, chert pebbles up to ¼" diameter; soft.
5.5	Conglomerate, buff-gray; varicolored chert pebbles up to 2" diameter; matrix coarse-grained sandstone; not well cemented.
17.9	Conglomerate, green; varicolored chert pebbles.

JURASSIC

	Morrison formation (486.2 feet)
188.5	Shales, green, prismatic fracture; some thin conglomerate lenses near top.
3.0	Shale, light to dark green, hard.
8.3	Shales, green, prismatic.
2.7	Limestone, light to dark green; red and white calcite-filled fractures.
4.5	Shale, green, soft.
2.0	Limestone, greenish-gray, clastic, hard.
28.1	Shales, green, prismatic fractures.
10.5	Shale, red, soft.
34.2	Sandstone, white, fine-grained; not well cemented; quartz grains rounded and frosted.
69.3	Sandstone, red and light green mottled at base, chiefly red above.
87.9	Sandstone, white, cross-bedded. Quartz grains rounded and frosted. Fine-medium grained; weathers yellow.
1.6	Sandstone, green, shaly.
2.0	Shale, red; lower 1' sandy.
42.6	Sandstone, white, cross-bedded. Quartz grains rounded and frosted; not well cemented. Weathers buff-yellow.
	Curtis formation (45 feet)
45.0	Shales, green sandy, hard, well-bedded. Top 4' composed of green, black, and red paper-thin shale. Some hard beds of limestone and sandy limestone, green and fossiliferous.
	Entrada-Navajo (244 feet, base not exposed)
244.0	Sandstone, buff, fine-medium-grained, sugary. Large and small scale cross-bedding. Glauconitic in upper part. Weathers brown to yellowish-brown. Quartz grains well-rounded and frosted.
	(Permian and Triassic beds not exposed, but present under surface cover and are approximately 1890' thick, by projection.)

LOWER PENNSYLVANIAN

	Morgan formation (1018.9 feet)
	(Measured west to east, from west entrance of canyon of the Yampa River.)
	Upper Morgan member (525.2 feet)
16.1	Limestone, bluish-gray, fossiliferous.
54.1	Covered; limestone float.
4.5	Sandstone, light gray-buff, fine-grained.
8.8	Limestone, gray; sandy at top. Weathers buff.
4.9	Limestone, gray, fossiliferous.
0.6	Shale, gray.
32.8	Limestone, gray, hard, well-bedded, fossiliferous.
29.1	Sandstone, light gray-buff; thinly laminated, with grain size varying from lamina to lamina. Weathers buff; limy in places.
2.1	Limestone, light gray, sandy, massive.
16.5	Sandstone, light gray, weathers buff-brown. Fine-medium-grained; cross-bedded.

THICKNESS (in feet)	DESCRIPTION
3.9	Limestone, gray, dense, sandy at base. Upper 1.3' bluish-gray, and contains weathered pyrite crystals.
7.3	Sandstone, gray, limy very fine-grained; well cemented.
6.4	Limestone, gray, clastic; fossiliferous.
2.0	Limestone, gray, crystalline; very fossiliferous.
26.5	Limestone, bluish-gray; well-bedded, dense; very fossiliferous, with some zones composed almost entirely of fusulinids. Weathers gray.
3.5	Limestone, light gray, sandy. Changes laterally to limy sandstone; weathers white. Irregular contact below.
4.5	Limestone, bluish-gray, massive. Weathers buff-gray; dense.
3.3	Sandstone, light bluish-gray, massive, dense.
3.7	Limestone, dark gray at top to bluish-gray below; dense, massive sandy.
2.0	Sandstone, light gray, limy, fine-grained; weathers buff; hard.
2.8	Limestone, light gray, sandy; weathers white. Calcite filled fractures.
8.0	Limestone, gray bluish-gray, very fossiliferous, with some zones composed almost entirely of fossil fragments. Red and black chert nodules. Irregular contacts above and below. Weathers light gray-white; dense, massive.
4.7	Limestone, dark gray, dense, fossiliferous. Gray chert lens 1½' from top, and thickness of lens varies from 1-3".
4.2	Alternation of nodular limestone, gray, and limy sandstone lenses. Weathers buff to light yellow.
7.0	Covered (nodular limestone float).
1.7	Limestone, dark gray; composed almost entirely of fossil fragments. Dense; weathers brown.
0.5	Conglomerate, gray, limestone matrix, with gray chert pebbles angular to sub-rounded.
8.5	Limestone, light bluish-gray, dense, crystalline, fossiliferous.
0.5	Chert lens, gray.
1.0	Limestone, light gray.
0.3	Chert lens, gray.
15.8	Limestone, gray, and alternating with layers of gray, nodular limestone; fossiliferous, sandy.
7.7	Sandstone, gray, fine-grained. Contains few scattered nodules of limestone. Weathers buff.
2.5	Limestone, gray, very fossiliferous, with many fusulinids; black chert nodules; calcite-filled fractures.
1.6	Sandstone, white, shaly.
1.0	Limestone, light gray, sandy, fossiliferous.
7.7	Marlstone, white, sandy at top; limestone nodules at base.
16.0	Limestone, light gray, nodular (many black and gray chert and limestone nodules). Weathers gray.
0.5	Limestone, gray-buff, sandy; weathers brown.
2.5	Limestone, dark gray, dense.
5.0	Sandstone, white-buff, fine-grained.
1.6	Limestone, gray, sandy.
3.0	Sandstone, white, fine-grained; limy at top. Weathers black.
3.7	Limestone, light gray-white, dense, weathers black.
2.7	Sandstone, buff-light pink, fine-grained, cross-bedded.
15.8	Limestone, light gray, nodular (both limestone and red chert nodules); very fossiliferous—many brachiopods. Some light blue intercalated shale. Fossils weather red, due to replaced red silica.
2.5	Sandstone, light gray-white, fine-grained, hard.
3.6	Limestone, gray, fossiliferous, gray chert nodules.
6.3	Covered, partly; fossiliferous, gray limestone, nodular at top; underlain by red and green shales. Red sandstone, fine-grained, at base.
9.0	Sandstone, white-light gray, quartzitic; weathers buff.
5.3	Limestone, light gray, crystalline, dense.
4.0	Sandstone, fine-grained, light gray; weathers buff.
10.0	Limestone, gray, nodular, very fossiliferous, and fossils replaced by red silica. Some gray chert.
4.0	Limestone, gray, very sandy at base.
3.8	Limestone, gray, dense; weathers rough.
5.0	Limestone, light gray-buff, sandy.
18.7	Sandstone, light gray-buff, fine-grained; limy at top. Weathers buff-brown.
0.8	Shale, light greenish-yellow; sandy.
8.1	Sandstone, light gray-buff, limy, fine-grained; weathers yellowish-brown, massive.
1.5	Shale, greenish-yellow, sandy.
1.2	Limestone, gray, nodular, sandy, fossiliferous, with red-colored fossils.
3.5	Limestone, light bluish-gray, crystalline, massive. Top 0.5' composed almost entirely of fossil fragments.
1.6	Limestone, gray, nodular; upper 0.7' gray shale.
4.5	Limestone, gray, fossiliferous, dense, crystalline.
11.3	Limestone, gray, nodular (black chert and limestone); weathers buff, fossiliferous.
7.0	Limestone, gray, very fossiliferous, dense.
5.1	Sandstone, light gray, fine-grained, limy.
6.0	Sandstone, light gray-white, fine-grained.
2.6	Limestone, grayish-purple; very fossiliferous.
3.0	Limestone, light purple, fossiliferous, shaly; many crinoid stems.
1.0	Limestone, gray, fossiliferous, dense.
5.0	Limestone, gray, nodular; fossiliferous, with many fusulinids. Weathers white.
1.8	Limestone, light gray, crystalline, dense.
4.1	Sandstone, white-light gray, fine-grained, top 1 foot thin-bedded.
2.0	Shale, yellowish-green.
5.6	Shale, red and purple.
1.0	Limestone, nodular, gray.
3.0	Sandstone, white (color varies from white-pink-purple-red), fine-grained.
2.2	Limestone, purple and green mottled, shaly.
3.0	Shale, red.
1.6	Sandstone, red and green mottled; fine-medium-grained.
3.6	Shale, red.
3.0	Conglomerate, limestone matrix, with limestone pebbles up to 1½".
5.0	Shale, brick red, green, and purple.
2.0	Limestone conglomerate, limestone matrix and pebbles.

Middle Morgan member (427.3 feet)

7.7	Limestone, nodular, with intercalated purple marl layers. Fossiliferous, and fossils weather red.
3.0	Limestone, gray, crystalline, dense.
1.0	Marlstone, grayish-purple, fractured.
0.4	Limestone, gray, composed almost entirely of fossil fragments.
1.2	Limestone, gray, crystalline. Red chert layers and nodules.
0.6	Limestone, gray, composed almost entirely of fossil fragments; red chert nodules.

THICKNESS (in feet)	DESCRIPTION
1.0	Limestone, gray, fossiliferous.
19.1	Limestone, gray, nodular, with intercalated red and purple shales, very thinly bedded.
1.5	Shale, purple; some limestone nodules.
3.1	Limestone, gray, massive. Very fossiliferous.
0.5	Shale, purple.
1.0	Limestone, gray, crystalline, dense.
0.8	Shale, green.
6.6	Shale, purple and green mottled.
3.9	Shale, red, green at base.
12.3	Limestone, gray, nodular, with thin, purple intercalated shales. Fossiliferous, and some fossils replaced with red silica.
1.2	Shale, purple.
0.5	Sandstone, gray, fine-grained, limy.
3.0	Shale, red and gray.
7.0	Limestone, gray, composed almost entirely of fossil fragments; dense.
11.8	Limestone, gray, nodular, with intercalated red and purple shale.
1.6	Limestone, gray, nodular, fossiliferous.
1.4	Shale, green.
1.8	Limestone, red and gray mottled, fossiliferous; weathers buff and red mottled.
5.5	Shale, grayish-purple, very soft.
3.5	Limestone, gray, composed almost entirely of fossil fragments; some limestone nodules.
6.8	Limestone, gray, nodular, many fusulinids.
2.0	Limestone, dark gray, dense, fossiliferous.
2.4	Sandstone, buff, fine-grained, limy; weathers pink-brown; some nodular limestone.
1.7	Sandstone, yellowish-brown, fine-grained, thin-bedded.
3.5	Shale, red and green.
3.2	Limestone, gray, fossiliferous, dense.
1.2	Limestone, gray, nodular, fossiliferous.
0.8	Limestone, gray fossiliferous.
0.8	Shale, purple and gray.
2.0	Limestone, gray, fossiliferous, nodular.
2.8	Limestone, gray, dense.
12.0	Limestone, black nodular, fossiliferous; bottom 3' alternating black shale and black nodular limestone; weathers light gray.
8.1	Shale, black and olive green.
0.6	Sandstone, yellowish-green, fine-grained, limy.
1.4	Limestone, dark gray, nodular, dense, fossiliferous.
2.0	Limestone, black, nodular, with intercalated purple shale; fossiliferous.
9.3	Limestone, gray, fossiliferous, nodular; bottom 2' bluish-gray.
3.0	Limestone, gray; composed almost entirely of fossil fragments.
0.6	Shale, gray.
5.2	Limestone, dark gray, nodular, fossiliferous.
7.2	Shale, green at top, red below; bottom 1.5' gray, nodular limestone.
2.7	Limestone, gray fossiliferous, hard; red chert lens near top, 0.5' thick.
5.8	Limestone, bluish-gray, nodular, fossiliferous.
7.7	Limestone, dark gray, massive.
6.8	Limestone, gray, nodular, with intercalated purple shale.
1.0	Shale, purple.
31.8	Limestone, gray, nodular, with intercalated purple shale beds.
18.1	Limestone, gray, dense, weathered surface rough.
0.6	Chert lens, black.
5.2	Limestone, gray, dense; weathered surface rough.
4.0	Limestone, bluish-gray; contains some brown chert.
6.6	Limestone, very dark gray, dense, massive.
2.0	Conglomerate, limestone pebbles and matrix.
3.0	Limestone, light gray, fractured.
6.5	Limestone, gray, weathers buff, highly fractured.
2.0	Limestone, light gray-pink, very dense, finely crystalline.
2.8	Limestone, gray; weathers buff-pink.
3.0	Limestone, brownish-gray; very fossiliferous.
5.2	Limestone, gray, fossiliferous; hackly fracture.
1.0	Limestone, bluish-gray, massive, fossiliferous.
0.6	Shale, purple.
1.4	Limestone, conglomeratic, dark gray, irregular contacts above and below.
12.0	Limestone, purplish-gray, soft; calcite-filled fractures.
4.5	Limestone, light gray-buff, very brittle; weathers yellow.
3.0	Limestone, gray, crystalline, dense.
6.0	Limestone, buff-gray, brecciated.
7.2	Limestone, buff-gray, lithographic at top; extremely dense.
3.7	Limestone, gray, conglomeratic-limestone pebbles and matrix; fractured.
10.0	Limestone, gray, fossiliferous; hackly fracture.
2.3	Limestone, gray, conglomeratic-limestone pebbles and matrix.
1.6	Limestone, gray, crystalline, dense.
2.8	Limestone, pink, crystalline.
5.0	Limestone, gray, crystalline, dense; cliff-forming.
3.0	Limestone, pink, finely crystalline.
2.0	Shale, green and yellow, soft.
5.0	Shale, purple and red.
6.4	Limestone, purplish-gray, fossiliferous, dense.
1.0	Shale, red, soft, well-bedded.
0.4	Shale, purple, well-bedded, soft.
3.5	Sandstone, fine-grained; facies changes to sandy limestone. Color varies from pink to gray.
1.5	Limestone, light gray, very dense; contains some purple limestone nodules.
2.0	Limestone, gray, coarsely crystalline; weathers black, resistant.
0.6	Shale, red, sandy.
3.3	Sandstone, fine-grained; quartz grains well-rounded and frosted; limy, cross-bedded.
5.0	Shale, purple, not well-bedded.
2.0	Limestone, gray; facies changes to sandstone, limy, and to red shale, sandy.
3.5	Shale, purple.
1.3	Limestone, gray, finely crystalline.
2.0	Sandstone, red and green mottled, very fine-grained.
1.8	Shale, purple, well-bedded.
2.0	Limestone, gray, nodular.
3.3	Limestone, gray, dense.
6.0	Limestone, white, finely crystalline; dense.

Lower Morgan member (66.4 feet)

13.0	Sandstone, white, fine-medium-grained; color varies laterally to pink; quartz grains well-rounded and frosted; weathers buff.
2.0	Limestone, black, sandy.

THICKNESS (in feet)	DESCRIPTION
22.9	Sandstone, white, coarse-grained, soft, cross-bedded; color changes to pink laterally; quartz grains sub-rounded; sugary at base.
3.0	Sandstone, yellow, medium-grained.
6.5	Sandstone, white; porous and permeable; poorly sorted.
2.0	Sandstone, yellow, fine-grained; well-bedded, shaly at base.
2.0	Limestone, buff, finely crystalline.
1.3	Sandstone, buff, sugary, coarse-grained.
4.7	Sandstone, buff, fine-grained.
1.5	Conglomerate, buff; matrix, quartz grains; limestone pebbles.
2.0	Limestone, dolomitic; gray-pink.
2.5	Limestone, dolomitic; brownish-gray, finely crystalline; weathers chocolate brown.
3.0	Sandstone, white, grains sub-rounded to sub-angular; limy at base.

MISSISSIPPIAN

(Measured in section 24, T. 6 N., R. 95 W.)
Madison formation (269.7 feet)
(Measured from top of erosion surface)

23.4	Dolomite, limy, light gray at top; pink to buff at base; weathered surface is rough; many small cavities.
20.5	Dolomite, buff, pink, and white mottled; weathers white; many small cavities.
22.8	Limestone, dolomitic, light gray, dense.
27.6	Limestone, light gray-buff; "boxwork structure" on weathered surface.
25.0	Limestone, pink, gray, and buff mottled; porous.
21.2	Dolomite, light gray-buff; finely crystalline, dense, massive; cliff-forming. Color varies to pink.
16.7	Dolomite, limy; light gray, coarsely crystalline, dense, massive.
20.0	Dolomite, light gray-pink; fractured; "boxwork structure" on weathered surface; some gray chert nodules; facies changes to limestone, dolomitic; light gray, dense; weathers white; well-bedded.
15.8	Limestone, dolomitic at top; pinkish-gray; grades into dolomite, below; gray, finely crystalline, calcite-filled fractures; weathers white.
16.0	Limestone, dark gray, finely crystalline; weathers light gray, dense; grades into dolomite below; gray, crystalline, dense.
11.3	Limestone, gray; weathers white.
21.2	Dolomite, limy; pinkish-gray to white; "boxwork structure" on weathered surface; very porous, crystalline.
21.5	Dolomite, pinkish-gray; coarsely crystalline; some calcite-filled cavities; weathers brown-black; grades into sandy dolomite at base, near "Cambrian (?) - Madison contact; quartz grains angular, and range from fine-coarse-grained.

CAMBRIAN-DEVONIAN

Beds of unknown age (Cambrian ?) (203.2 feet)
(Measured in section 23, T. 6 N., R. 95 W.)

4.0	Shale, brick red, sandy; asymmetrical ripple-marks.
199.2	Conglomerate, white-buff, quartz pebbles up to one centimeter in diameter; not well cemented at base, and degree of cementation varies. The coarse conglomerate at the base grades upward into well-cemented quartzite; buff, light gray, and pink. Contains some thin conglomerate lenses. Both conglomerate and quartzite are cross-bedded, unfossiliferous, resistant, and cliff-forming.

PRE-CAMBRIAN

Uinta Mountain Group (base not exposed)
Conglomeratic sandstone, quartzite, and red, sandy shale.

H. R. Ritzma

Rapids in Whirlpool Canyon, Dinosaur National Monument.
Cambrian - pre-Cambrian contact above and to right of boat.

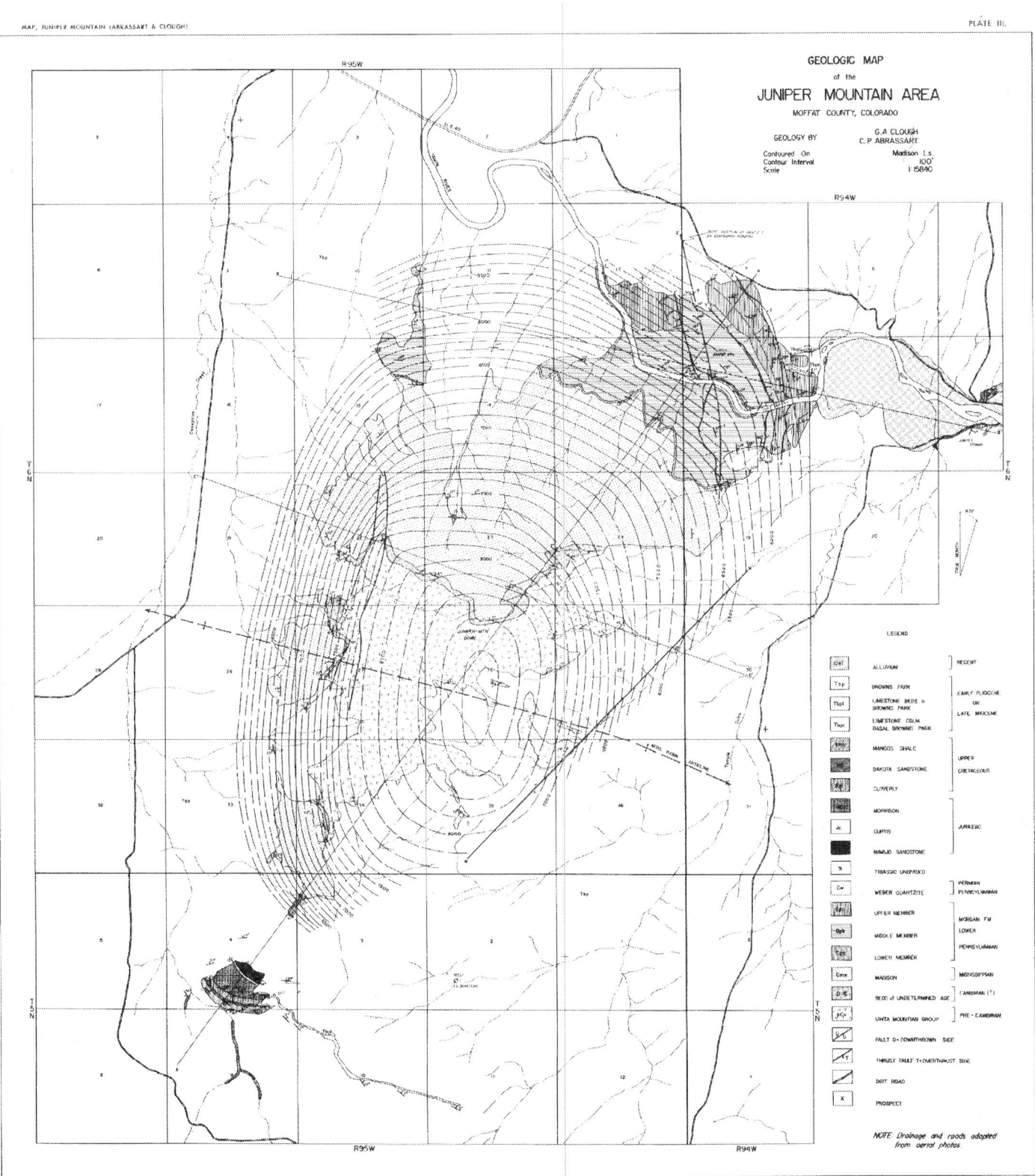

GEOLOGY OF THE NORTH HAHNS PEAK AREA, ROUTT COUNTY, COLORADO

By JAMES M. HUNTER
University of Wyoming, Laramie, Wyoming

INTRODUCTION

The author is greatly indebted to the faculty at the University of Wyoming and to all others who made this work possible. This paper is a condensation of a Master of Arts thesis in geology submitted to the University of Wyoming.

The North Hahns Peak area includes 45 sections in and adjacent to the south half of T. 11 N., R. 85 W., Routt County, Colorado, on the western slope of the Park Range and east of the Elk Head Mountains. The elevation of the mapped area varies from 7,900 feet in the northwest corner to 10,150 feet at Dome Peak on the eastern margin. An areal geologic map is enclosed in the pocket at the back of this guidebook.

STRATIGRAPHY

The exposed sedimentary sequence in the North Hahns Peak area includes somewhat less than 3,000 feet of sedimentary rocks which range from Permian or Triassic to Middle Miocene (?) in age. The generalized stratigraphic section of the North Hahns Peak area is very similar to that described by Barnwell (1955) in the South Hahns Peak District. Thickness of units is as follows:

```
MIOCENE (?)
    Browns Park formation.............. 500'+
    Conglomerate ...................... 0-40'
CRETACEOUS
    Mancos shale
        Upper shale unit............... 600'±
        Middle sandstone unit.......... 55'
        (Frontier equivalent)
        Lower black shale unit......... 250'
    Cloverly formation ................ 90-120'
JURASSIC
    Morrison formation ................ 267'
    Curtis formation ...............125-137'
    Entrada or Nugget formation........ 58'
TRIASSIC
    Jelm formation .................... 227'
PERMIAN (?) - TRIASSIC
    Chugwater formation ............... 285'+
        FAULT CONTACT WITH PRE-CAMBRIAN
PRE-CAMBRIAN
```

STRUCTURE

The north Hahns Peak area is located in a zone of adjustment related to the change in direction of the axes of two mountain ranges. The Park Range axis trends generally north, but to the east of the mapped area the axis swings northwest into the Sierra Madre Range. This change in direction of the axis of the Park Range appears to have been associated with a zone of weakness, part of which is represented by the King Solomon Creek fault. The sedimentary rocks which form a reentrant into the pre-Cambrian core of the Park Range suggest persistence of this fault zone. Beekley (1915, p. 24) reported an occurrence of infaulted redbeds, which generally aligns with the King Solomon Creek fault, in the col between Flattop Mountain and Mt. Zirkel. This indicates that a possible zone of structural weakness may extend completely across the Park Range.

The structural features of the north Hahns Peak area are the result of two principal periods of deformation; late Cretaceous-Eocene (Laramide), and the post-Middle Miocene. Laramide structure consists of uplift with extensive folding and faulting, whereas post-Middle Miocene structure is characterized by intrusion and uplift with associated faulting.

Laramide Structure

The overall Laramide structure of the mapped area south of the King Solomon Creek fault varies in detail from that to the north. The southern part of the north Hahns Peak area is the northward extension of the structure mapped by Barnwell (this guidebook) which includes two major structural elements: Farwell Mountain anticline, and Hahns Peak syncline. This is in contrast to four structural elements to the north which consist of Dome Peak anticline, a biaxial synclinal basin, a low anticlinal arch, and a shallow synclinal basin (map, in pocket). The contrast is quite striking since the axes of these folds are not continuous across the east-trending King Solomon Creek fault.

The line of weakness, King Solomon Creek fault, between the two structural divisions probably extends across the Park Range manifesting itself in the north Hahns Peak area as a scissors type fault. This fault is considered to be related to the change in direction of the axis of the Park Range from north to northwest during Laramide time. Movement on the fault was perhaps renewed during a later period because Mesozoic sediments are infaulted and form a prominent hogback

at 10,200 feet in elevation on the divide between Dome Peak and Farwell Mountain.

Post-Middle Miocene Structure

The post-Middle Miocene structure is characterized by abundant intrusives. There are two phases, a rhyolite porphyry phase and an olivine basalt phase. The rhyolite porphyry phase is considered to be the earliest, but the evidence is vague. The porphyry is discussed by Barnwell in this guidebook and is therefore not discussed in this paper. However, it is believed that trapdoor faulting accompanied the porphyry intrusions in several localities within the mapped area.

The post-Middle Miocene olivine basalt surrounds Circle Bar Basin and has the circular outcrop pattern of a cone-sheet or a ring-dike. Its thickness ranges from 10 to 30 or more feet. Columnar jointing and small vesicles up to 1 mm. in diameter are commonly found.

Possible origins for such a structure are a cone-sheet, ring-dike, folded lava flow, or a folded sill. The dips on the main body of the basalt range from 40 degrees on the north side to 80 degrees on the west and south sides. The observed dips are toward the center of the structure, but accurate dip measurements are difficult to obtain.

The inward dips omit the possibility of a ring-dike since a ring-dike is a subsidence feature, and dips are vertical or outward on such a structure. The steepness of the dips on the basalt in a bed of post-Laramide age seems to indicate that the structure is discordant. It is possible that dips were increased by intrusion of magma later than the basalt. However, this possibility is eliminated since the porphyry is considered older than the basalt. A microscopic thin-section cut from a hand specimen taken at the upper contact of the basalt and the Browns Park (?) formation strongly suggests that the basalt is an intrusive rather than an extrusive igneous body. Therefore, the structure is probably not a ring-dike, a folded sill, or a folded lava flow but is probably a cone-sheet from the available evidence.

Cone-sheets are not known to occur in sedimentary rocks, but it is likely that the sedimentary cover overlying the pre-Cambrian basement is thin. Thus, the structure is probably closely associated with crystalline rock rather than with the overlying thin sedimentary sequence.

SUMMARY

A zone of weakness appears to extend across the Park Range east of the mapped area. During the Laramide orogeny the Mesozoic rocks were strongly folded and westward thrusting occurred. The post-Middle Miocene orogeny was characterized by two intrusive phases; a rhyolite porphyry phase, and an olivine basalt phase. The basalt is intruded as a cone-sheet whereas the porphyry is intruded as stocks.

REFERENCES

Barnwell, W. W. (1955), "Geology of the South Hahns Peak District, Routt County, Colorado," This guidebook.

Beekley, A. L. (1915), "Geology and Coal Resources of North Park, Colorado," U. S. Geol. Survey Bull. 596, 121 pp.

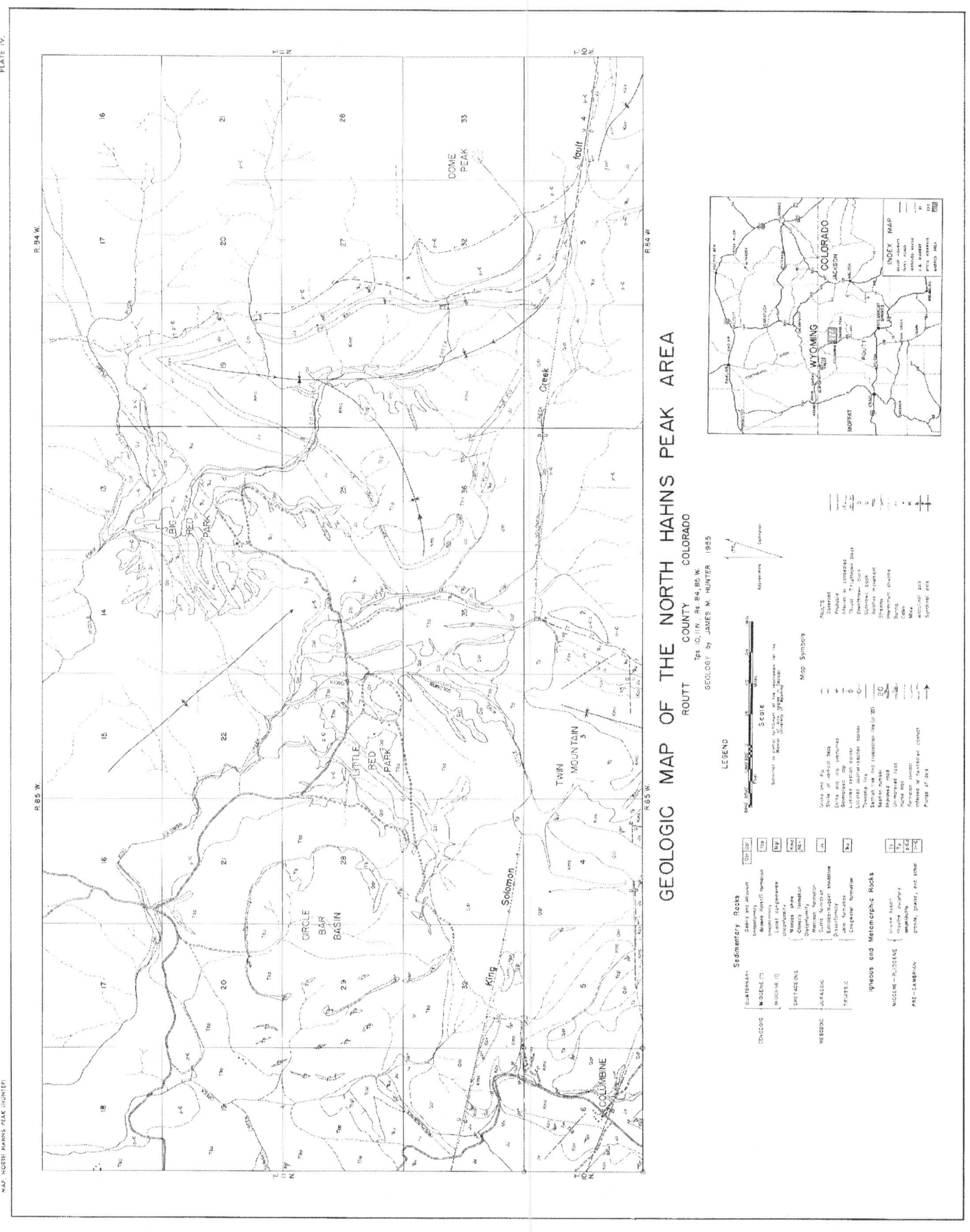

THE GEOLOGY OF THE SOUTH HAHNS PEAK DISTRICT, ROUTT COUNTY, COLORADO

By WILLIAM W. BARNWELL
Standard Oil Company of California, Seattle, Washington

INTRODUCTION

The south Hahns Peak district is located in T. 10 N., Rs. 85 and 86 W., Routt County, Colorado. The Hahns Peak district was first described by George and Crawford (1909) in a 33 page reconnaissance report. The present paper is a condensation of a masters thesis at the University of Wyoming. The writer wishes to acknowledge the help of Dr. Brainard Mears, Jr., Dr. D. L. Blackstone, Jr., Dr. Robert Houston and Mr. James M. Hunter during its preparation. An areal geologic map is included in the pocket at the back of this guidebook.

STRATIGRAPHY

Chugwater Formation (Lower Triassic)

The Chugwater formation, 600 to 800 feet thick, consists of irregular redbeds of mudstone, siltstone and fine-grained sandstone. The formation was deposited on a Permo-Pennsylvanian erosion surface cut on pre-Cambrian rocks.

Jelm Formation (Upper Triassic)

The Jelm formation, 227 feet thick, overlies the Chugwater formation and consists of four lithologic units. The basal unit, 25 feet thick, consists of a cross-laminated clay-pebble conglomerate. The basal unit is overlain by 42 feet of purple, medium- to coarse-grained, slabby, often calcareous sandstone which is in turn overlain by 75 feet of cream-colored, thin-bedded, medium-grained sandstone interbedded with red claystone. The upper unit, 85 feet thick, is mostly covered but from isolated exposures appears to consist of red siltstone and mudstone.

Entrada Sandstone (Upper Jurassic)

The Entrada sandstone 50 to 60 feet thick, overlies the Jelm formation and consists of white to buff massive well-sorted sandstone composed of medium frosted quartz grains.

Curtis Formation (Middle Upper Jurassic)

The Curtis formation, 110 feet thick, overlies the Entrada sandstone. The formation consists of littoral deposits of black shale; calcareous, tan, ripple-marked, fucoidal shale; glauconitic, calcareous, often fossiliferous sandstone and arenaceous limestone.

Morrison Formation (Upper Jurassic)

The Morrison formation, 267 feet thick, overlies the Curtis formation and consists of 75 feet of tan, silty, often cross-laminated, fine-grained sandstone which is overlain by 192 feet of dense grey to green thin-bedded limestone of fresh water origin and interbedded variegated claystone.

Cloverly Formation (Lower (?) and Upper Cretaceous)

The Cloverly formation, 90 to 110 feet thick, overlies the Morrison formation and consists of three units. The basal unit is composed of cross-laminated chert and quartz pebble conglomerate. The middle unit consists of 25 to 35 feet of soft medium-grained often iron-stained sandstone. The upper unit is composed of 20 to 25 feet of resistant, fine- to medium-grained, well-bedded, rectangularly jointed, ferruginous, tan sandstone which is interbedded with thin black beds of shale.

Mancos Shale (Upper Cretaceous)

The Mancos shale, 900 feet thick, overlies the Cloverly formation. The basal unit, 200 to 300 feet thick, is composed of ferruginous, rectangularly jointed, black shale interbedded with thin beds of bentonite. The middle unit, 53 feet thick, consists of highly arenaceous, oil-stained, grey limestone and calcareous sandstone interbedded with bituminous black shale. Thin beds of bentonite are present at the top and bottom of the middle unit. Hunter (1955) identifies *Inoceramus perplexus* and *Prionocyclus wyomingensis* (?) of Frontier age in the limestone of the middle unit. The upper unit consists of 576 feet of grey to black shale. Hunter (1955) identifies *Inoceramus* (Haploscapha) *grandis* (Conrad, 1875) and *Ostrea congesta* (Conrad, 1843) of Niobrara age in a thin limestone bed located 875 feet above the top of the Cloverly formation and near the top of the upper unit.

Miocene (?) Conglomerate

Isolated exposures of a locally derived stream-laid conglomerate composed of large sub-angular to sub-rounded cobbles of pre-Cambrian debris are present in

the district. The conglomerate overlies the Mancos shale with angular unconformity, has a maximum thickness of 40 feet, and is thought to have been deposited as a mountainward phase of early Browns Park deposition.

Browns Park (?) Formation (Middle Miocene)

The Browns Park formation, estimated to be 1200 feet thick, overlies the Mesozoic sediments with angular unconformity and is composed of medium-hard to soft tuffaceous calcareous fine-grained quartzitic cream-colored sandstone. The sandstone contains occasional shards and rare sub-angular fine-grained fragments of sodic plagioclase. The formation is overlain by recent alluvium and debris.

STRUCTURAL GEOLOGY

The south Hahns Peak district is bounded on its east by the exposed metamorphic core of the north trending Park Range and on its west by the volcanic region known as the Elkhead mountains. The structural geology of the district is related to three periods of orogeny:

1. Permo-Pennsylvanian uplift
2. Laramide folding and faulting
3. Post Middle Miocene igneous intrusion

Permo-Pennsylvanian Uplift

The south Hahns Peak district was part of the Front Range Highland and was a source area throughout Permo-Pennsylvanian time. It is believed that the development of Laramide structure in the district was controlled by zones of crustal weakness which were inherited from this uplift.

Laramide Folding and Faulting

The Laramide orogeny formed en echelon northward trending folds in the south Hahns Peak district. One such fold is a northward trending anticline, the Farwell anticline, which bounds the east side of the south Hahns Peak district. A syncline, the Hahns Peak syncline, is adjacent to the Farwell anticline. Its steep east limb, in which Mancos shale is overturned, has been overthrust by the metamorphic core of the Farwell anticline. The syncline's west limb relatively gently rises toward a second northward trending anticline, the Elkhead anticline, which is located to the west of the south Hahns Peak district in R. 86 W. The writer believes that the Elkhead anticline, whose west limb bounds California Park, may have been thrust westward. Evidence for this suggestion lies in the presence of Lancian sediments in California Park (Fenneman and Gale, 1906).

The regularity of the northward trending folds of the region is disrupted by complex faulting in northerly, northeasterly, northwesterly and westerly directions. The northerly trend of faulting is emphasized by the Farwell thrust zone, located on the east flank of the Hahns Peak syncline. The Farwell zone can be traced 10 miles southward from the south Hahns Peak district to the Elk river, and Hunter (1955) has traced a similar zone six miles northward from the south Hahns Peak district along the west flank of Dome mountain.

The northward trend of folding and thrusting in the south Hahns Peak district is broken by northeastward and northwestward trending cross-faults. South of the Hahns Peak stock, movement on the northeastward trending faults caused the south blocks to rise or move westward and therefore resulted in a progressive westward displacement of the Farwell thrust zone in a southerly direction.

The general result of faulting has been to cut the synclinal structure west of the Farwell thrust zone into graben and horst blocks. For example, the Hahns Peak syncline is bounded on its north and south by major normal faults. The syncline is bounded on its south by a northwestward trending fault, the Hahns Peak fault, which places the Chugwater formation in fault contact with Mancos shale. The syncline is bounded on its north by the King Solomon fault (Hunter, 1955). Between the limits of these two major faults are numerous secondary cross-faults. The form of the syncline is thus a highly faulted graben bounded on its east by the Farwell thrust zone and on its north and south by upthrown blocks of normal faults.

Post Middle Miocene Igneous Intrusion

Following deposition of the Browns Park formation the Hahns Peak stock and the other intrusive bodies of the district were emplaced. The intrusive bodies take many forms but for the most part they are discordant in respect to the sedimentary formations into which they were intruded.

REFERENCES

Fenneman, N. M., and Gale, H. S. (1906), "The Yampa Coal Field, Routt County, Colorado," U. S. Geol. Survey Bull. 297, 96 pp.

George, R. D., and Crawford, R. D. (1909), "The Hahns Peak Region, Routt County, Colorado," Colo. Geol. Survey, First Report, 1908, pp. 189-229.

Hunter, J. M. (1955), "The Geology of the North Hahns Peak District," Unpublished Masters Thesis, Univ. of Wyoming, Laramie, Wyoming.

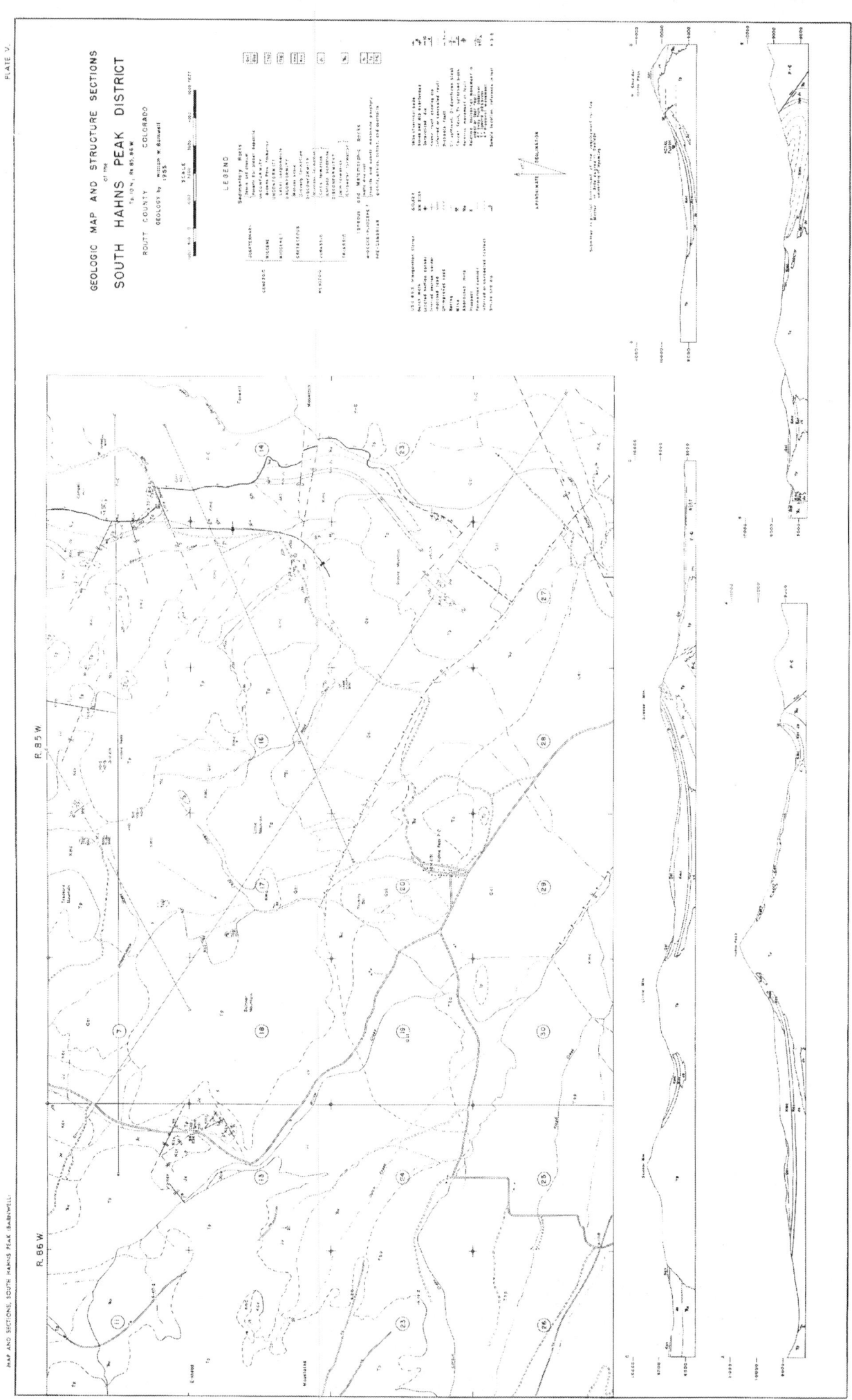

Hahns Peak in an Autumn setting.

Wiggins Studio, Craig

ECONOMIC GEOLOGY

Many limitations have prevented presentation of papers on all of the oil and gas fields of Northwest Colorado. For some fields too little new significant information is available to warrant publication.

The volume "Oil and Gas Fields of Colorado," published by the Rocky Mountain Association of Geologists in 1954, is an excellent source of information on the fields of this area. Some of this data has been repeated in this guidebook along with considerable supplemental information.

Uranium has made its mark on Northwest Colorado. Although not yet a major producer, the region has experienced a rash of exploratory activity. Papers are included on three types of uranium ore deposits: in pre-Cambrian rocks, in Paleozoic and Mesozoic sediments and in late Tertiary sediments.

Short discussions of Piceance Creek and Pagoda gas fields are included in the third day's road log where the route of the caravan nears these structures.

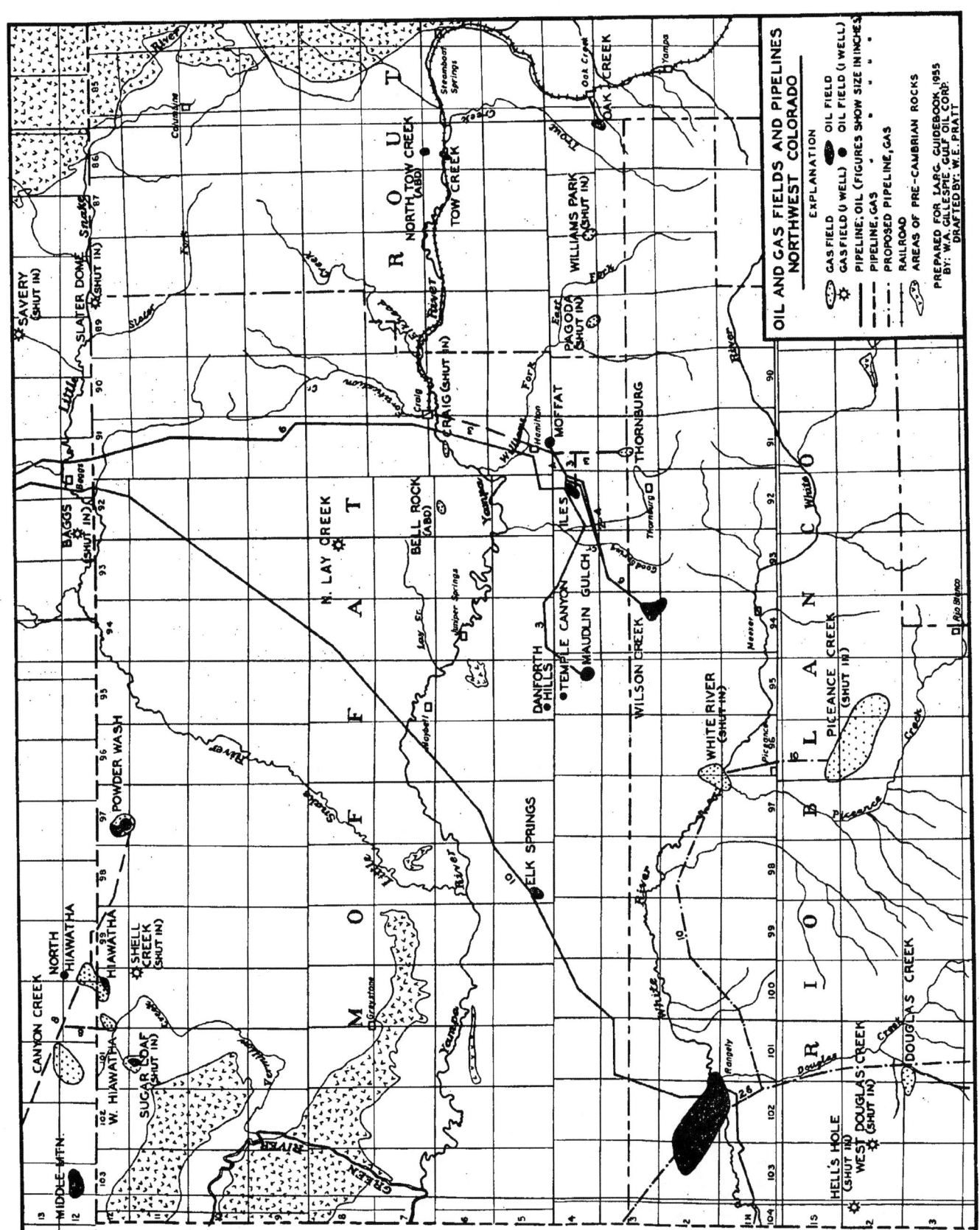

VERMILION CREEK BASIN AREA*
Sweetwater County, Wyoming and Moffat County, Colorado

By VICTOR B. GRAS

Mountain Fuel Supply Company, Rock Springs, Wyoming

LOCATION

The Vermilion Creek Basin area is located in Townships 11, 12, and 13 North, Ranges 99, 100, and 101 West of the Sixth Principal Meridian in northern Moffat County, Colorado, and southern Sweetwater County, Wyoming. The nearest town is Rock Springs, Wyoming, 65 miles northwest of the area via oil-surfaced Highway No. 430.

TOPOGRAPHY AND DRAINAGE

Vermilion Creek Basin is a topographic depression bounded on the northeast by Kinney Rim, a prominent Green River-capped escarpment rising some 1,000 feet above the basin. On the southwest and west, this basin is bounded by Cold Spring Mountain and Diamond Peak which lie at the eastern end of the Uinta Mountain range. The southeastern flank of the Rock Springs uplift defines the boundary of the Vermilion Creek Basin on the northwest. Sea level elevations average 6900 feet in the basin. It is an area of rolling hills with deeply eroded badlands along streams and washes. All drainage is into Vermilion Creek and its tributaries. Vermilion Creek in turn flows out the southern end of the basin to its confluence with the Green River in Browns Park. Exposed at the surface in Vermilion Creek Basin are the Tipton tongue of the Green River formation, and the Cathedral Bluffs tongue and the Hiawatha member of the Wasatch formation. All are of Tertiary (Eocene) age. The Mesozoic and Paleozoic formations are exposed south of the Sparks fault.

STRUCTURE

As shown on the accompanying map, there are six anticlines within the Vermilion Creek Basin. These are Hiawatha, Sugar Loaf, Canyon Creek, Alkali Creek, Haymower, and Shell Creek. These structures were all located by surface structural mapping. Trail Unit structure is located in Townships 13 and 14 North Range 100 West. Although it was outlined by seismic work, there is a slight surface suggestion of its presence.

Detailed surface structural mapping was started in 1928 with the mapping of Canyon Creek anticline by The Ohio Oil Company. Mountain Fuel Supply Company mapped the remainder of the Vermilion Creek Basin. Hiawatha, Alkali Creek, Sugar Loaf, and Shell Creek were mapped in 1929. Haymower structure was mapped in the summer of 1930.

The major structural feature of the Vermilion Creek Basin is the line of uplift along which lie the Sugar Loaf, Hiawatha, and Alkali Creek anticlines. This major line of uplift extends completely across the basin. From Sugar Loaf to Hiawatha, the trend of this uplift is southwest-northeast. At Hiawatha the trend changes to a north-south direction through the Alkali Creek anticline. Canyon Creek and Shell Creek anticlinal axes parallel the northeast-southwest trend of this major line of uplift between Hiawatha and Sugar Loaf anticlines. Of interest is the fact that the main axis of folding in the Vermilion Creek Basin is normal to the northwest-southeast axis of folding of the Uinta uplift in this area, and that it also lies at an angle of 45 degrees to the north-south axis of folding of the Rock Springs uplift.

Faulting is present on some of the anticlines in the area, and two major faults cut the area. The Sparks fault, on the southwest edge of the area, is a reverse fault downthrown on the north side and is probably an eastward continuation of the Uinta fault. Sparks fault cuts out 10,000 feet of sediments but may not have a displacement of that magnitude. The Shell Creek fault extends northeastward from the Sparks fault, south of Sugar Loaf field to the north of the east end of the Shell Creek anticline. It is a normal "scissors" type fault. At the east end the north side is downthrown 100 feet, and at the west end the south side is downthrown 300 feet. Folding and faulting followed the deposition of Eocene beds in this area. The Tipton shale tongue of the Green River formation is involved in faulting. The Cathedral Bluffs member of the Wasatch formation and the Laney shale member of the Green River formation are involved in folding.

The Hiawatha anticline is an unfaulted, symmetrical anticline with a northeast-southwest trending axis. It is composed of two minor highs separated by a very shallow syncline but both minor highs lie within a common lowest closing contour. These minor highs are designated as East Hiawatha dome and West Hiawatha dome.

Haymower dome is a symmetrical, slightly elongated anticline with the axis of folding trending north-south. Three north-south trending normal faults at the north

---
*Published by permission of Mountain Fuel Supply Company.

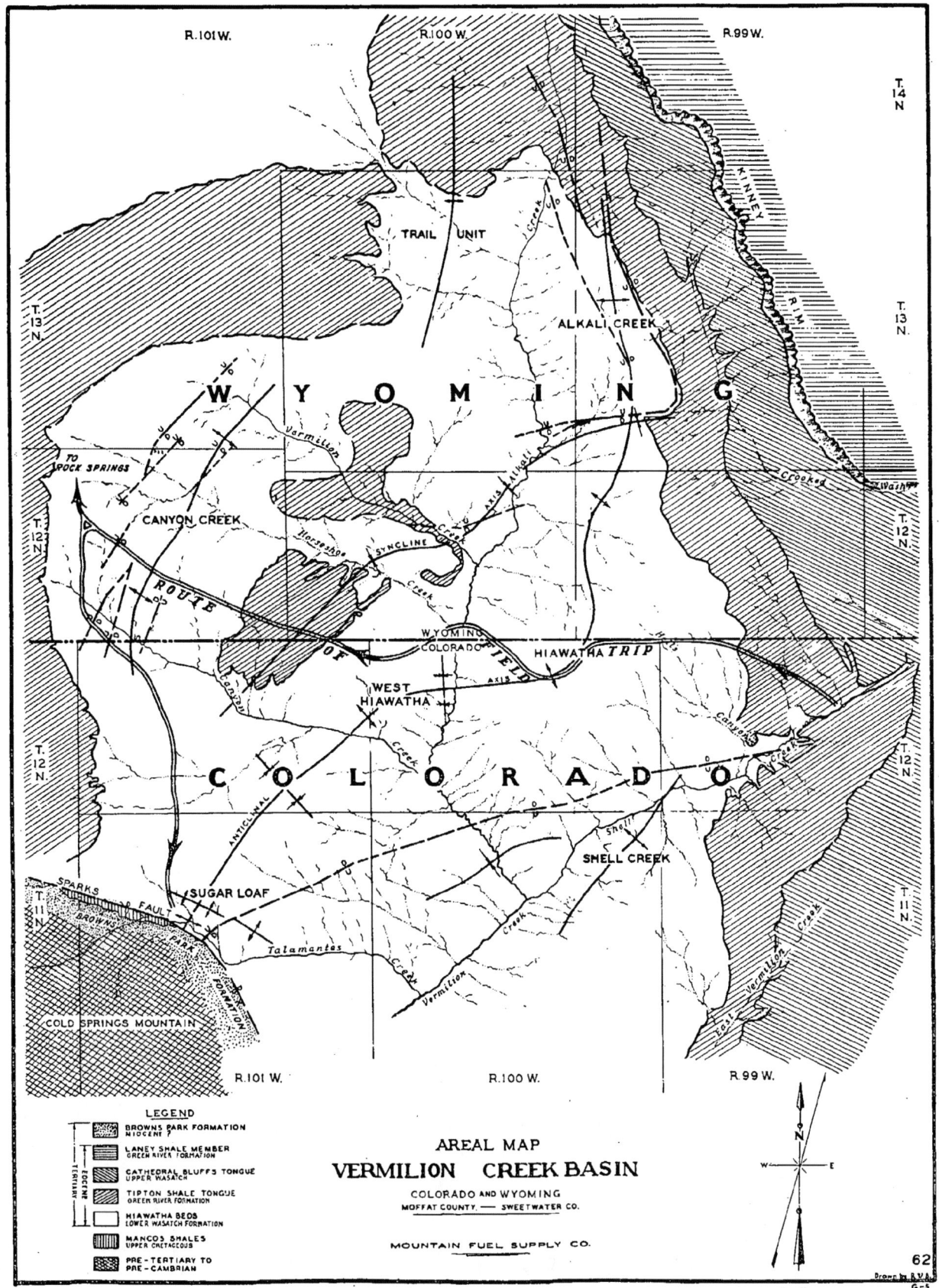

Figure 1. Areal map showing route of field trip.

end of Haymower all have their downthrown block on the side nearest the axis of folding.

Sugar Loaf dome is a symmetrical, slightly elongated, northwest-southeast trending anticline. Shell Creek fault cuts the southeast flank of Sugar Loaf dome and puts the dome on the upthrown block. Sugar Loaf dome also closes against the Sparks fault on the downthrown side. Closure is dependent upon a combination of folding and faulting.

Canyon Creek anticline is a symmetrical, elongated, highly faulted, northeast-southwest trending anticline. There are seven minor faults cutting Canyon Creek anticline. All are normal faults of small displacement and none cuts completely across the anticline. All faults are approximately parallel to the axis of the anticline and have their downthrown block on the side nearest the anticlinal axis. Closure is entirely the result of folding.

Alkali Creek anticline is an asymmetrical anticline with the steeper side being the east flank. Alkali Creek anticline is cut by three normal faults. The fault with a throw of 150 feet at the north end of the anticline, and the fault with a throw of 40 feet on the west flank are approximately parallel to the anticlinal axis with the downthrown blocks on the side nearer the anticlinal axis. The third fault extends the length of the anticline on the east flank until it reaches the south end of the anticline, thence it turns 90 degrees to the west cutting across the anticline into the adjoining syncline on the west where it disappears. Along this fault, the block nearest the axis of folding is upthrown with a maximum displacement of 775 feet. Closure at Alkali Creek is dependent upon folding and faulting.

Shell Creek dome is a symmetrical, slightly elongated, northeast-southwest trending anticline. Shell Creek dome is on the upthrown block of the Shell Creek fault. The fault cuts across north of the east end of Shell Creek dome. Closure at Shell Creek is probably all due to folding. Additional closure may be established if the Shell Creek fault is an effective seal.

STRATIGRAPHY

In the Vermilion Creek Basin area, the surface formation is mainly the Hiawatha member of the Wasatch formation. In the syncline between Hiawatha and Canyon Creek anticlines, the Tipton shale tongue of the Green River formation is exposed. Vermilion Creek Basin is rimmed by Tipton shale except along the southwest margin where Browns Park (Miocene?) lies in fault contact with Hiawatha beds along the Sparks fault. The intertongued Wasatch and Green River formations are exposed in Kinney Rim along the eastern edge of the area. The Hiawatha member (Wasatch formation) is overlain by the Tipton tongue (Green River formation), which is in turn overlain by the Cathedral Bluffs tongue (Wasatch formation), which is in turn overlain by the Laney shale (Green River formation).

The accompanying generalized stratigraphic section lists formations exposed and drilled in Vermilion Creek Basin area. All formation thicknesses given for formations older than Wasatch were taken from Hiawatha Unit Well No. 1 which is the deepest test in the area. This test penetrated 230 feet of Nugget sand at a total depth of 15,041 feet.

Commercial gas production in the Vermilion Creek Basin area is from Wasatch, Fort Union, and Mesaverde sands. All oil is produced from the Wasatch formation. Wasatch and Fort Union production is from sandstone lenses. Accumulation results from facies change and from closed structure. Due to the small areal extent of individual sandstone lenses, each well drilled to the Wasatch or Fort Union is practically a wildcat well. Mesaverde production at Sugar Loaf and Canyon Creek comes from different zones within the Mesaverde. Top of production at Sugar Loaf is 120 feet below the top of the Mesaverde group, while top of production at Canyon Creek is 800 feet below the top of the Mesaverde group.

HISTORY

Gas in the Vermilion Creek Basin area was discovered at the Hiawatha field in the Florence Wilson Well No. 1 in 1927. This well was drilled by Sormir Petroleum Company to 2220 feet in the Hiawatha member of the Wasatch for an open flow of 45,000,000 cubic feet of sweet gas a day. This was the first commercial production of gas from the Tertiary in the Rocky Mountain area. In 1933, the Florence Wilson Well No. 2 bottomed at a total depth of 7577 feet after penetrating 1677 feet of the Mesaverde group. This test discovered no production in the Fort Union, Lance, Lewis, or Mesaverde; and was completed as a gas well in a Wasatch sand topped at 2260 feet. In February, 1952, Hiawatha Unit Well No. 1 bottomed at 15,041 feet after penetrating 230 feet of Nugget sand. This is the deepest test to date in the Vermilion Creek Basin area. In Hiawatha Unit Well No. 1, the Frontier and Dakota sands were shaled out and the Nugget sand was never tested when mechanical difficulties junked the hole. Unit Well No. 1 was plugged back and completed as the Fort Union gas discovery in a basal Fort Union sand topped at 4880 feet. This sand made 11,000,000 cubic feet of sweet gas a day with a surface shut-in pressure of 1750 psi. Oil production was first discovered in 1935 when Florence Wilson Well No. 1, the discovery well, began producing 12 barrels of oil per million cubic feet of gas. In October, 1935, G. Kuykendall Well No. 1 was deepened and completed as

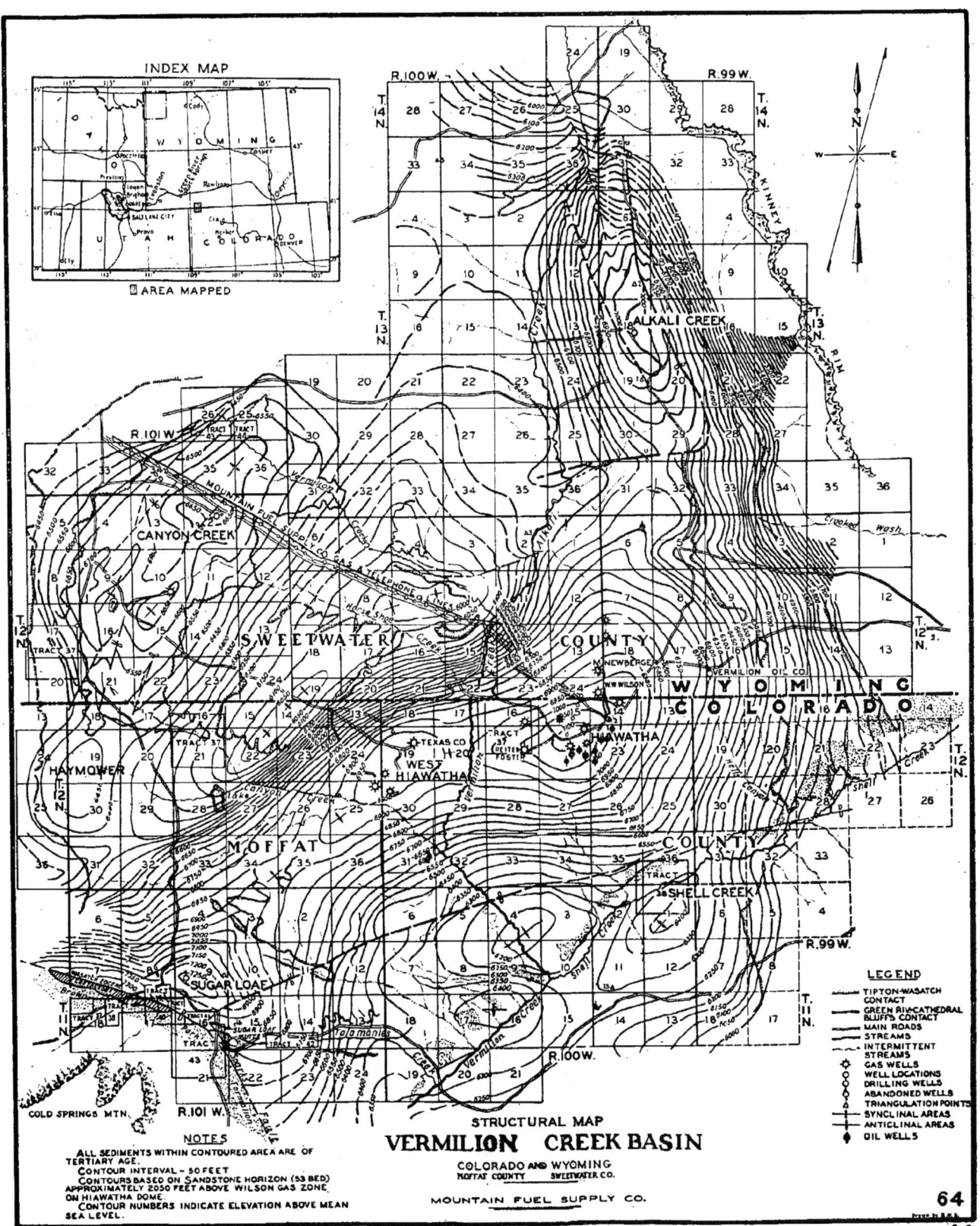

Figure 2. Structure contour map.

the first oil well for 38 barrels a day at 2468 feet in the Hiawatha member of the Wasatch formation.

The Sugar Loaf structure was first tested in R. T. Hatch Well No. 1, drilled by Mountain Fuel Supply Company in 1933. This test bottomed at 3518 feet in the Wasatch as a dry hole. In 1953, Mountain Fuel Supply Company drilled Sugar Loaf Government Well No. 1, a gas discovery well in the Mesaverde formation at 5145 feet. Government Well No. 1 bottomed at 7658 feet in Baxter shale and is the deepest test in the field. The Mesaverde discovery at Sugar Loaf produced 4,670,000 cubic feet of sweet gas a day with a bottom hole shut-in pressure of 1960 psi on drill stem test. There have been some shows of gas in the Fort Union at Sugar Loaf, but no production.

Haymower dome was tested in Haymower Unit Well No. 1, drilled by Pacific Western and Frontier in 1945. The well bottomed at 4030 feet in the Wasatch formation as a dry test.

Canyon Creek anticline was drilled five times before commercial production was obtained. In 1929, W. H. Humphreys Well No. 1, drilled by Mountain Fuel Supply Company, bottomed at 4231 feet in the Wasatch formation as a dry hole. J. G. Horne Well No. 1 which was drilled by Westates Petroleum Corporation in 1941 bottomed at 3955 feet in the Wasatch formation. Horne No. 1 was completed in a sand at 2680 feet for an initial production of 750,000 cubic feet of sweet gas a day with a shut-in casing pressure of 1075 psi. This well was not commercial as the well would not produce against pipeline pressure. In 1943, J. G. Horne Well No. 2, drilled by Westates Petroleum Corporation, bottomed at 2787 feet in the Wasatch formation as a dry hole. In 1994, Stanolind Oil and Gas Company drilled Canyon Creek Unit Well No. 1 to a total depth of 13,322 feet in the Dakota formation. The Dakota sand made 1,200,000 cubic feet of sweet gas on test, but pressures did not justify commercial production. Unit Well No. 1 had shows in the Mesaverde at 5445 and 6450 feet. In 1952, Canyon Creek Unit Well No. 2, drilled by Mountain Fuel Supply Company, bottomed at 7376 feet in the Mesaverde. In 1954, Canyon Creek Unit Well No. 3, drilled by Mountain Fuel Supply Company, bottomed at 6549 feet in the Mesaverde group as the Mesaverde gas discovery. Unit Well No. 3 made 4,500,000 cubic feet of sweet gas a day from 5460 to 5544 feet with a shut-in surface casing pressure of 1815 psi. This well was drilled in with gas to avoid water-blocking of Mesaverde sands, suspected in Canyon Creek Unit Well No. 1.

No production has been discovered at Alkali Creek anticline although four tests have been drilled. In 1954, Government Foreman Well No. 1, drilled by The Chicago Corporation and Republic Natural Gas, bottomed at 7751 in the Baxter formation as a dry test. This is the deepest test to date on the Alkali Creek structure.

Shell Creek dome was tested in S. A. Duncan Well No. 1, drilled by Mountain Fuel Supply Company, in 1937. This well bottomed at 3762 feet in the Wasatch formation as a dry test. Fort Union gas was discovered in Shell Creek Government Well No. 1, drilled by Mountain Fuel Supply Company, in 1955. This well bottomed at 7948 feet in the Mesaverde group. A Fort Union sand topped at 5350 feet was good for 7,450,000 cubic feet of sweet gas a way with a shut-in casing pressure of 1845 psi.

The Trail anticline was tested in Trail Unit Well No. 1, drilled by Mountain Fuel Supply Company and Stanolind Oil and Gas Company, in 1952. Unit Well No. 1 bottomed at 7731 feet in Baxter shale. A Baxter sand at 7530 feet made 3,000,000 cubic feet of gas a day following Hydrafrac treatment. The well would not produce again after it was killed to run tubing, and it was subsequently plugged and abandoned.

PRODUCTION AND DEVELOPMENT

In 1929, The Western Public Service Corporation constructed a 330 mile pipeline from Hiawatha to Salt Lake City, Utah, providing a market for gas from the Vermilion Creek Basin area.

At the Hiawatha field, production is from the Wasatch and Fort Union formations. Oil production is found in only one area on the southeast flank of the East Hiawatha dome. The trap is formed by sand lenses on an anticlinal closure. The type of drive is gas expansion. Hiawatha gas is sweet with a specific gravity of .65 and a heating value of 1075 BTU. Oil is a green paraffin base with an A.P.I. gravity of 40 degrees and pour point of 70°F. On December 31, 1954, there were 32 gas wells, 12 oil wells, 3 dry holes, and 2 depleted and abandoned wells in the Hiawatha field. The annual production of gas for 1954 was 6,383,709 M cubic feet, and the total cumulative production to December 31, 1954 was 92,768,004 M cubic feet. The annual oil production in 1954 was 158,859 barrels, and cumulative production to December 31, 1954 was 1,658,479 barrels.

Sugar Loaf field production is dry sweet gas from the Mesaverde group. Trap type is anticlinal closure with the possibility of some permeability barrier. The type of drive is gas expansion. The specific gravity is .67 and the heating value is 1087 BTU. The Sugar Loaf field was produced only during the latter part of 1954. Its cumulative production to December 31, 1954, was 430,606 M cubic feet. There are five producing wells and two dry holes at Sugar Loaf field.

The Canyon Creek field produces dry sweet gas from the Mesaverde group. Trap type is anticlinal closure with possibly some permeability barrier. A gathering system is being completed at the present time and some wells have been producing in 1955. For some weeks, the Canyon Creek field has been producing 4,000,000 cubic feet a day on test runs. In the Canyon Creek field, there are six producing wells, five dry holes, and one well drilling as of April 20, 1955. Drive is by gas expansion. The heating value is 1115 BTU and specific gravity is .66.

The Shell Creek field is an anticlinal closure type trap and is probably gas expansion drive. With only the one well in the field, it cannot be determined whether or not the producing sand is a lense. Production is from Fort Union and could easily be from lens-type sands as is the Fort Union production at Hiawatha.

GENERALIZED SECTION OF GEOLOGIC FORMATIONS EXPOSED AND DRILLED IN VERMILION CREEK BASIN

System	Series	Group or Formation	Member	Thickness	Character and Remarks
Tertiary	Miocene	Browns Park formation		200'	Soft, chalk-white sandstone with calcareous cement
		UNCONFORMITY			
	Eocene	Green River formation	Laney shale	500'	Grey fissile shale and oil shale; grey clay shale; grey and buff fresh water limestone; oolitic limestone and algal reefs
		Wasatch formation	Cathedral Bluffs tongue	1750'	Red and green clay shales with some thin beds of red to brown sandstone
			Tipton tongue of Green River form.	367'	Grey fissile shale with few thin-bedded red to brown sandstones
			Hiawatha member	5000'	Grey clay shale; grey shale; grey to buff sandstone lenses and coal beds; GAS AND OIL SANDS
	Paleocene	Fort Union formation		747'	Grey to greenish-grey shales; hard grey sandstones; coal beds; GAS SANDS
		UNCONFORMITY			
Cretaceous	Upper Cretaceous	Lance		255'	Grey shales; white fine grained sandstones; coal
		Lewis		795'	Grey marine shales
		Mesaverde group		2610'	Grey, white and brown sandstones; some grey shales; some coal beds in upper part; GAS SANDS
		Baxter		5200'	Grey marine shale, some sandy
		Frontier		125'	25' grey, fine grained, hard, tite sandstone; remainder dark grey to black, hard, sandy shale
		Aspen		183'	Black, hard, siliceous shale
		Dakota		137'	15' sandstone, white, coarse grained, hard and siltstone, dark grey, very hard and light grey claystone; SHOW GAS
		UNCONFORMITY			
Jurassic		Morrison		385'	Claystones, light green, hard; maroon shales; some light grey, hard, tite sandstones with calcareous cement
		Curtis		105'	Grey, glauconitic limestone; some oolitic limestone; dark grey to black shale with thin grey sandstones and limestones
		Entrada		70'	Light brownish-grey, very fine to coarse grained, quartzitic sandstone
		Carmel		16'	Red shale and anhydrite
		Nugget		230'	Light grey, fine to medium grained sandstone with scattered coarse, rounded grains

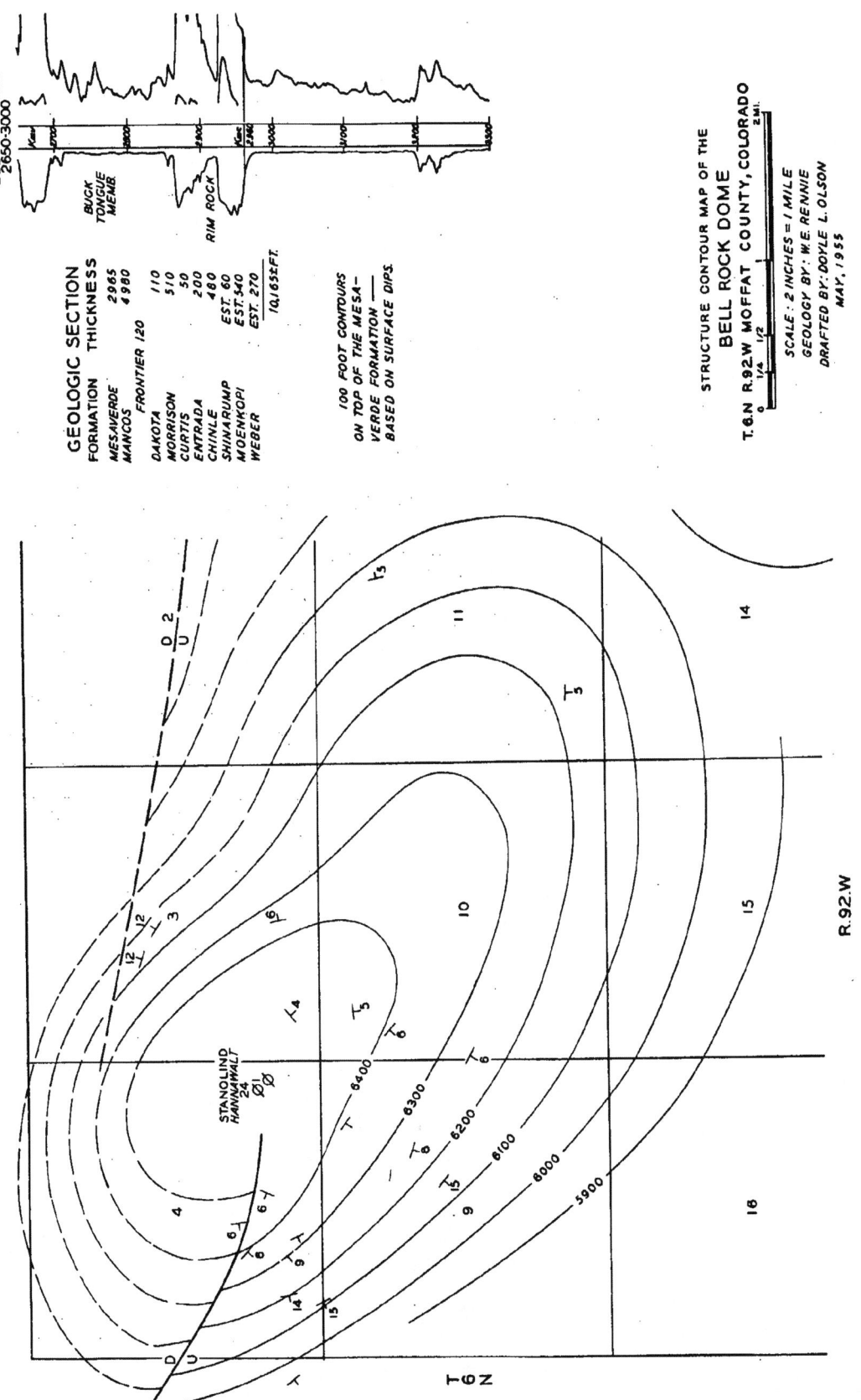

BELL ROCK DOME
Moffat County, Colorado

By JOHN C. WYETH

Continental Oil Company, Salt Lake City, Utah

The Bell Rock Dome is a partially tested separate high on the northwest end of the Moffat Dome anticlinal trend, and is about eight miles southwest of Craig in T. 6 N., R. 92 W., Moffat County, Colorado.

Only two wells have been drilled on the structure. In 1931, the Midwest Refinery Company drilled the No. 24 Hannawalt in NW NE SE SE, Section 4, T. 6 N., R. 92 W. The hole was drilled with much difficulty and was abandoned with junk in the hole at 3,041 feet. Gas was encountered in the Rim Rock member of the Mesaverde (basal sand). The estimated amount of flow is unknown, but one source of information suggests that it was approximately 3,000 ? MCF/D from the interval 2840 to 2861 feet.

In 1944, the Stanolind Oil and Gas Company drilled their No. 1 Hannawalt in NE SE SE Section 4, T. 6 N., R. 92 W. This well was drilled to a total depth of 9,084 feet, where it was abandoned in the Entrada sandstone. Several D.S.T.'s recovered various amounts of gas and/or oil from the Mesaverde, Frontier, Dakota and Entrada sands. One of the better tests was taken in the Rim Rock member, (interval 2860-2887). This test recovered 2,250 MCF/D.

The Bell Rock Dome has at least 500 feet of northwest-southeast trending closure which covers about 2700 acres. The anticline is expressed in the surface rocks, which belong to Williams Fork member of the Mesaverde formation. The presence of two faults is expressed by surface mapping, neither of which shows over 100 feet of throw on the surface.

The Dome lies on the south flank of the Sand Wash Basin (southern portion of the Washakie Basin). The structure was folded in Laramide time (Eocene ?).

GEOLOGIC SECTION

FORMATION	THICKNESS
MESAVERDE	2800
MANCOS	4980
FRONTIER 120	
DAKOTA	120
MORRISON	500
CURTIS	50
ENTRADA	200
CHINLE	480
SHINARUMP	60
MOENKOPI	540
WEBER	270–
	10 000 FT.

GEN. PET. 88-9-P ELECTRIC LOG 2600-300

100 FOOT CONTOURS ON TOP OF THE MESA-VERDE FORMATION BASED ON SURFACE DIPS.

STRUCTURE CONTOUR MAP OF THE

CRAIG DOME

T.6.N R.92.W MOFFAT COUNTY, COLORADO

0 1/4 1/2 1 2 MI.

SCALE: 2 INCHES = 1 MILE

GEOLOGY BY: W.E. RENNIE

DRAFTED BY: DOYLE L. OLSON

MAY, 1955

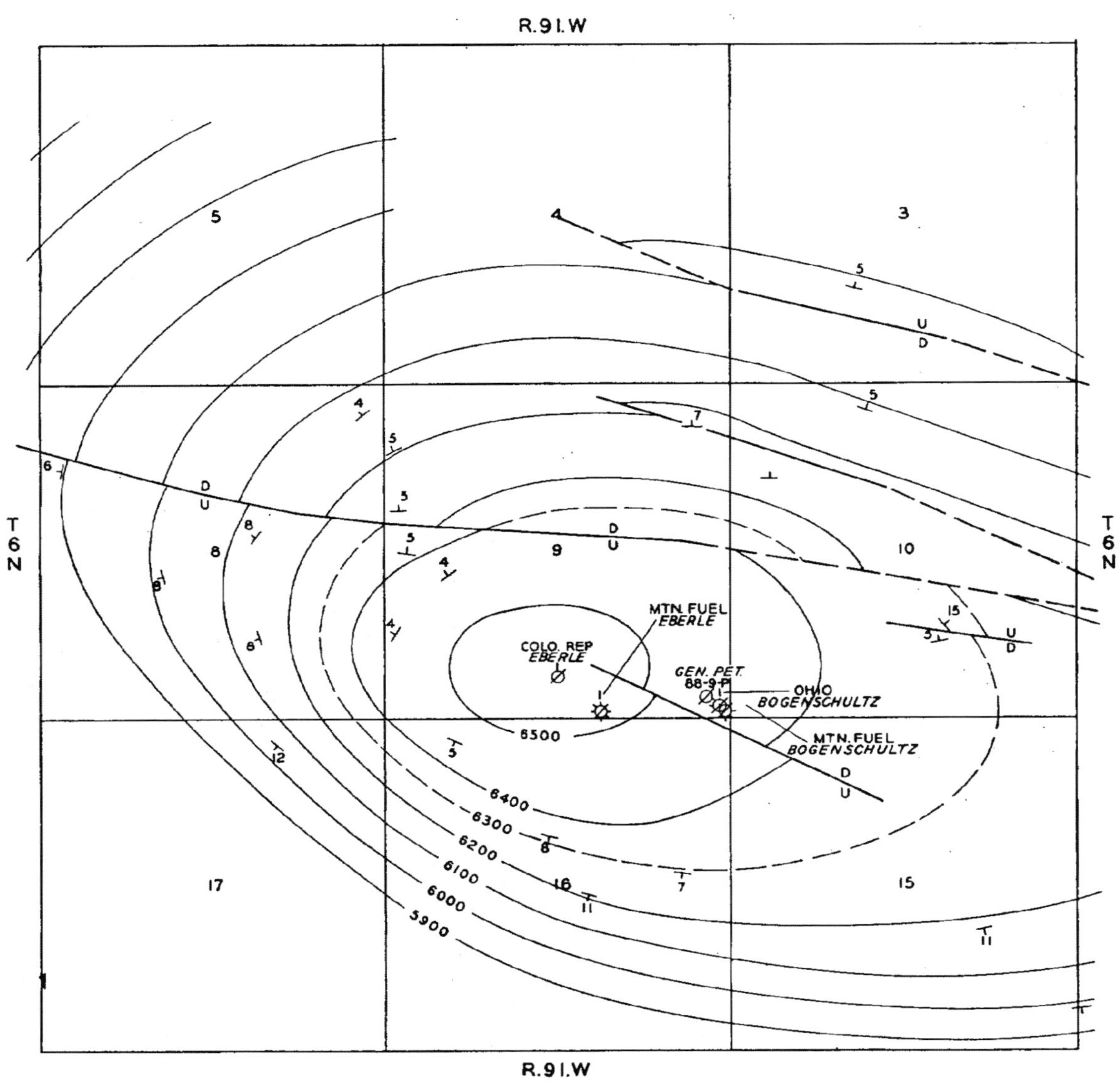

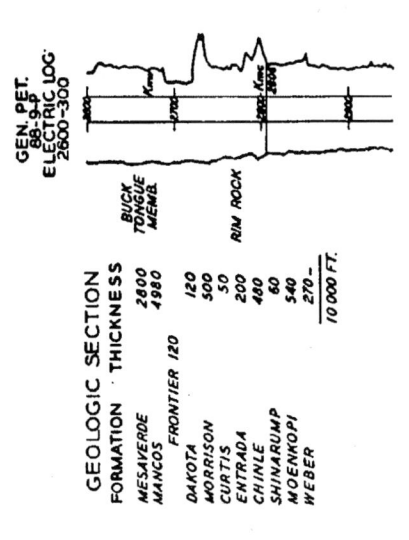

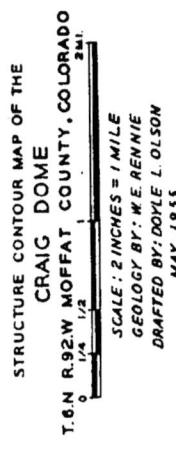

CRAIG DOME
Moffat County, Colorado

By JOHN C. WYETH

Continental Oil Company, Salt Lake City, Utah

The Craig Dome is located about three miles south of Craig, Colorado in the Williams Fork Mountains. The area is just north of the Iles and Moffat fields in Axial Basin. The structure is classed as an abandoned gas field that was depleted in 1940.

Five wells have been drilled on the structural high, but only two produced gas. The discovery well, Bogenschultz No. 1, was drilled by the Ohio Oil Company in 1932 in the SE¼ of Section 9. It had an initial flow of 11,000 MCF/D from the Rim Rock member of the Mesaverde formation (basal sand).

The second well to be put on production was put down by Mountain Fuel Supply Company. This well, the Eberle No. 1 in SE¼ of Section 9, was completed for an initial flow of 300 MCF/D from the Morapos sand member of the Mancos shale. Production from both of these wells fell off rather rapidly.

A deep test was drilled by General Petroleum in 1951. This test, G.P. 89-9-P, had several small shows of gas in the Frontier, Entrada, and Shinarump formations. About 10 feet of spotty stain was cored in the Entrada. The well was bottomed in Pennsylvanian sediments at 10,003 feet. Some geologists feel that the Weber formation was never reached, but the writer believes that the Weber or its equivalent was topped at 9,727 feet. The Weber facies loses its identity as it is traced eastward across Moffat County. Thus, the gray quartz sands exposed at Split Mountain thin and turn to pinkish colored silts and sands, that are interbedded with shales, some of which contain anhydrite. Thus, it appears that the line where the Weber facies grades into the Morgan sediment is fast being approached in the Bell Rock-Craig Dome vicinity.

The Craig structure, which has an east west orientation, lies on the south fringe of the Sand Wash Basin. This anticline was folded during Laramide time (probably Eocene).

The structure has several faults, all of which are parallel to the axis of the fold. The throw of these faults is under 100 feet where mapped on the surface. The mapped surface closure is between 150 and 200 feet.

The Mesaverde pay zone is typical of the Mesaverde formation. The sands are light gray to yellowish white. They are fine grained and often grade to siltstone. Permeability is low and porosity irregular. The high initial gas flow at the Craig discovery well may have been helped by having a fracture-system present in the reservoir.

SUMMARY OF WELLS DRILLED

Company and Lease	Location	Found or Tested Gas	Producing Interval
*Ohio Oil — Bogenschultz No. 1	C SE SE SE, 9	11,000 MCF/D	2769-81
Mt. Fuel — Bogenschultz No. 2	SE SE SE, 9	1,580 MCF/D	2750-52
*Mt. Fuel — Eberle No. 1	C SL, SW SE, 9	300 MCF/D	3560
Colo. Rep. — Eberle No. 1	C EL, SE SW, 9	No shows	
Gen. Pet. — 88-9-P	SE SE SE, 9	750 MCF/D	7120

*Wells put on production.

GEOLOGIC SECTION

	FORMATION	THICKNESS
CRET.	MANCOS	5000+ WELLS SPUD 1000 BELOW Kmc TOP
	FRONTIER	240
	MOWRY	133
	DAKOTA	76
JUR.	MORRISON	564
	ENTRADA NAVAJO	635
TRI.	CHINLE	227
	SHINARUMP	20
	MOENKOPI	515
PER.	PHOSPHORIA	153
PENN.	WEBER	MAXIMUM PENETRATION 180' EST. 600' TOTAL

100 FOOT CONTOURS ON TOP OF THE WEBER FORMATION — WEBER SAND PRODUCTION

TYPICAL ELECTRIC LOG
CONTINENTAL OIL CO.
2-41 M.V. SMITH

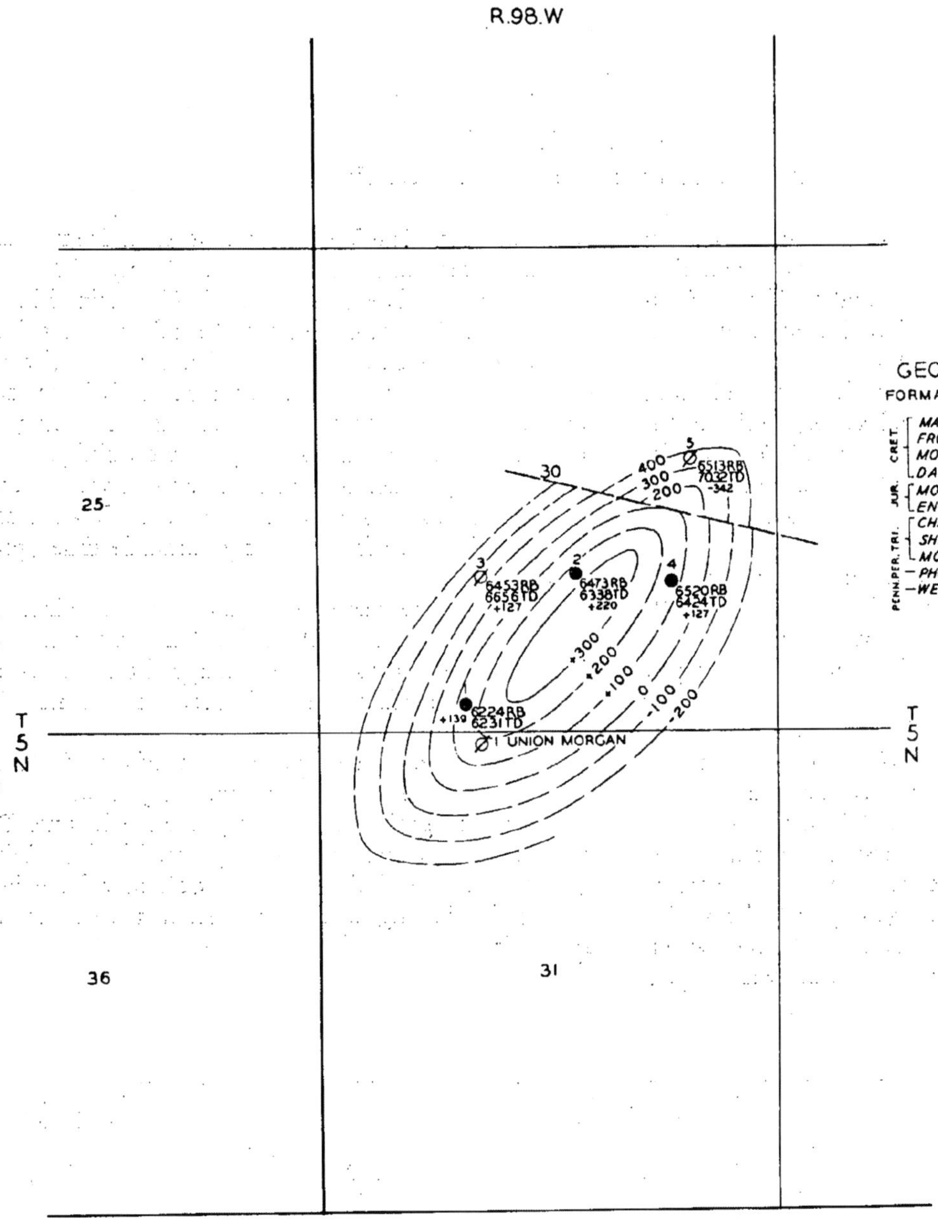

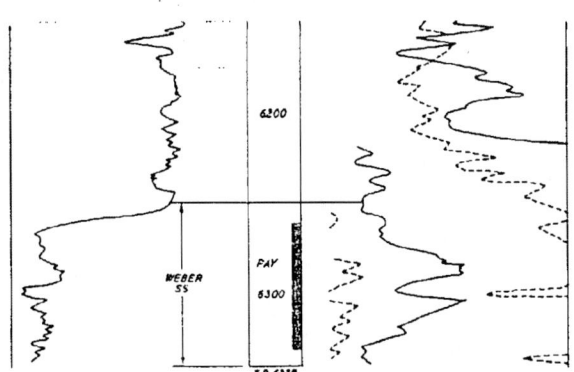

STRUCTURE CONTOUR MAP OF THE
ELK SPRINGS FIELD.
T. 5 N. R. 98 W. MOFFAT COUNTY, COLORADO.

0 500 1000 2000 3000 4000 FT.

SCALE: 1 INCH = 2000 FEET
GEOLOGY BY: W. D. FENEX
DRAFTED BY: DOYLE L. OLSON
CONTINENTAL OIL COMPANY
MAY, 6. 1955

ELK SPRINGS OIL FIELD
Moffat County, Colorado

By WILLIAM D. FENEX
Continental Oil Company, Salt Lake City, Utah

The Elk Springs Field is located approximately midway between the towns of Craig, Colorado and Vernal, Utah on U.S. Highway 40. The field lies in Section 30, T. 5 N., R. 98 W., Moffat County, Colorado.

The surface structure of Elk Springs was determined from strike and dip measurements taken in the Cretaceous Mancos shale. Most of these dips were obtained in deeply dug pits. The surface picture obtained from this type of mapping served as the basis for locating the well sites on the Elk Springs structure.

The first well of importance on the structure was drilled by Union Oil Company in 1925. This well, the Morgan No. 1 (NE NE NW 31-5N-98W), was drilled to a total depth of 3,973 feet. The hole was bottomed in the variegated shales of the Jurrasic Morrison formation. Considerable gas, estimated at over three million cubic feet per day, was recovered on tests of the Cretaceous Dakota sandstones in this well.

In November, 1946, commercial oil was recovered from the Pennsylvanian Weber sandstone in Continental's M. V. Smith No. 1. The initial production from the Smith well was estimated at 200 BOPD. Following the discovery of oil in the Smith No. 1, four more wells were drilled on the structure. The Smith No. 2 and the Smith No. 4 were completed as Weber producers, with an estimated initial production of 300 BOPD and 276 BOPD, respectively. The Smith No. 3 and Smith No. 5 were drilled through the producing horizon, but recovered only salt water on tests of the Weber.

Regionally, the Elk Springs structure is located on the southeast plunge of the Yampa Plateau uplift, which is a large horst bounded north and south by faults of considerable magnitude. This horst is an associated feature of the Uinta Mountain uplift.

Locally, the structure, as shown by shale dips, is a northeast-southwest trending anticline with about 500 feet of surface closure. The Weber structure, which is shown on the accompanying map, has been drawn by projecting surface contours down to the Weber datum. It is possible that the Weber structure is decidedly more complex than the picture herein presented.

At the surface on the southern portion of the structure, the Mancos shale of Cretaceous age is exposed. The Brown's Park formation of Miocene age overlaps the Mancos on the northern portion of the structure, making surface mapping impossible in this area.

The Weber pay zone consists of sandstone, white to pink, medium-grained, sub-rounded to sub-angular, slightly porous to tight, with fractures. The maximum penetration into the Weber is about 180 feet.

Other pertinent information such as nature of oil and statistics on production may be found in the recent Rocky Mountain Association of Geologists publication entitled "Oil and Gas Fields of Colorado."

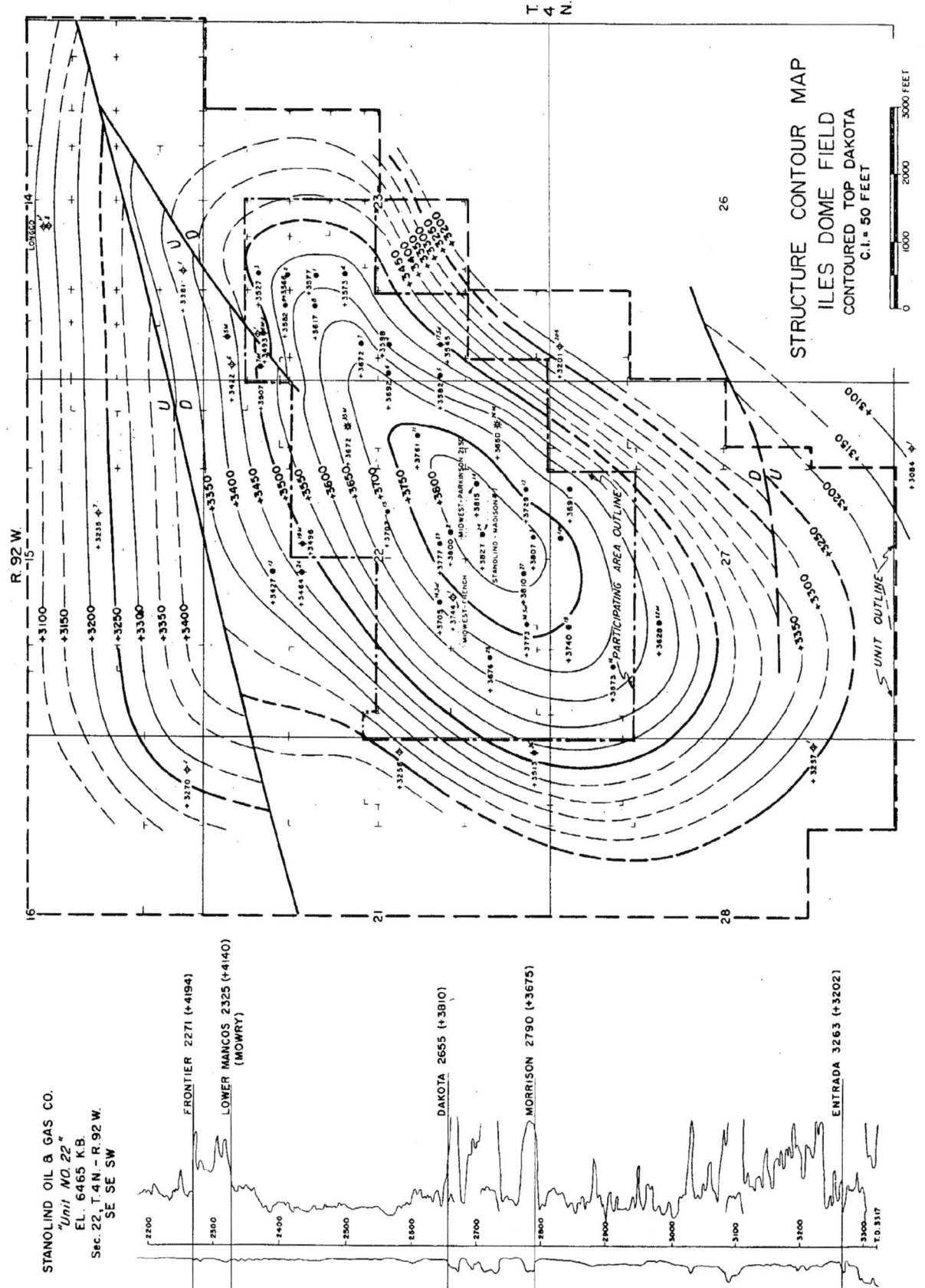

ILES DOME
Moffat County, Colorado

By ERIK NELSON
Stanolind Oil and Gas Company, Salt Lake City, Utah

Iles Dome oil field is located in Sections 22, 23, and 27, T. 4 N., R. 92 W., Moffat County, Colorado. The anticline is well expressed in the surface outcrops, and as a result, was recognized as a possible oil producing structure during the early period of oil exploration in the Rocky Mountains. Most of the wells were drilled from 1924 to 1947 with sufficient production established to make Iles Dome the largest producing field in northwestern Colorado until the discovery of Weber oil at Rangely.

The first test in this area, a dry hole drilled by Longco Oil in 1918 on the north flank of the structure in NE SW Section 14, T. 4 N., R. 92 W., encountered water-bearing Frontier sand and shows of gas in the Mancos shale section. Six years later, the Midwest Oil Company drilled the No. 1 French, in the NE SW of Section 22, T. 4 N., R. 92 W. This test, located high on the structure, was completed September 8, 1924, as an oil well with an estimated I.P. of 1100 BOPD from the lower Mancos (Mowry) formation. This well quickly went to water. Three additional shale wells were completed in 1925 and 1926.

On April 1, 1927, the Midwest Parkinson No. 4 (2150) located in SW SE of Section 22, was completed for 1900 BOPD from the Entrada (Sundance) formation. This well also encountered gas and oil shows in the Morrison formation and is credited as being the discovery well for both horizons. From 1927 to 1936 twenty-two successful wells were completed and the yearly production increased to 1,164,540 barrels in 1936.

In 1946, Stanolind Oil and Gas Company drilled the No. 1 Madison, located in NW SW SE Section 22, T. 4 N., R. 92 W., to a total depth of 7487 feet in the Madison formation of Mississippian age. A drill stem test taken in the Pennsylvanian Weber formation recovered 270 feet of 32.2 degree gravity oil and 480 feet of drilling mud, but attempts to complete in this formation were unsuccessful. A drill stem test in the Shinarump formation (Triassic) recovered only fresh water and mud, although staining was noted in samples.

Iles Dome is a closed anticline with 500 feet of structural closure expressed in the surface outcrops. The trend of the structure is NE - SW with the high point, as mapped on the Dakota, in the south-central part of Section 22, T. 4 N., R. 92 W. The dome is located near the southeast end of Axial Basin anticline where the axis of this major trend bends abruptly to the south. Transverse faulting cutting across the Axial Basin anticline was probably responsible, in part at least, for the formation of the Iles Dome structure. The largest fault is south of the producing area and has a maximum displacement of approximately 1000 feet (as measured east of the mapped area) downthrown to the north. Another fault has been mapped on the surface to the north of Iles Dome field. This fault is upthrown to the north, which places the closed portion of the structure in a graben between two upthrown blocks.

The surface of the dome is eroded deep in the Mancos shale (upper Cretaceous) with upper Mancos sands forming high ridges on the east and northeast sides of the dome. With the exception of these sands, the upper Mancos is a thick mass of homogeneous gray shale which develops steeply sloping hummocky topography covered by vegetation and detritus.

The Frontier sands at the base of the upper Mancos section are gray, fine-grained, hard, quartzitic, interbedded with sand shale, and carried gas shows in a few of the wells. The lower Mancos (Mowry) shale section underlies the Frontier sands. This 300 foot interval of gray and black shales is the productive portion of the Mancos formation at Iles Dome. The gray to white, fine- to medium-grained sands of the Dakota formation (upper Cretaceous) are water-bearing at Iles Dome.

The Morrison formation (Jurassic) is composed of variegated shales which are interbedded with sands in the lower half of the formation. The sands are fine- to medium-grained, white to light gray to gray-green, with sub-angular to sub-rounded fair- to well-sorted grains, slightly micaceous, and calcareous.

The Entrada formation (Jurassic) is a massive white sandstone with fine, sub-rounded, well-sorted grains, clean, friable with good porosity.

The oil produced from the Morrison and Entrada is 31.5 degree gravity with a pourpoint of 30 degrees Fahrenheit.

CUMULATIVE PRODUCTION
ILES DOME FIELD
To January 1, 1955

Formation	Barrels of Oil
Mancos	147,470
Morrison	1,210,462
Entrada	12,770,252
TOTAL	14,128,184 Barrels

MOFFAT DOME
Moffat County, Colorado

By DON VIEAUX and E. R. HAYMAKER
Phillips Petroleum Company, Denver, Colorado

LOCATION

Moffat Dome is a closed anticline in Ts. 4 & 5 N., R. 91 W., Moffat County, Colorado on the southern terminus of the regional northwest-southeast trending Williams Fork Anticline. The post-office and store of Hamilton are approximately three miles north of the Moffat field and Craig, Colorado is 12 miles northeast, straight line distance. The Iles field is approximately four miles southwest, being separated by a syncline.

The structure was originally named Hamilton Dome for the ranch and post-office on the north flank. Confusion with the Hamilton field in Wyoming caused the name to be changed to Moffat Dome about one year after discovery in 1924.

HISTORY AND DEVELOPMENT

The discovery well, No. 1 B. Knowlton, was drilled jointly by the Texas Production Company (The Texas Company) and the Transcontinental Oil Co. It was spudded June 19, 1923 with cable tools, in the SE¼, section 34, T. 5 N., R. 91 W. The first show of oil was encountered in the Mancos shale at 2,045'. A test of this zone filled the hole with 1,200' of 36° gravity oil and produced 41 BOPD. At 2,145' a second zone in the Mancos tested 125 bbls of oil the first day, however, production declined to 61 BOPD by the sixth day, and the test was terminated because of tank shortage. The shale oil was cased off at 2,364'. On January 15, 1924 the top of the Dakota sand was encountered at 3800'. Five feet of the sand was drilled which tested 1000 BOPD, and at a T.D. of 3820' the production increased to 4,500 BOPD. The No. 1 B. Knowlton was completed February 29, 1924. Development history for this and other wells completed in the Dakota formation indicates that early flowing production was maintained by water drive. Shut in well head pressure of the discovery well was 400 psi. The initial production of the discovery well averaged 4580 BOPD during the month of March, 1924. It was shut-in at the end of March of that year and not produced again until July, 1924, due to lack of market. At sometime during production in July, water encroachment became evident and the producing rate had declined to 3600 B.F.P.D. with 20% water. The amount of water produced continued to increase with time, and by November, 1925 the well had died. All attempts to revive the well were unsuccessful, and the hole was abandoned. Total production from the well was approximately 429,000 bbls. of oil.

Subsequently, commercial quantities of oil have been discovered in the year indicated in the following formations: Mancos, 1953; Morrison, 1948; Entrada, 1928; and Shinarump, 1954.

A total of 18 commercial wells were drilled in the field and at the present time, there are 11 producing, one from Mancos, three from Dakota, six from Entrada and one from Shinarump. No wells are producing from the Morrison at present. Total cumulative production from all pays in the field was 7,245,921 bbls. of oil to the end of 1953. Annual field production for 1953 had declined to 61,556 bbls. of oil from all zones.

STRUCTURE

Moffat Dome is a closed anticlinal structure. The domal symmetry undoubtedly suggested a portion of its name. Surface contours show the south and west flanks to be steep, while the north and east are relatively gentle. The steep flanks have dips exposed up to 25 degrees, while the dips on the north and east flanks are 4 degrees to 11 degrees. Approximately 1000' of closure has been mapped on the surface.

The subsurface structural picture indicates a much more gently dipping structure than does the surface. This contrast is no doubt due in part to lack of structural control on the flanks. All of the production from Moffat Dome Field is restricted to an area of approximately one half square mile, while the surface closure includes approximately six square miles.

STRATIGRAPHY

The following are the geologic formations and approximate thicknesses encountered in the Texas Company's No. 14 B. Knowlton deep test in Sec. 19, T. 4 N., R. 91 W.:

	Drilling Depth	Thickness
Mancos	Surface	3948'
Dakota	3928'	97'
Morrison	4045'	461'
Curtis	4506'	87'
Entrada	4593'	242'
Chinle	4835'	445'
Shinarump	5280'	304'
Moenkopi	5584'	403'
Weber	5987'	206'
Morgan	6193'	TD 6193' in Morgan

The surface rocks on the anticline have been eroded approximately 1300' into the Mancos formation. An irregular escarpment of Morapos sandstone surrounds the center of the dome and is 900'± below the top of the Mancos. The Mesaverde formation forms steep cliffs at the perimeter of the Mancos outcrop on three sides of the structure. The syncline on the southwest side, where no Mesaverde is present, dips down and rises immediately to form the northeast flank of the Iles Dome.

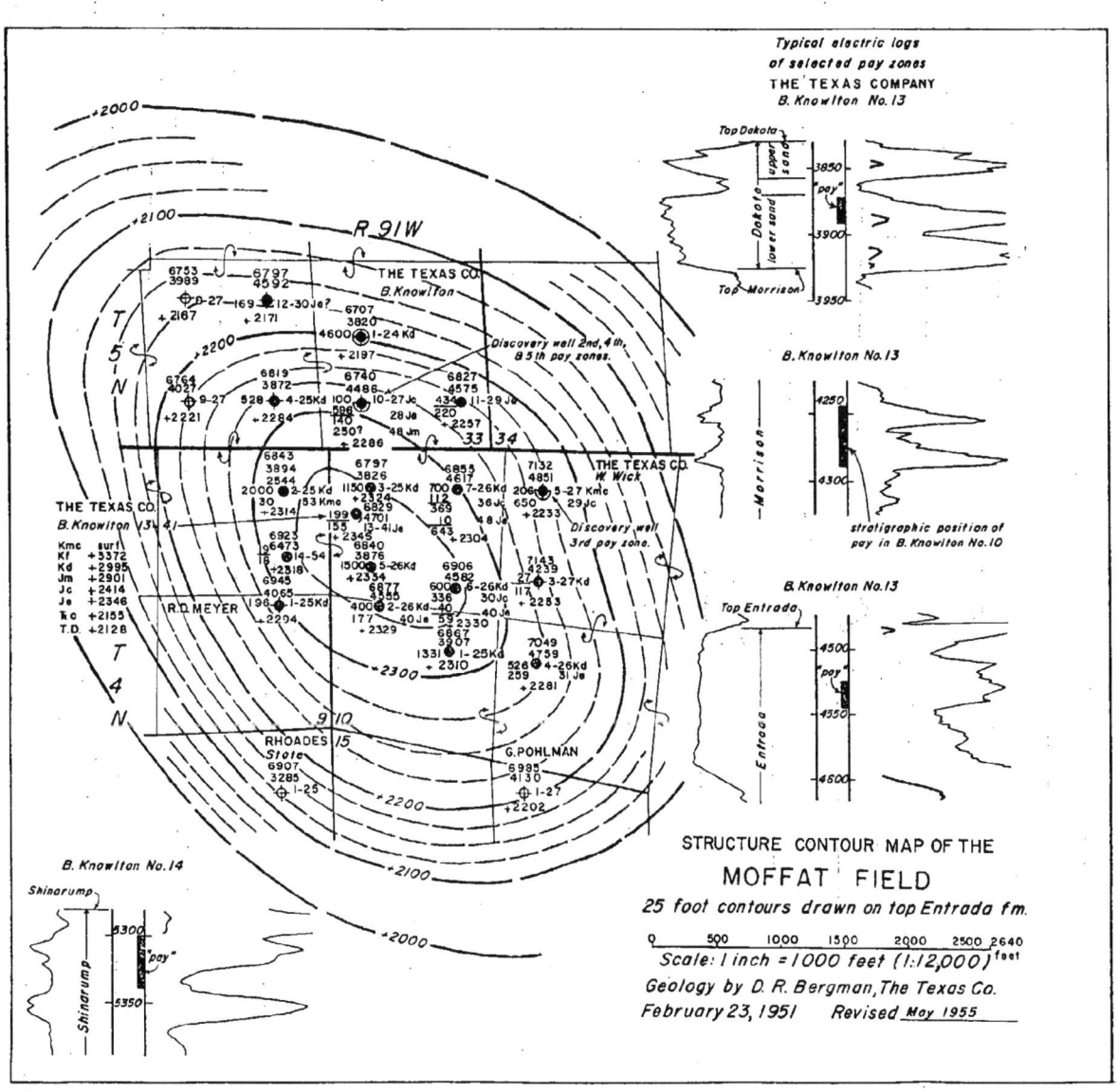

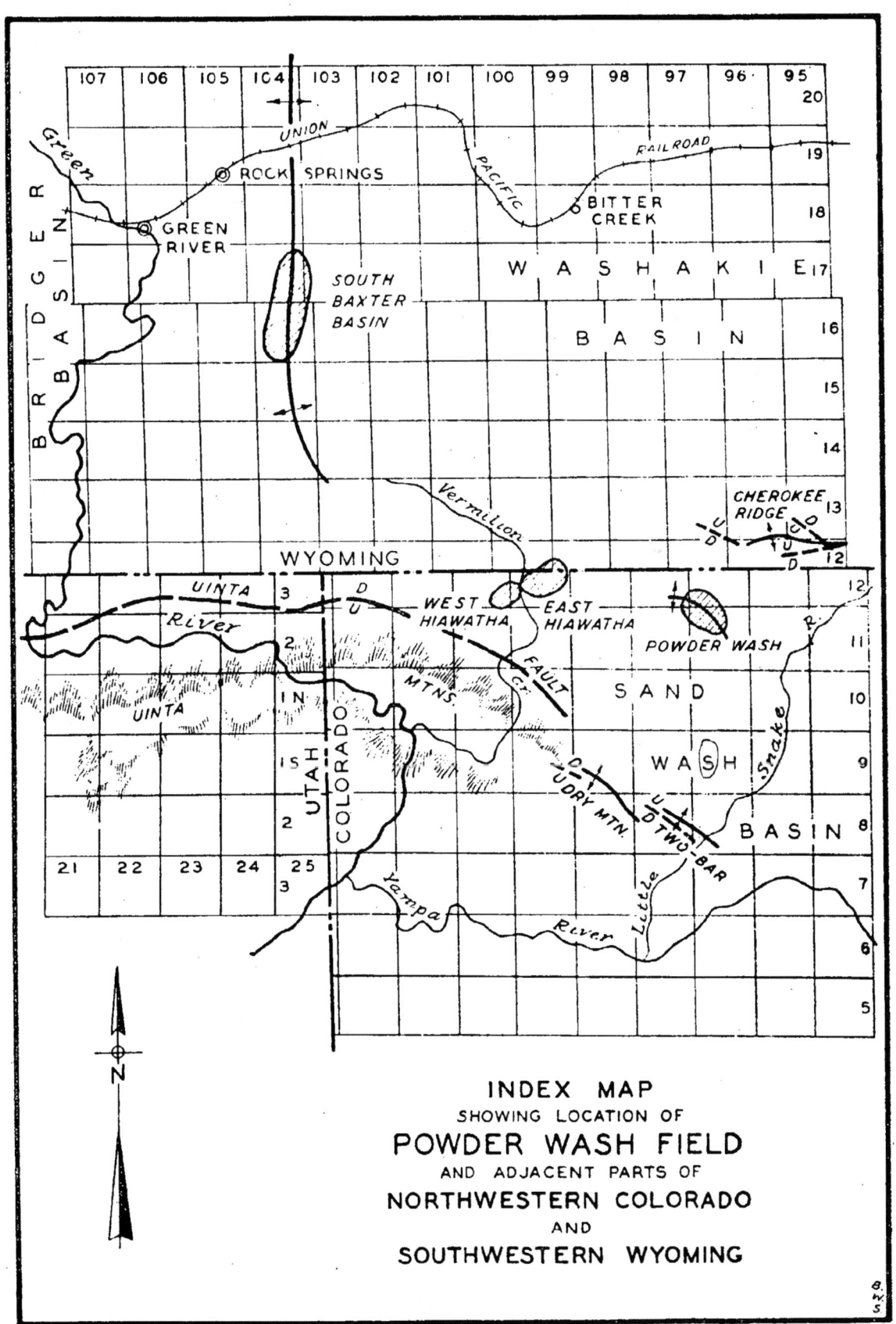

INDEX MAP
SHOWING LOCATION OF
POWDER WASH FIELD
AND ADJACENT PARTS OF
NORTHWESTERN COLORADO
AND
SOUTHWESTERN WYOMING

POWDER WASH - ACE FIELD*
Moffat County, Colorado

By L. W. FOLSOM
Mountain Fuel Supply Company, Rock Springs, Wyoming

INTRODUCTION

Thirty-four wells have been drilled in the Powder Wash-Ace field including the original discovery well, B. W. Musser Well No. 1, which was completed in 1931. Development through the 1930's and early 1940's was relatively slow. All but one of the early wells were completed in sands of the Wasatch formation between depths of 2100 feet and 5100 feet. The J. C. Donnell Well No. 5 was completed, however, in a sand topped at 5362 feet, questionably near the top of the Fort Union formation. Since 1950 development has been greatly accelerated, especially since 1952 when deeper Fort Union production was encountered in the Powder Wash field. The eight Ace Unit wells and Carl Allen Wells Nos. 9, 10, and 11 have all been completed since early 1954.

LOCATION

The Powder Wash-Ace field is located within Townships 11 and 12 North, Ranges 97 and 98 West of the Sixth Principal Meridian, Moffat County, Colorado, about three miles south of the Wyoming-Colorado state line (Fig. 1). It lies within the tributary system of the Green River, being located within the drainage area of the Little Snake River which drains into the Yampa River of northwest Colorado. The Yampa River is, in turn, a tributary of the Green River. It is outside the limits of the Green River structural basin, however, which is located to the northwest of this particular area.

HISTORY AND DEVELOPMENT

The Powder Wash field was discovered by surface geological mapping. As previously mentioned, the discovery well was drilled in April, 1931. This well, B. W. Musser Well No. 1, drilled by Mountain Fuel Supply Company, was completed in a nonmarine sand of the Wasatch formation between 2152 feet and 2182 feet. Initial open flow was 34,118,000 cubic feet of natural gas a way. Development of the field was slow because of the absence of pipeline outlet. In November, 1936, Carl Allen Well No. 1 was completed in a Wasatch sand at 5014 feet for an indicated initial production of 45 barrels of 37.4 degree A.P.I. gravity oil an hour with about 3,000,000 cubic feet of natural gas a day. In 1937 two wells were completed, Hal W. Stewart Well No. 1 and J. C. Donnell No. 1. Initial production in these wells was 33 barrels of oil an hour and 4,000,000 cubic feet of natural gas a day respectively from sands of the Wasatch formation. A period of about four years elapsed before fairly persistent development began in 1941 with completion of the B. W. Musser Wells Nos. 2, 3, and 4. B. W. Musser Well No. 1 was drilled deeper in 1943 to a new total depth of 5509 feet. At this new total depth the well penetrated about 384 feet of sediments considered to be of Fort Union age. This well was recompleted dually from two Wasatch sands at 2870 feet to 2901 feet and 5109 feet to 5125 feet. No new drilling was done in the Powder Wash field in 1943, 1946, or 1947, but beginning in 1948, development became fairly persistent. Carl Allen Well No. 6 was drilled in 1952 and 1953 to a total depth of 8407 feet. The well encountered natural gas and distillate in a Fort Union sand between 8085 feet and 8130 feet, and was plugged back and completed for an initial production of 8,770,000 cubic feet of natural gas and 74 barrels of distillate a day.

The discovery of Fort Union production in Carl Allen Well No. 6 encouraged drilling of Carl Allen Wells Nos. 7 and 8. Subsurface information gained by the drilling of these wells indicated the possibility of an eastward extension of Fort Union production, and drilling was started in the Ace Unit area. Ace Unit Wells Nos. 1, 2, and 3 were drilled and completed in 1953 and 1954 as commercial gas wells in Fort Union sands. A shallow oil sand in the Wasatch formation encouraged further drilling in the Ace Unit area and, to date, ten wells have been completed on the east side of the Powder Wash field in the Ace Unit area. Of these ten wells, only one, Ace Unit Well No. 4 has been dry.

Interest in the possibility of Mesaverde production resulted in the drilling of Carl Allen Well No. 11. No commercial production was encountered in the Mesaverde formation, however, and the well was plugged back and completed in a Fort Union sand at 7992 feet to 8007 feet after having reached a total depth of 12,985 feet.

---

*Published by permission of Mountain Fuel Supply Company.

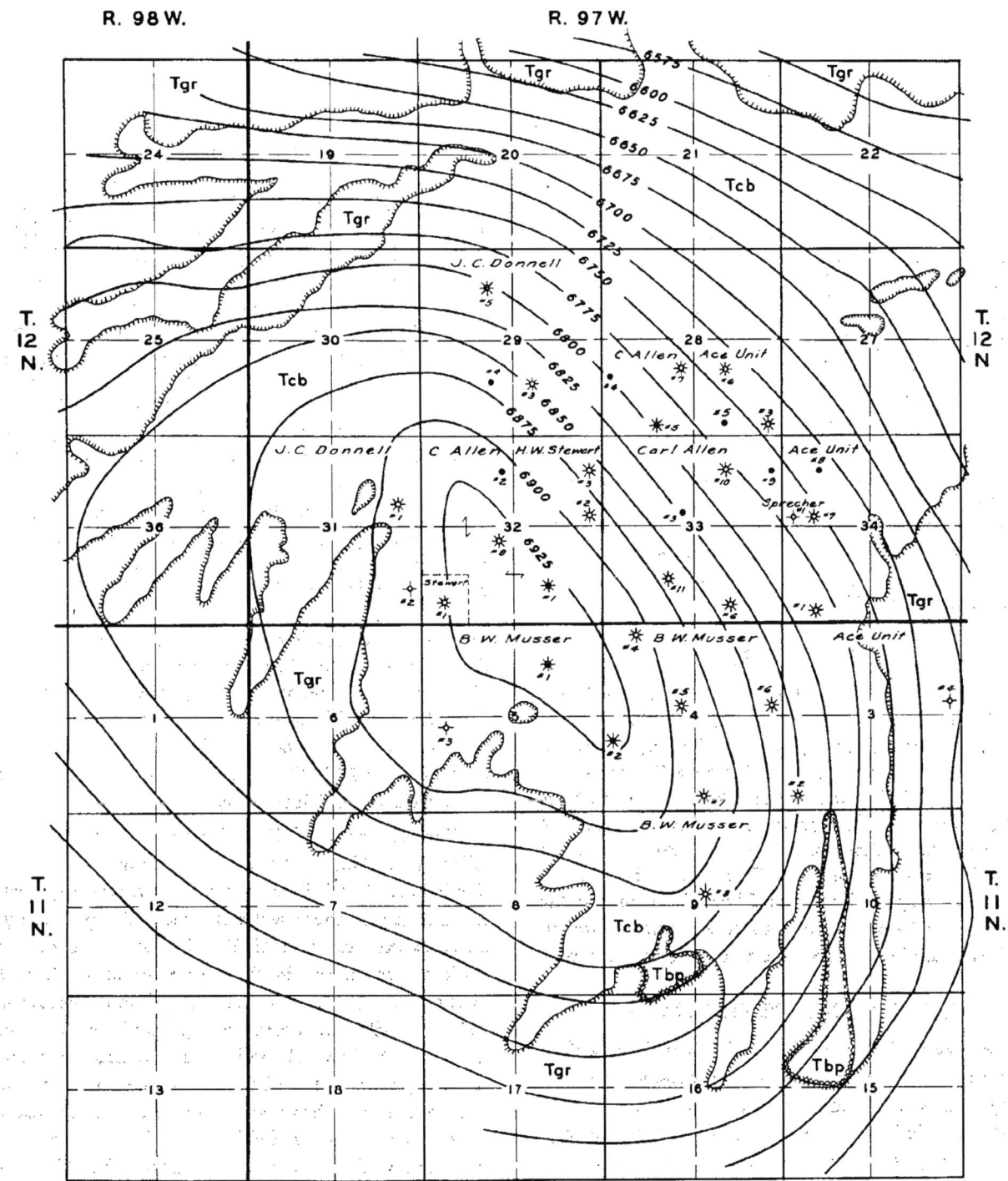

LEGEND
- Tbp Browns Park formation
- Tgr Green River formation
- Tcb Cathedral Bluffs tongue

GEOLOGICAL MAP
OF
POWDER WASH STRUCTURE
MOFFAT CO., COLORADO

STRATIGRAPHY

Rocks exposed at Powder Wash and in the adjacent area belong principally to the Upper Wasatch and Green River formations of Eocene age. A few remnants of the Browns Park formation of Miocene (?) age occur around the borders of the Powder Wash structure. The conspicuous rimrock surrounding the Powder Wash field is composed of the Laney shale member of the upper Green River formation. As exposed in the vicinity, the thickness of the Laney shale member is about 1200 feet. Erosion has exposed the upper part of the Cathedral Bluffs tongue of the upper Wasatch near the crest of the structure. The thickness of the Cathedral Bluffs tongue is about 1850 feet, most of which is unexposed.

The deepest well drilled in the Powder Wash-Ace field, Carl Allen Well No. 11, bottomed in the Mesaverde at a total depth of 12,985 feet. Following is a general description of the stratigraphic section penetrated in the field:

System	Series	Formation		Thickness	Lithology
Tertiary	Eocene	Cathedral Bluffs tongue (Upper Wasatch formation)		1850±	Reddish clay shale with coarse sandstones. Continental.
		Tipton shale tongue (Lower Green River formation)		75±	Thin-bedded, fissile shale. Oil shale. Lacustrine.
		Hiawatha member (Lower Wasatch formation)		3100±	Gray to variegated clay shale; lenticular sandstones. Coal. Continental.
	Paleocene	Fort Union formation	Undifferentiated	4100±	Dark shales; gray to brown sandstones. Some thin coals.
Cretaceous	Upper Cretaceous	Lance			
		Montana Group	Lewis shale	1400±	Gray marine shales.
			Mesaverde formation	2515+	Gray sandstones, thin to massive with gray shales and coal.

The producing formations of the Powder Wash-Ace field, the Hiawatha member of the Wasatch and the Paleocene Fort Union formation, are typically variable. The sediments of the Hiawatha member show paludal and lacustrine depositional environments, but are probably predominately fluviatile. The sands are extremely lenticular, but some more persistent sands occur which in general are water bearing.

The Fort Union formation consists of dark carbonaceous shales with well developed gray sandstones. Numerous thin carbonaceous partings and thin coals are present. The sands are lenticular in character to nearly as great an extent as those of the Wasatch. Fluviatile, lacustrine, and paludal depositional environments are indicated by the Fort Union sediments as well as by those of the Hiawatha member.

The Lance formation cannot be differentiated from the Fort Union on the basis of samples or Schlumberger logs in the Powder Wash field. It contains lenticular sandstones, dark carbonaceous shales, and thin carbonaceous zones indicative of depositional environments similar to those of the Fort Union and Hiawatha member of the Wasatch. The Lance formation may represent a transitional zone between the underlying marine Lewis shale and the overlying nonmarine Tertiary formations. A thickness of about 4100 feet has been assigned to the undifferentiated Fort Union formation and the Lance. A thickness of about 800 feet can possibly be assigned to the Lance formation at Powder Wash and about 3300 feet to the overlying Fort Union formation.

STRUCTURE

The structure of the Powder Wash-Ace field was first mapped in detail by W. T. Nightingale of Mountain Fuel Supply Company, prior to 1930. Previously, Julian D. Sears, in United States Geological Survey Bulletin 751, had indicated in a general way the presence and location of the Powder Wash structure.

Regionally, the Powder Wash dome forms a structural divide between the central Washakie Basin of southwestern Wyoming and the subsidiary Sand Wash

Basin of northwestern Colorado. Powder Wash is located along a trend which lies between the Cherokee Ridge line of folding and faulting to the northeast, and the Dry Mountain-Two Bar line of folding and faulting to the southwest. The origin of the Powder Wash dome was probably the result of Tertiary (Eocene) deformation along these older lines of weakness which probably had their inception in late Cretaceous orogeny.

The surface structure of Powder Wash is that of a broad, gentle dome, relatively regular and unfaulted. The axial trend is northwest and in an alignment between that of the Cherokee Ridge trend on the northeast and the Dry Mountain-Two Bar trend on the southwest. Surface closure is about 100 feet. Dip on the northeast flank averages 125 feet a mile. Dip on the southwest flank is somewhat less, averaging about 75 feet a mile (Fig. 2). An area of about 5000 acres is included within the area of surface closure.

There is no definite evidence of subsurface faulting, although there is some indication of weak faulting between Carl Allen Well No. 6 and Carl Allen Well No. 11. This faulting, if present, is indistinct on the Schlumberger logs and would not have more than about 25 feet of displacement.

The accumulation of oil and gas in the Powder Wash-Ace field is controlled primarily by structure, but obviously is modified by the lateral extent of the lenticular sands and by porosity and permeability barriers within these sands. The hydrocarbons in the Wasatch and Fort Union formations are undoubtedly indigenous to the group of sediments in which they occur. Shallow lakes or barred basins were obviously present at intervals during deposition of the Fort Union and lower Wasatch sediments. Such basins could have served as traps for large quantities of organic material which, through decomposition in the protected basins, were transformed into petroleum and natural gas. The protected nature of these catchment basins is evidenced by the saline connate waters associated with some of the sandstone horizons. There is no evidence of extensive faulting which could have served as migratory conduits in the Powder Wash field, although some minor faulting may be present. Because of the extreme lenticularity of the lower Wasatch and Fort Union sands, any vertical or long range horizontal migration of hydrocarbons is almost inconceivable. These facts lend support to the idea of the indigenous nature of the petroleum and natural gas in the Powder Wash field.

REFERENCES

Nightingale, W. T. (1930), "Geology of Vermilion Creek Gas Area in Southwest Wyoming and Northwest Colorado," Amer. Assoc. Petrol. Geol. Bull., vol. XIV, no. 8, pp. 1013-1040.

─────────── (1938), "Petroleum and Natural Gas in Nonmarine Sediments of Powder Wash Field in Northwest Colorado," Amer. Assoc. Petrol. Geol. Bull., vol. XXII, no. 8, pp. 1020-1047.

Sears, J. D. (1924), "Geology and Oil and Gas Prospects of Part of Moffat County, Colorado, and Southern Sweetwater County, Wyoming," U. S. Geol. Survey Bull. 751, pp. 269-319.

WEBER POOL OF RANGELY FIELD, COLORADO

By GRAHAM S. CAMPBELL
Consultant, Salt Lake City, Utah

LOCATION

The north trending Douglas Creek Arch is situated between the Uncompahgre Uplift and the Uinta Mountain fold and has been the principal tectonic factor in the estrangement of the Piceance Basin of Colorado and the Uinta Basin of Utah. Rangely Field lies on the north end of the Douglas Creek Arch in Rio Blanco County of Northwest Colorado. The field is 12 miles south from U. S. Highway 40 between Denver and Salt Lake City.

HISTORY

Reported oil seeps motivated drilling and the establishment of small shallow production from fault generated fractures within the Mancos shale in 1902 in SW SW NE section 33, T. 2 N., R. 102 W. The next notable development began in 1919 when another Mancos well was completed for 130 barrels oil per day at 520 feet depth. The Mancos shale yields about 800 barrels of 52° gravity oil per day at present.

Intermittent drilling continued until 1933 when the California Company drilled the Raven 1-A (see field map in pocket) into the Weber at 5700 feet and completed a flowing well with an initial production of 300 barrels of 31° gravity oil per day. The well was produced for a few months, then shut in until 1943 when it was reopened to produce over 200 barrels oil per day. Circumstances, presumably economic, prompted the "Rangely Boom" from 1945-49 during which time a total of 480 Weber completions were made. A ten-inch pipeline was completed in 1948 from Rangely to Salt Lake City and to Wamsutter, Wyoming. Deeper possibilities had always been considered and in 1953 the Texas Company-Union Pacific Railroad well was drilled to the Cambrian quartzite at 9360 feet. No commercial oil was found below the Weber.

STRUCTURE (See map in pocket)

The Rangely anticlinal dome is about 10 miles long trending north 60 degrees west. Though essentially a mild fold, the southwest flank attains dips of 30 degrees, suggestive of severe subsurface crowding beneath that flank. The crestal area, once called Raven Park, is a topographic basin surfaced with soft Mancos shale and rimmed by the Cretaceous Iles (lower Mesaverde) formation. There is nearly 2000 feet of surface closure and subsurface data are in reasonable conformity. There is some small scale normal faulting on the south flank of the structure. The faults trend north 60 degrees east and are closely related to the Mancos shale production.

STRATIGRAPHY (With related comments)

With slight variations, the drilling data shown on the map are adequate and representative for formation thicknesses. The Mancos shale is 3000 feet thick at the structural crest. In descending order the Dakota, Morrison, Curtis, Entrada, Carmel, Navajo and Chinle total about 1800 feet. The Shinarump, which now produces 30 barrels oil per day, from one well, is from 40-70 feet thick and the Moenkopi is about 700 feet in thickness. The Permian Park City formation ranges up to 130 feet in thickness. The total average drill depth to the Weber is 5700 feet.

The Weber sandstone is the principal producing formation at Rangely, yielding about 98% of the total field production. The oil is mixed base, averages 35 degrees A.P.I. gravity, with 40°F. pourpoint. The Weber is 1200 feet thick, containing effective pay zones in the upper 400 feet. The pay interval is very fine grained, light colored calcareous sandstone. The formation gains red coloring downward and has a few shale breaks. The lower third is arkosic, characterized by tight clayey, micaceous matrix. In the pay zone the permeability varies from good on the west end of the field to very poor on the east. The increase in argillaceous red beds is inversely proportionate with porosity and permeability to the east. Vertical fracturing is an important production and permeability factor, particularly where primary porosity is low.

The low millidarcy character of the east end of Rangely Field has led to the conclusion that the field is first a stratigraphic trap and secondly, a domal entrapment. It is interesting to speculate on when the Rangely oil pool might have been discovered, had there been no discernible east flank.

The type Weber quartzite of Weber Canyon, Utah is Pennsylvanian in age and is probably older than the upper part of the Rangely Weber sandstone. The upper third of the formation at Rangely is now regarded as of Permian age by most stratigraphers who also match the upper Weber (?) sandstone with the Coconino and regard the Uncompahgre highland as source area for Rangely Weber sediments.

The Pennsylvanian Morgan formation is about 1600 feet thick, containing an interbedded series of reddish

sandstones, siltstones, limestones, and shales. Indicated permeabilities are very low.

The Mississippian Madison consists of 500 feet of limestone and dolomite, the medial part of which is apparently quite permeable, judging from the vugular character of the cores.

The Devonian Chaffee formation is made up of gray and tan carbonate rocks with purple, red, and green staining in a series about 200 feet thick. The lower part of the formation contains a quartzitic sandstone tentatively regarded as the Parting quartzite member of the Chaffee. A tenable correlation also exists between the Rangely Devonian section and the Ouray and Elbert formations of southwestern Colorado.

A notable similarity exists between the pre-Devonian section at Rangely and the same section at Glenwood Canyon, about 70 miles southeast. The lower Paleozoic series consists of 400 feet of reworked dolomites and dolomitic sandstones correlative with the Ordovician Manitou, and the upper Cambrian Dotsero and Sawatch formations.

FUTURE CONSIDERATIONS

Rangely field is the most prolific field in the Rocky Mountain Region. It has produced a cumulative total of about 150 million barrels of oil resulting in a gross income of about 300 million dollars. The structure has received only one deep test and further testing is feasible if not imminent. The asymmetric nature of the Rangely dome suggests a shifting axial plane and perhaps a subsurface flank fault. The apparent permeability in the Madison along with oil shows in the Morgan are encouraging for future considerations. Notwithstanding its economically gratifying past performance, it would seem that the Rangely structure still has unexplored possibilities worthy of future attention.

THE RANGELY BOOM

Rangely Anticline was known as an anticlinal structure from the days of the Hayden Survey. In 1902, shallow production from fractured Mancos shale was discovered, and more than two million barrels of high gravity crude oil was produced from a myriad of wells before the deep Weber reservoir was tapped forty years later. Thus Rangely, in addition to being a local trading center for stockmen, was a small oil camp from the early years of the century.

Discovery of the Weber reservoir at Rangely was made in 1933 by the California Company with its Raven No. 1 well. However, remoteness of the area, transportation problems and market conditions did not encourage development of the field. It was not until 1943 that World War II triggered the explosion of activity that was the Rangely Boom.

The 1933 discovery well was put on production in September 1943; and the second deep test, California Emerald No. 1, was spudded in April 1944. Rigs converged from all directions, and drilling was commenced hastily by half a dozen or so companies and individuals whose leases blanketed the sprawling structure. Through 1944 and 1945, the two wells increased to 182; and at one time during 1946, 54 rigs were running at once in the field. By the end of 1948, Rangely was delimited by 473 wells.

Every oil boom has a history of unbelievable happenings, colorful characters, fortunes made and lost, hardships and heartbreak. The "Oil Basin", as Rangely was known, was remote, 100 miles over crumbling blacktop, through mud and dust from the nearest railroad. It was war time with its tangle of allocations, rationing and shortages of everything but red tape. Often a bankroll couldn't buy a hamburger for there was none to be had. Flattened oil cans, packing cases and used and re-used tarpaper were fought over for building material. Re-capped tires failed, gas coupons ran out and the milk of human kindness and decency soured in the frenzy and frustration. At times the difficulties seemed insurmountable; but bits went down, gathering lines spread over the basin, and two major pipelines snaked across country to carry the surge of crude oil northeast and west to refining centers.

Rangely shared its boom with the whole region. Craig burst into activity as the 100 mile distant railhead; and Vernal, nearest sizable town, strained at the seams with population that could not be accommodated at the field 40 miles away. A rash of exploration activity spread over northwest Colorado and eastern Utah as new Rangelys were eagerly sought after. The Weber, however, proved to be fickle and there is just one Rangely today.

When veterans of the boom meet, there is the usual swapping of yarns, the recalling of legend and lies and some incredible truths. There are stories of fun and frustration, exhilaration and exasperation. Most swear they wouldn't have missed it for the world; but, like beachhead veterans, they hope it never happens to them again.

Howard R. Ritzma

FRACTURE PRODUCTION FROM THE MANCOS SHALE, RANGELY FIELD, RIO BLANCO COUNTY, COLORADO

By V. E. PETERSON

Equity Oil Company, Salt Lake City, Utah

With few exceptions, geologists of the Rocky Mountain area are acquainted with the Rangely Field and know it for the prolific production and reserves found there in the Weber sandstone of Permian and Pennsylvanian age. The structure associated with this field is one of the most outstanding surface anticlines occurring in northwestern Colorado. Because of its lesser economic significance and confinement essentially to operations by small independent producers, very few geologists have delved into the geologic relationships associated with the occurrence and production of oil from the shallow Cretaceous Mancos shale of this field.

Structure

The Rangely Field is located on the Raven Park Anticline which is a large asymmetrical and somewhat arcuate fold approximately 20 miles long and 8 miles wide and trending in a northwest-southeast direction. (See generalized geologic map and oil and gas field map, this guidebook.) It is situated intermediate between the Douglas Creek arch and the Blue Mountain uplift. Except for recent sediments, surface rocks cropping out in the central part of the fold are all of Cretaceous age. The periphery of the fold is outlined almost completely by a hogback formed by the sandstones of the Mesaverde formation. The Mesaverde sandstones have been eroded from the top of the structure to form Raven Park throughout which the underlying Mancos shale section is exposed. Production from the Rangely Field is within the confines of the Mancos shale exposure and it is within this area where production from the Mancos shale has been found and is being developed.

Shale Reservoir

The Mancos shale section of this area consists of a monotonous series of dark gray to black, somewhat calcareous and bentonitic shales 4,200 feet in thickness. A few silty members occur within the section, but these are rare and confined mainly to the upper part of the formation. Electric log characteristics of the shale section are consistent, and it is possible to correlate readily equivalent horizons in the section throughout the field. On the basis of electrical log correlation, it has been attempted unsuccessfully to tie the production from the Mancos shale to particular horizons within the section. For the present consideration, there is one outstanding characteristic of this shale section, that is that it does not contain an effective reservoir within itself. The shale section is all of very low permeability and porosity and possesses none of the usual reservoir characteristics. Such reservoir characteristics as the shale section does possess are due to retained openings in secondary fracturing associated with the formation of the fold.

Orientation of Fractures

At an early period in the development of the Mancos pool of this field, operators noted that the appearance of calcite in the well cuttings had a definite correlation with the occurrence of the oil in the shale section. They were quick to recognize that this occurrence of calcite in the samples came from calcite incrustations or calcite veins associated with fractures within the shale section. Fracture calcite veins are exposed abundantly at the surface at several localities in the field and most of these fracture veins have been noted and mapped by the shallow oil operators of the field. The majority of the fracture calcite veins that have been mapped are shown on Figure 1A, which represents the fractures thus far found in search of the western two-thirds of the Rangely Field. From this map it will be readily recognized that the fracture veins fall into two groups, of which one is by far the more important. Fracture veins trending in a northeast-southwest direction across the field are relatively much more abundant than those occuring south of and parallel to the axis of the fold. The northeast-southwest trending fracture veins are observed to hade predominantly to the southeastward, though quite a number of these fracture veins hade to the northwestward. It has been found through drilling that the fracture veins having a hade to the southeastward are more persistent with depth and have consequently been found to be of most importance in the location of oil. Fracture veins paralleling the axis of the fold have also been observed to hade in both directions, but more commonly they have been found to hade to the northeastward or toward the axis of the fold. The hade of the fracture veins in either group averages between 5 degrees and 15 degrees from the vertical, though wide variations from these figures have been observed. From information obtained through drilling it has been observed that fracture veins having high angles of hade

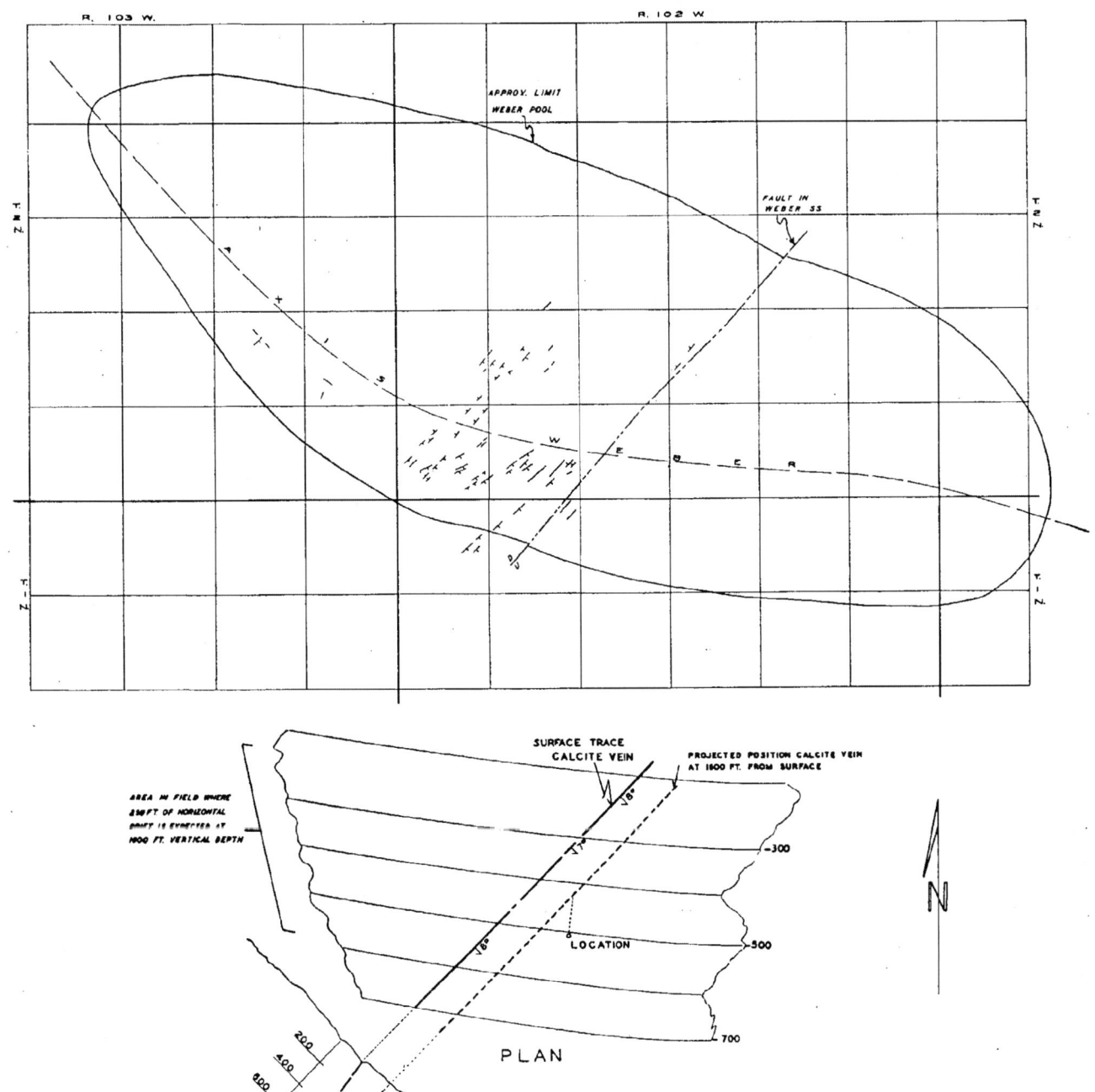

Figure 1. A. (Above) Fracture pattern in Rangely Field.
B. Typical location problem.

are inclined to be associated with near-surface fractures or local surface slumping and do not persist with depth.

From further observation of Figure 1A it will be noted that the preponderance of the northeast-southwest trending fracture calcite veins are confined to essentially one area within the field. While it must be recognized that the information shown on this map is incomplete in so far as the southeastern part of the field is concerned, this concentration of fracture veins is representative of observed conditions. This concentration of fracture calcite veins is located within the area of greatest arcuate deviation of the axis of the fold and within the area of greatest tangential tension. It should be noted that this group of fracture calcite veins essentially dies out on reaching the apex of the fold and does not extend into the northeastward flank of the anticline. Forces creating tension on the south side of the axis of the fold in this area of greatest arcuate deviation conversely would exert compression on the north side of the axis. This relationship is believed to be responsible for the dying out of the fracture calcite veins to the northeastward. An alternate hypothesis for the localization of this group of fracture calcite veins might be that they are related to the fault that crosses the field in about this same area. This fault has only 50 feet of displacement, however, and it appears questionable to the writer that a fault of this magnitude would possess the degree of associated fracturing found in this wide belt. In addition, if the fracturing is secondary to the formation of the fault, the question may be asked as to why the fracture calcite veins are not found all the way across the field as is the fault. It is the writers opinion that the fault is also related to this zone of relatively greater readjustment in the fold and that the fractures and fault have a consanguineous relationship to each other.

Nature of Fractures

Calcite marking the fractures at the surface may be in veins from a fraction of an inch up to as much as a foot in thickness and usually displays great variation in thickness along the vein. In nearly all cases, the calcite does not weather out into relief on the surface, but occurs as a scattered rubble above the buried vein. For this reason, an excavation of one kind or another is required to reveal the hade and exact trend of the vein.

The mode of surface weathering associated with the fractures and related calcite veins clouds the nature of the reservoir characteristics within the fractures. Where observable, the character of the calcite veins within themselves appears to offer only limited possibilities for the development of substantial reservoirs in which oil might accumulate. Openings have been observed within the calcite veins, but generally the veins themselves are tight. The relatively great local variation in the thickness of the veins on the surface appears to indicate a possible interpretation of the character of the reservoirs. It is believed that the fractures originally did not possess this great variation in width, and that the variation in the thickness of the calcite veins found in the fractures is due to localized variation in the rate of deposition not necessarily related to the width of the fracture opening. Thus, the function of the calcite occurring in the fracture may commonly be that of a bridge or a support sustaining the fracture opening. Under this interpretation, calcite-barren or deficient areas within a given fracture could form a substantial reservoir in which oil might accumulate. So far as surface information is concerned, there is little evidence to support this theory. However, information gained through drilling is at least limitedly corroborative of this interpretation. It has been noted that there is no direct relationship between the amount of calcite recovered in the well samples and the potential and recovery of oil from the well. It is not uncommon to drill through extensive calcite at an appropriate depth and not obtain oil. On the other hand, some of the best wells in the field have been obtained from holes in which no calcite was found. If the calcite veins themselves were responsible for the principal reservoir, a direct relationship between well potential and the thickness of the calcite should be expected.

That the fractures are not uniformly open and freely interconnected is shown by the way in which the oil occurs within them. It is not uncommon to find perched pockets of oil in fractures which are known to be either drained or barren at lower levels. Likewise, pockets of gas or heavily gas-cut oil are quite commonly found in the fractures — sometimes with sufficient contained pressures to cause the wells to flow for short intervals.

Another consideration in the nature of the porosity of these fractures is the size of the openings that may be found in some of them. That some of the openings in these fractures reached a considerable size is indicated by experiences associated with lost circulation in drilling with rotary tools through this section. Though all manner and kinds of lost circulation materials were used in some of the wells, it became impossible to cure the lost circulation problems enough to continue drilling and several wells within the field had to be abandoned before reaching the Dakota sandstone. In some instances gunny sacks were pumped down the wells without material effect. In one instance, willows about one-half inch in diameter and 6 inches long were pumped down a rotary hole to create a bridge for lost circulation material, and a portion of these willows was recovered along with some oil and mud from a nearby cable tool

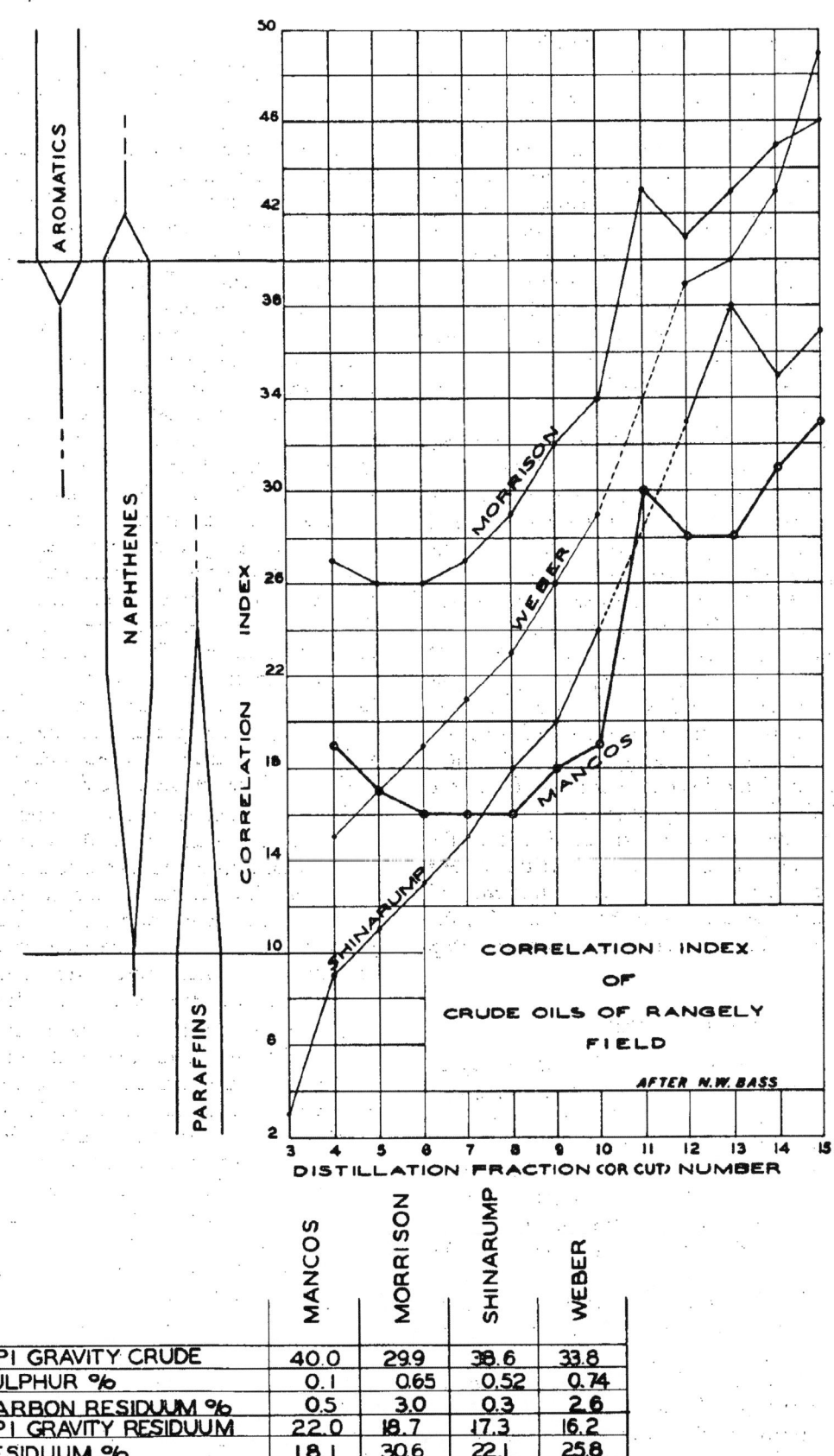

Figure 2. Analyses of oils.

hole to the same fracture. The intersections of the fracture by the two holes are known to be at least 175 feet apart.

Production and Drilling

With all of the divergence in the size and extent of the fracture openings found in the Mancos shale, there is obviously a very great divergence in the rates of production and ultimate recoveries from the wells. With few exceptions, the wells drilled in the Rangely Mancos pool are produced by pumping. Production varies from a few barrels per month to several hundred barrels per day, or up to the capacity of the pumping equipment. Ultimate recoveries from the wells completed as producers may be from a few hundred barrels up to several tens of thousands of barrels. A few exceptional wells within the Mancos pool have made in excess of 100,000 barrels of oil before abandoning. Most wells in the Mancos pool of the field are short lived and produce an ultimate amount of oil measured by a few thousand barrels.

Because of the short lived character of the wells, continuous drilling in the field must be maintained in order to sustain production. Unlike most development programs where it is possible to drill up a pool completely this oil pool is still being drilled after having been discovered 52 years ago and having been actively produced for more than 35 years. The number of wells that have been drilled per year has not varied greatly during the past 15 years.

Locating Wells

With the scattered character of the fractures found in the field, it is readily apparent that locating wells in a proper position to intersect these openings at appropriate depths is quite a problem and results in numerous failures. Over the years it is estimated that the success in drilling for oil in the Rangely Mancos pool has not averaged more than one oil well for each three wells drilled. This relationship would be a very disturbing element if it were not for the low cost of drilling and completion of the wells. The average total depth of the wells drilled for Mancos shale production in the pool is probably somewhere between cost of drilling wells is less than $3,000 per well. To date all Mancos wells have been drilled with cable tools, though it is expected that air rotary drilling will soon be tried. No pipe is set in the cable tool holes unless production is obtained, in which case it is customary to set one joint of 8 inch conductor pipe. The remainder of the hole is left open. Two-inch line pipe is customarily used for tubing and in most cases this pipe is salvageable and may be used in several wells. Thus, except for one joint of conductor pipe, all of the equipment placed on a completed oil well is salcageable. (It should be pointed out here that the ability of these holes to stay open without casing is an index to the reason open fractures are found within the shale section.)

It is of interest to outline the procedure by which a location for a Mancos shale well is customarily made. An example of this procedure is given in Figure 1B which might apply to any part of the south side of the pool. Most locations are made on the basis of calcite veins located on the surface, a study of which has revealed the average hade and bearing of the associated fracture. For this case it is assumed that the approximate level at which oil might be expected in the fracture is 1,400 feet. It is likewise assumed that in this case the potential location is within an area of the field where approximately 250 feet of horizontal drift is expected in drilling to a vertical depth of 1,600 feet. It has been found that cable tool holes in this field always drift up dip at nearly right angles to the strikes of the beds. In this example, a location is made to theoretically intersect the fracture 200 feet below the level of the oil. The many variables make this procedure only partially successful, however, it has met with more success than any other method yet devised.

Nature of Oil

Oil found in the Mancos shale fractures is a very good grade with excellent refining qualities. In Figure 2, an analysis of this oil is shown with other oils found in the field. The nature of the occurrence of this Mancos oil strongly suggests the possibility that it might be migratory from a lower horizon. While it is not known what effects migration might have on an oil, at least as it is found today, this oil is entirely different from the other oils of the field and does not suggest a common origin. It would appear that this oil is indigenous to the Mancos shale section.

Reserves and Production

To date, the area developed is only a small percentage of the field; and as yet, no well drilled in the Mancos shale section in the field has encountered water natural to the formation. Thus the limits of the potential producing area on the structure are not yet known.

By 1954, Petroleum Information reported 4,727,139 bbls. cumulative production from the Rangely Mancos pool, 75% of this production being from less than two square miles in the south central part of the field. Localized areas have produced in excess of 10,000 bbls. per acre. At present, the production from the Rangely Mancos pool is approximately 25,000 bbls. per month.

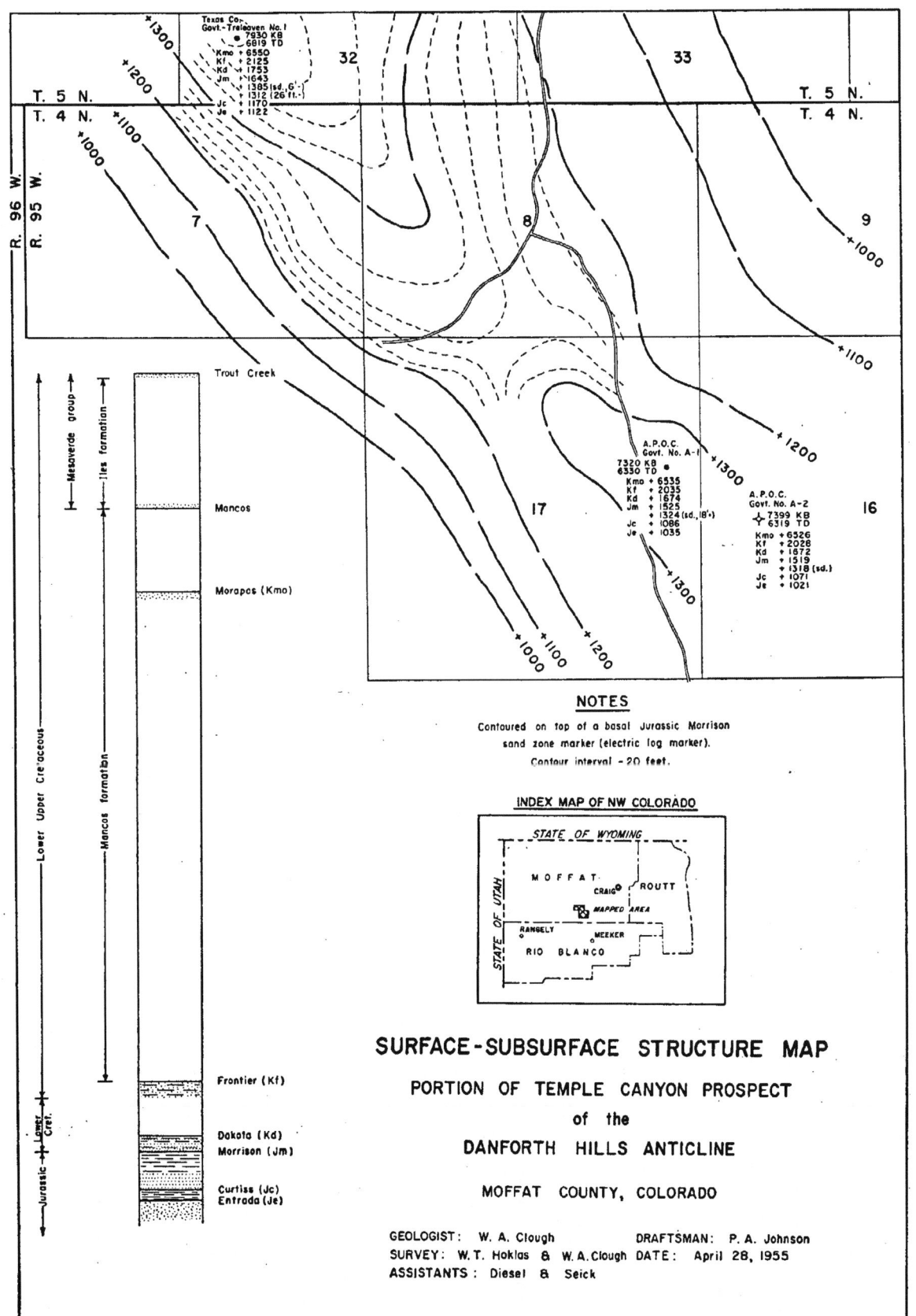

TEMPLE CANYON OIL FIELD
Moffat County, Colorado

By W. A. CLOUGH
Anderson-Prichard Oil Corporation, Denver, Colorado

INTRODUCTION

The Temple Canyon Oil Field of Stanolind Oil & Gas Company and Anderson-Prichard Oil Corporation is located in the heart of the Danforth Hills in southern Moffat County, Colorado, approximately 45 miles southwest of Craig. This part of the Danforth Hills is reasonably rugged, mountainous country with elevations in the area ranging from 6200 feet along the southern margin of the Axial Basin, four miles to the north of the producing lease, to 8259 feet at Coal Mountain, about one mile east of the discovery well.

HISTORY AND DEVELOPMENT

The field was discovered in 1953 as a result of combined detailed surface and subsurface mapping.

The No. 1 Government-"A" was spudded on July 14, 1953 and completed on September 10, 1953, for an initial gauge of 173 BOPD flowing, cut 0.6 of 1% BS, no water. A pump was soon set for an initial pump gauge of 237 BOPD. Production is from the upper bench of the basal Morrison sand series through performations from 6000 to 6008 feet. A second test of the structure was dry because the producing sand in the initial well was absent. Lower sand zones in the basal Morrison sand series in the second test carried shows of oil but were found to be practically impermeable.

Aside from the basal Morrison sand series, shows of oil were encountered in the top of the Entrada and in the upper benches of the Dakota in the initial test and in the same benches of the Dakota in the second test of the structure. The Entrada show was insignificant but the Dakota shows were of some importance. Despite the recovery of a small amount of high gravity oil on a drillstem test, pipe was not set in the second well because only a relatively thin, tight zone of possible productivity was encountered above the probable oil-water contact.

STRUCTURE

The Temple Canyon structure is a small, low-relief closure on the apex of and near the northern end of the Danforth Hills anticline. The latter feature is a large northwesterly trending asymmetric anticlinal fold of some 45 miles length on whose crest also rests the prolific Wilson Creek oil field as well as the Maudlin Gulch and Danforth Hills pools.

The Temple Canyon anticlinal closure is not fully expressed on the surface but can be confirmed by a combination of surface and subsurface geology utilizing nearby well control.

STRATIGRAPHY

The generalized columnar section and structural elevations on the map indicate the section involved in the area.

Surface beds in the area are assigned to the Iles formation and consist of several massive sands separated by gray shale and coal. Thickness of the Iles formation in the area is approximately 1240 feet.

Only the upper few feet of the Mancos formation is exposed on the surface anywhere in the immediate area. The Mancos formation, which consists primarily of dark gray to black siliceous shale with two recognizeable sand zones (Morapos and Meeker sands), is the only unit that shows any appreciable change in thickness locally. This factor limits the direct application of surface geology in the area. Average thickness of the Mancos is approximately 5200 feet.

The Frontier sandstone consists of gray to dark gray tight shaley sand and sandy shale zones separated by gray shales. The Frontier zone is approximately 140 feet thick locally.

Underlying the Frontier sandstone is the Mowry shale, locally about 220 feet thick. It is a dark gray to black fissle shale and carries abundant fish scales and some ammonites.

The Dakota group includes several sand lenses of variable number and development. Locally the term Dakota group has been used to include the persistent basal sandstone member which probably correlates with the Lakota of southwestern Wyoming. The overall thickness of the Dakota group approximates 150 feet in this locale.

The Morrison formation which is approximately 440 feet thick in this area is broken into two major zones. The upper 200 feet of the Morrison is characterized by varicolored shales with a few thin sands; the lower 240 feet, by primarily lenticular sands separated by varicolored shales.

The Curtis formation is approximately 45 feet thick locally and can be best described as it is termed in this area, "the Jurassic black shale".

The Entrada sandstone has not been penetrated in this immediate area below its upper 30 feet. However, it is recognized as a blanket-type, light gray to white sand with excellent porosity and permeability locally. It has been found productive along the Danforth Hills trend only where a fairly sizeable amount of structural closure is present.

The Entrada overlies a fairly thick section of Mesozoic and Paleozoic sedimentary rocks whose potentialities along this portion of the Danforth Hills anticline have not been explored.

POOL INFORMATION

The producing mechanism at Temple Canyon is a combination gas expansion and water drive from a closed anticlinal trap. Average porosity in the pay zone is approximately 18%, and the average permeability of the sand analyzed in the pay zone is approximately 60 md. The pay sand is 22 feet thick but to date only the upper 8 feet have been perforated. Gravity of the oil is 35.8° A.P.I.

THORNBURG DOME
Moffat County, Colorado

Contributed

The following is largely adapted from Ross L. Heaton's paper, "Relation of Accumulation to Structure in Northwestern Colorado," pp. 103-105, *Structure of Typical American Oil Fields,* Vol. II., A.A.P.G., 1929.

Thornburg dome (Morapos), T. 3 N., R. 91 W., has a surface closure of approximately 900 feet, long axis trending northwest and southeast. It is on a southeast continuation of the Axial Basin anticline and has its steeper dips on the southwest. From a syncline which marks the limit of the closure on the southwest, all of the formations rise rapidly toward the Saw Mill anticline, an extension of the main White River uplift in which all of the older Paleozoic formations are exposed.

Due to the absence of key beds in the Mancos shale, surface mapping was accomplished with shale dips. Actual faulting on the outcrops was not seen except for several small faults on the north end of the structure. Minor faulting on most of the structure, especially along its crest, is evidenced by the plentiful occurrence of calcite slabs which have been weathered from faulted fissures.

Except for small amounts of oil from the Frontier in the northernmost well and from the "Sundance" in the southern well, nothing except gas has been produced from this structure. The gas is confined to the top of the structure, as shown by a water well in the Dakota approximately 300 feet below the top contour. The Dakota sand on top of the structure was reached at a depth of 1,950 feet.

After the discovery of oil in the "Sundance" on the Iles dome, the southern well on the Thornburg dome was deepened to that formation, but encountered only small amounts of oil.

EDITOR'S NOTE: In 1951, the No. 1 Lewin, SE NE SE 17-3N-91W, elevation 7030 RB, was completed for 11,800 MCF/D from the Weber open hole interval 4005-4326. The Weber sand cored in this well is gray, fine grained, subround, well sorted, quartzitic (4.4% porosity), massive, and has very low permeability (0.5 milidarcys). The well was shut in with an initial field pressure of 1,650 p.s.i. Gas from Thornburg dome is piped north to supply Craig, Colorado.

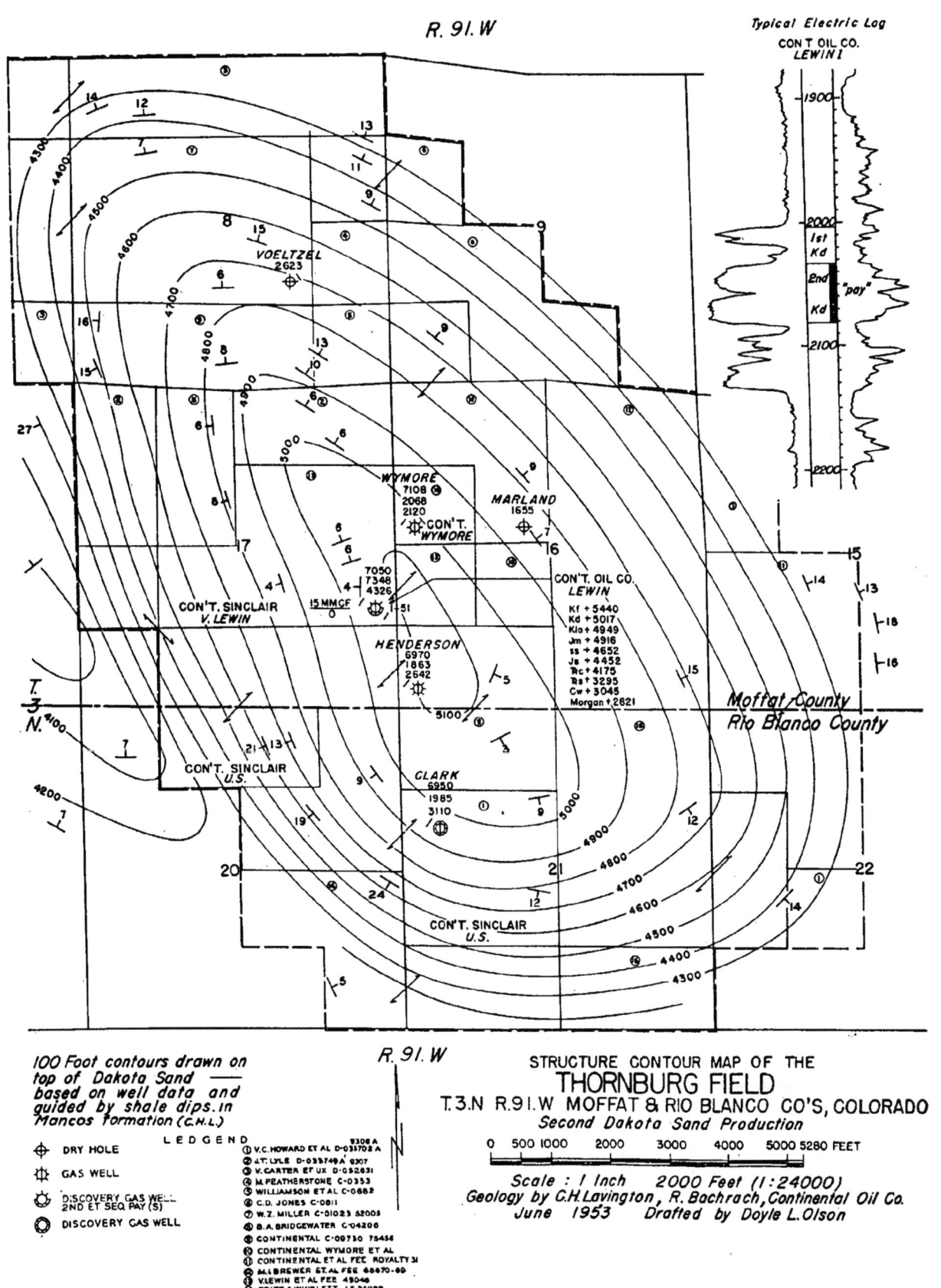

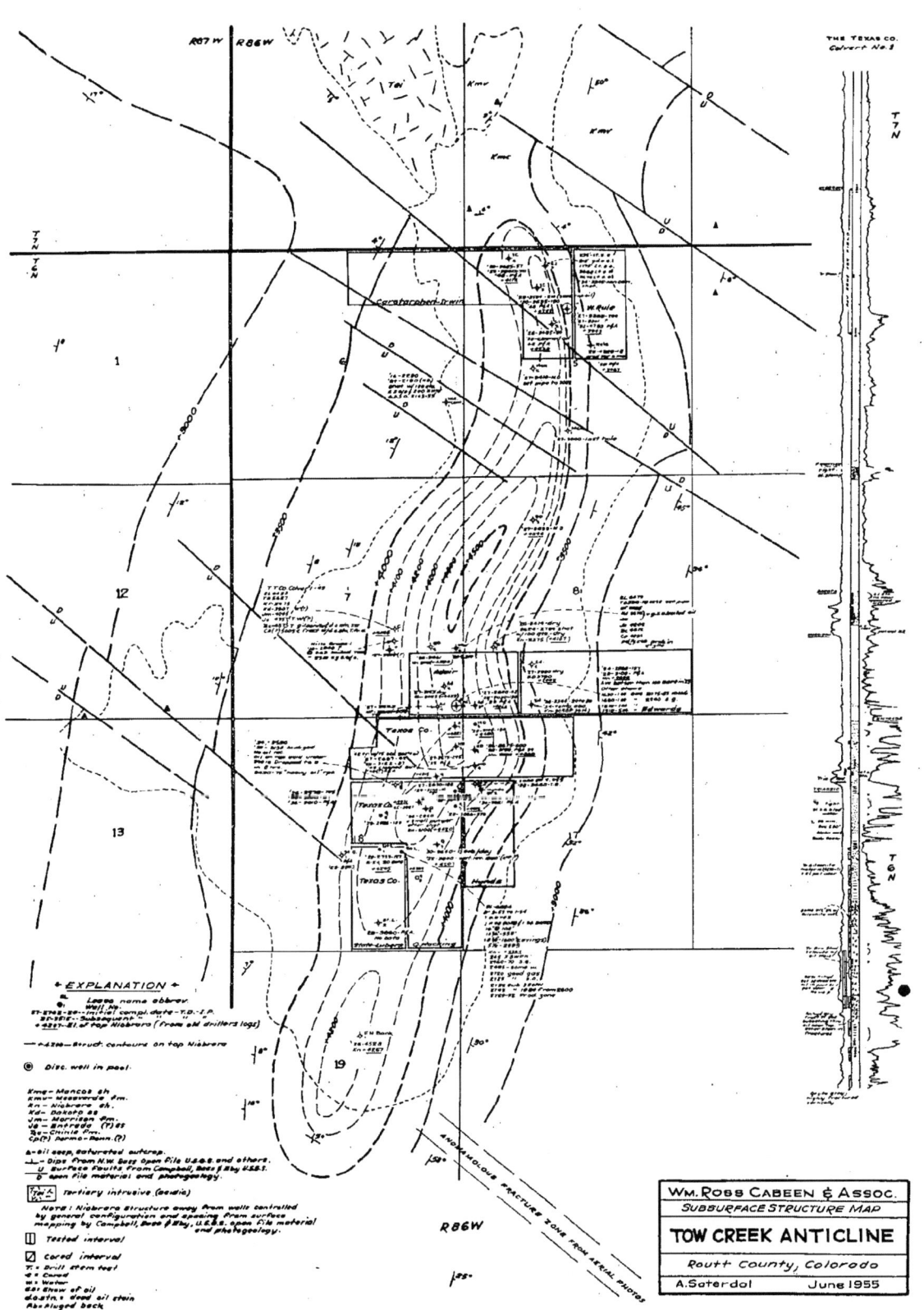

TOW CREEK OIL FIELD
Routt County, Colorado

By A. SATERDAL
Wm. Ross Cabeen & Associates, Denver, Colorado

LOCATION

The Tow Creek Anticline is located in Routt County, Colorado, approximately twenty miles west of Steamboat Springs. The axis is traversed by the Yampa River and by U.S. 40. The structure is plainly visible from this highway. Mancos shale is exposed in the core of the anticline and this part of the structure is quite accessible. Mesaverde sandstones, outcropping on the flanks, form more mountainous terrain which is characterized by incised gulches. This area is accessible mainly by trails and small subsidiary roads.

HISTORY AND DEVELOPMENT

The presence of an anticline was recognized as early as 1906 (Fenneman and Gale, U.S.G.S. Bull. 297). Willson and Perini mapped the structure and their map appears in Colo. Geol. Bull. 23, 1920, along with a fairly complete report. Open file U.S.G.S. material prepared by Campbell, Bass and Eby released in 1954 is probably the most complete available surface work in the area. However, it does not appear to cover details of structure along the axis. Mr. John E. Frey, with The Texas Company, has published a Niobrara map and presented considerable statistical data in "Oil and Gas Fields of Colorado" which is current through 1954.

The first exploratory test of record was the Hills-Gross No. 1, SE NW SE section 7, T. 6 N., R. 86 W., which was drilled in 1910 to T.D. of 2540 feet. Shows of oil were reported in shale but it appears that the well did not fully penetrate the Niobrara section. A second well, reportedly drilled by Mid Continent Oil Co. in 1916 to T.D. 2290 feet, probably did not reach the Niobrara. This well reportedly had shows of oil in the shale at 2145 to 2155 feet. It was plugged back to 2190 feet and shot with resultant production of 5 BO and 200 BWPD. Records on this well, as on all wells drilled prior to 1949, are very poor.

The discovery well in the Tow Creek field was located on the basis of surface geology and was drilled by The Texas Production Co. on the Adair Lease in 1924. It had an I.P. of 131 BOPD from fractures in Niobrara shale. This well averaged better than 100 BOPD of 35.4° gravity oil during 1925. It was deepened to 3108 feet in 1928 and abandoned.

Discovery well at the North Tow Creek pool was The Texas Production Co. Carstarphen-Irwin No. 1 in NE SE NW, section 5. This was completed in 1927 at T.D. 3288 feet for 200 BOPD. It was deepened to 3301 feet later in 1927 and produced until 1932, when it was deepened to 4793 feet in the Dakota (?) and abandoned.

Development subsequent to the two discoveries resulted in a total of 24 significant tests on the South Tow Creek pool of which 18 produced varying amounts of oil with I.P.s up to 600 BOPD. Cumulative production in one case, The Texas Co. State Lubers No. 1, is over 305,000 BO and the well still produces some 20 BOPD relatively free of water.

In the North Tow Creek pool there have been a total of seven wells drilled of which six were producers with I.P.s up to 210 BOPD.

In addition to the pool wells and stepouts, 11 dry tests have been drilled between the pools on the axis and on the flanks for a total of 42 significant tests drilled on the structure.

The only cable tool well drilled in recent years is The Texas Co. State Lubers No. 3, NE NW SE, section 18. This well is a 10 acre offset to the State Lubers No. 1 which has produced over 305,000 BO and is still pumping. At this writing The Texas Co. is drilling Quaintance Hocking No. 9, SW NE SE, section 18, T. 6 N., R. 86 W.

All production has been from fractured Niobrara shale. Significant shows found in deeper formations are as follows:

1. The Texas Co. Belle Dennis No. 4 (first rotary well drilled in Colorado) encountered a very good show of oil either in the "Dakota group" or in the Upper Morrison. Casing was run to test this show but considerable water along with oil was swabbed and the well was abandoned.

2. The Texas Co. Colvert No. 1 was drilled with rotary tools in 1949 and found slight stain in tight-appearing Frontier sand at 3475 feet. Also some 90 feet of lightly stained sand was found in the "Dakota" which appeared wet in samples and had very little odor. Porosity was generally low with exception of the lower bench which appeared wet and had no visible stain. This sand was not tested.

The Entrada also showed light oil stain in this well, but tested 630 feet of clear water. Considerable dead oil stain and gilsonitic material was found in the lower part of the Triassic section and throughout a highly

anhydritic section tentatively called Permo-Pennsylvanian. Shows were in fractures for the main part. Approximately 50% of the anhydritic section was cored and two tests were run. Considerable fresh water was recovered on one (probably from fractures) and on the second, only drilling mud was recovered.

STRUCTURE

Regionally, the Tow Creek Structure is located on the southeastern flank of the Sand Wash Basin of northwestern Colorado. Local structure appears to be controlled by the Park Range uplift, a major tectonic feature to the east. Local structure is that of a N-S trending anticline with a long, south-plunging nose. Some 500 feet of critical surface closure can be mapped. Structure is probably the result of very late Cretaceous movement modified by Paleocene and Eocene movements. The structure has been altered by a system of NW trending cross faults which can be mapped at the surface. Their presence at depth is conjectural.

An acidic igneous body, either a plug or a very steep sided laccolith, is intruded into the upper part of the Mancos shale immediately north of the Tow Creek closure. This can be dated with certainty only as post-Cretaceous but is probably post-Browns Park (Miocene-Pliocene). Mancos shale in contact with the acid rock does not appear to have been seriously altered for more than a few feet from the contact. Oil saturation has been observed in the basal Mesaverde sands on what could very possibly have been the flank of a closed dome caused by intrusion. This saturation is within a quarter of a mile of the intrusive body, and other saturated sand is reported to have been found on top of the igneous body. The igneous activity does not appear to have altered the structure of the Tow Creek field appreciably. There does appear to be a relationship, however, between the rather extensive fault and fracture system on the north end of Tow Creek and the trend of basic igneous dikes north of the area which probably took advantage of pre-existing lines of structural weakness.

FACTORS CONTROLLING ACCUMULATION

Accumulation of oil at Tow Creek appears to be caused by the coincidence of anticlinal structure with an upward loss of fracture permeability in the stratigraphic section. All production found to date is in the Niobrara shale and the "caprock" appears to be the relatively plastic overlying Mancos shale. Numerous cores in The Texas Co. Colvert No. 1 indicated extensive open vertical fractures lined with quartz and gypsum crystals which may be a deterrent to large accumulations of oil in lower horizons, at least in the highly fractured parts of the anticline.

It is considered significant that current drilling activtiy at Tow Creek is by the extension of a producing area along the trend of the fractures instead of along the structural axis. The logic for this development can be seen by constructing an iso-initial production map which defines the better producing areas as being parallel to the fracture trends.

Erratic permeability in the Dakota, Morrison and Shinarump, coupled with a high frequency of oil shows in the area in these formations indicate that productive possibilities are good on untested parts of the anticline but are not specifically defined at present. U. S. Geological Survey mapping and photogeologic work indicate that additional fault or fracture zones, prospective for additional Niobrara pools are present on the anticline, but detailed surface work and further wells are necessary to fully evaluate these possibilities.

The entire prospective section is considered to have been tested by the Belle Dennis No. 4 and the Colvert No. 1. The former reportedly bottomed in "Granite" and the latter bottomed in highly quartzitic, pyritic, gray, fine-grained sandstone which was interbedded with green waxy shale. No definite age correlation has been made.

ACKNOWLEDGEMENTS

The writer has drawn heavily on the excellent published and open file works by the authors listed in the references below and wishes to acknowledge the use of much of their material.

Campbell, M. R., Bass, N. W., and Eby, J. B.; U.S.G.S. open file material.

Coffin, R. C., Perini, V. C., Jr., and Collins, M. J. (1920), "Some Anticlines of Western Colorado," Colo. Geol. Surv. Bull. 24.

Crawford, R. D., Willson, K. M., and Perini, V. C., Jr. (1920), "Some Anticlines of Routt County," Colo. Geol. Surv. Bull. 23.

Fenneman, N. M., and Gale, Hoyt S. (1906), "The Yampa Coal Field, Routt County, Colorado," U. S. Geol. Survey Bull. 297.

Frey, John E. (1954), "Oil and Gas Fields of Colorado," Rocky Mtn. Assoc. of Geol.

WHITE RIVER DOME
Rio Blanco County, Colorado

By GALEN L. HELMKE
Consultant, Casper, Wyoming

LOCATION

The White River Dome is located in Ts. 1 and 2 N., Rs. 96 and 97 W., Rio Blanco County, Colorado. The field, 40 miles east of Rangely, has approximately 400 feet of structural relief and 13,000 acres within surface closure.

HISTORY AND DEVELOPMENT

Shallow gas was discovered in 1890 when two 500 foot$\pm$ tests were drilled on a gas seep near the axial crest of the structure. Intermittent drilling has been carried on in 1919-1923, 1930, 1944-1948 and 1951-1953. Most of the wells have had substantial shows of gas in the Wasatch. Some of the earlier cable tool wells reported initial potentials of 15,000,000 cubic feet of gas per day.

The deepest test to date, drilled in 1951, was the Union Oil of California-Frontier Refining Company No. 1, SW¼SE¼SW¼, section 30, T. 2 N., R. 96 W. This test reached the Morapos Sandstone at 8480 feet after penetrating the Wasatch, Mesaverde and upper 800 feet of Mancos Shale.

The Wasatch formation has made some gas and occasional shows of oil in all tests drilled.
66—IAPG

Type of Trap

Gas and shows of oil have been found from 500 to 8500 feet in the test wells. There seems to be no blanket sand that produces gas, which indicates that most of the accumulation is in stratigraphic traps. (The section between the base of the Green River formation and the top of the Mesaverde formation thins from 5000' at White River Dome to less than 1000' on the south flank of the Rangely structure.) It is possible that a structure contour map on the top of the Mesaverde would show little or no closure due to the westward thinning of the lower Wasatch and Paleocene sections.

Production

Most of the oil recovered from the Wasatch has been high gravity and high pourpoint. There have been several operators in this field and production histories are difficult to obtain or evaluate. Incomplete records indicate that 520,000,000 cubic feet of gas has been produced from this gas field since its discovery in 1890. There is no reliable estimate of oil production.

The building of the Pacific Northwest pipeline through Rangely should stimulate additional drilling in the field.

STRUCTURE

The White River Field is on a major northwest-southeast trending fold on the north flank of the Piceance Creek basin. The fold extends from near the Colorado-Utah border and ends abruptly at the Grand Hogback just west of the town of Meeker, T. 1 N., R. 94 W. This general line of folding parallels the Rangely-Yellow Creek and Piceance Creek line of structural folding to the south.

The White River Dome, with surface elevation of 6000 feet, is an east-west trending asymmetrical anticline with steep dips on the south flank of 5° to 10° and 2° to 3° dips on the north flank. The dome has approximately 400 feet of closure on the surface and the area of closure has some 13,000 acres. The crest of the anticline is covered by Wasatch formation, about 1000 feet of which has been removed by erosion. The younger Green River formation rings the structure on all but the northwest side. Surface faults are present on the White River Dome. The lenticular characteristics of the sands and wide spacing of the deeper wells make correlation of logs difficult and subsurface faults difficult to determine. However, it appears that the subsurface axis migrates toward the steep dip (south) side of the structure.

STRATIGRAPHY

White River Dome was originally mapped in the Wasatch formation (Eocene), and the deepest test has penetrated the upper 1000 feet of Mancos Shale. Operators in the field have released only Mesaverde and Mancos tops for the tests. However, it is probable that there is a section of Paleocene sediments in the White River area. Megafossils from the Wasatch-Mesaverde boundary as shown on the state geologic map were identified by J. B. Reeside as Fort Union type.

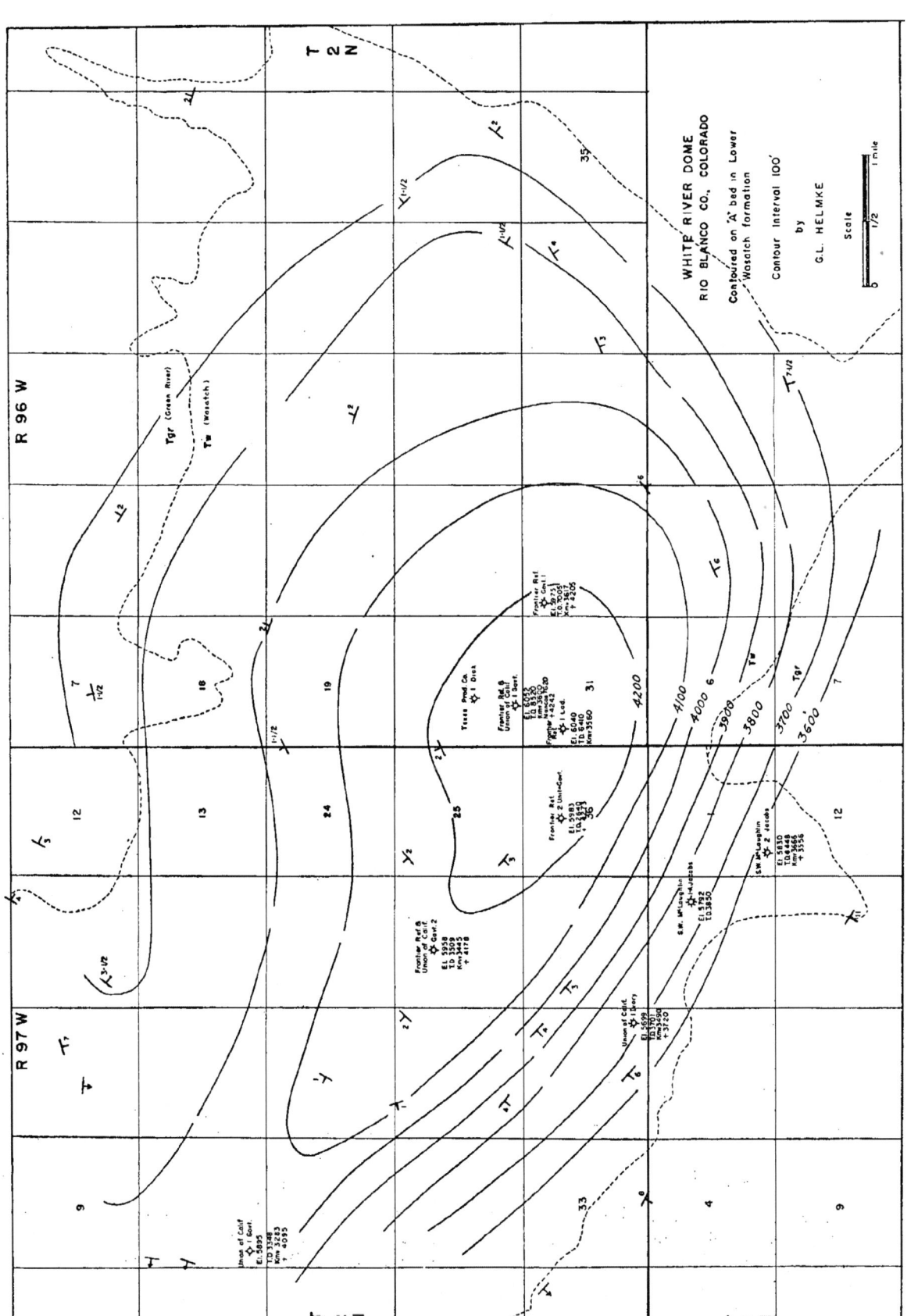

Wells drilled on White River Dome.

Company	Well No.	Location Sec. Twp. Rge.	Year	Total Depth	Bottomed
Unknown		31 2N 96W	1890	700±	Wasatch
Unknown				700±	Wasatch
Beaver Oil & Gas Company		31 2N 96W	1918	1250	Wasatch
White River Development Company		31 2N 96W SE¼SW¼	1919	1250	Wasatch
White River Oil Company	No. 2	25 2N 97W SE¼SE¼	1919	530	Wasatch
White River Oil Company	No. 3	31 2N 96W NW¼SW¼	1919	955	Wasatch
White River Oil Company	No. 4	36 2N 97W NE¼NE¼	1919	808	Wasatch
Inland Oil Company		31 2N 96W SW¼NW¼	1922	1150	Wasatch
White River Development Company	No. 2	31 2N 96W SW¼NW¼	1923	2790	Wasatch
White River Oil Company		30 2N 96W SW¼NW¼	1923 (?)	1500	Wasatch
Texas Production Company	No. 1	30 2N 96W NW¼SE¼	1929	850	Wasatch
Texas Production Company		30 2N 96W NE¼SW¼	1930	5987	Mesaverde
Frontier Refining Company	No. 1	32 2N 96W NW¼NW¼	1944	7005	Mesaverde
Frontier Refining Company	No. 1 Lad	31 2N 96W CWL NW¼	1948	6410	Mesaverde
Union of California and Frontier Refining Company	No. 1 Unit	31 2N 96W SW¼SE¼SW¼	1951	8520	Mancos
Union of California and Frontier Refining Company	No. 2 Govt.	26 2N 97W NE¼SE¼NW¼	1952	3509	Mesaverde
Union of California	No. 1 Govt.	16 2N 97W C SW¼SW¼	1952	3348	Mesaverde
Union of California	No. 1 Ivory	34 2N 97W C SE¼SE¼	1952	3701	Mesaverde
Frontier Refining Company	No. 2 Govt.	36 2N 97W NE¼SE¼NW¼	1953	2460	Wasatch
S. W. McLaughlin	No. 1 Jacobs	2 1N 97W NW¼SE¼NE¼	1953	3805	Mesaverde
S. W. McLaughlin	No. 2 Jacobs	1 1N 97W SE¼SW¼SW¼	1953	4428	Mesaverde

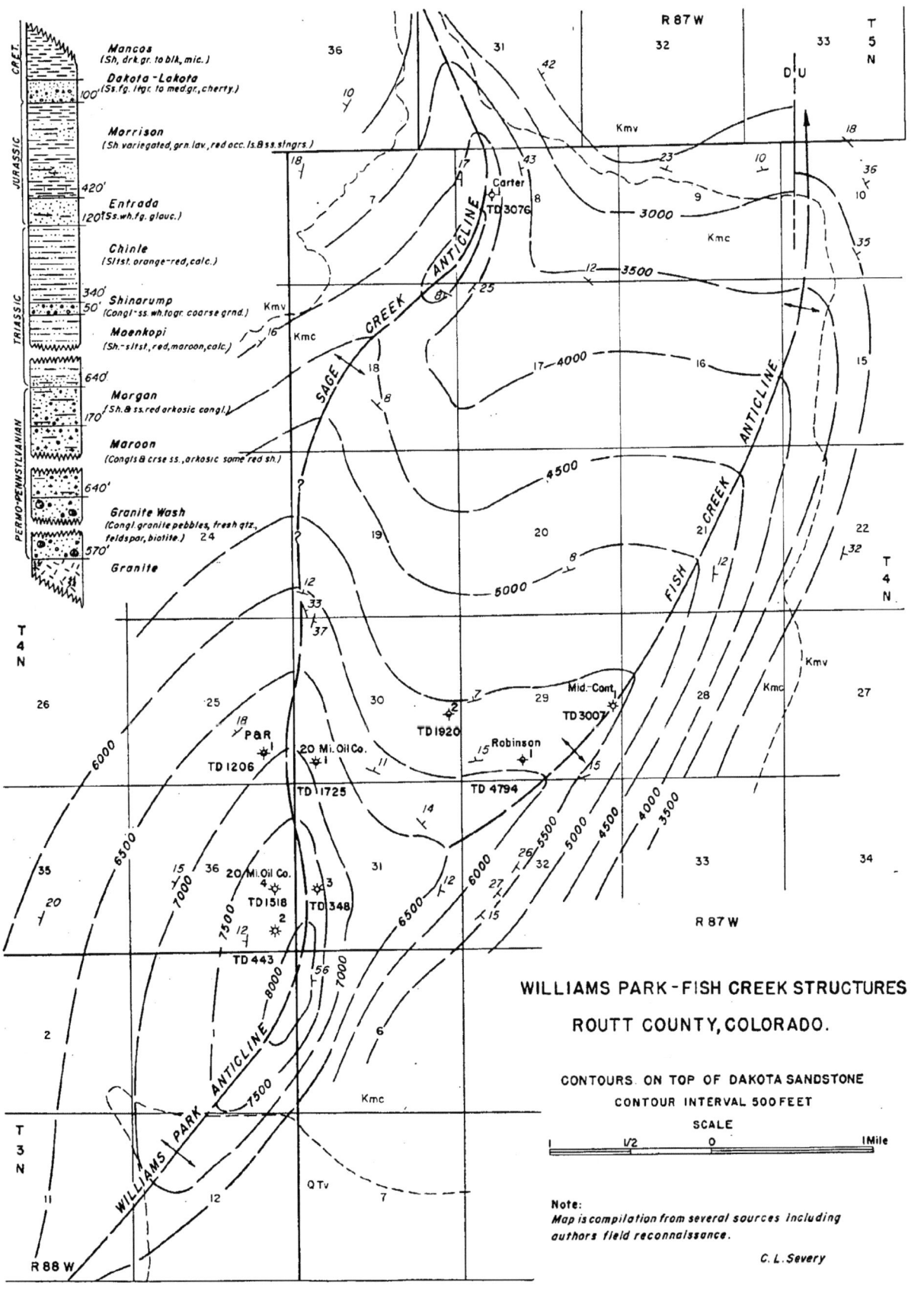

WILLIAMS PARK - FISH CREEK STRUCTURES
ROUTT COUNTY, COLORADO.

CONTOURS ON TOP OF DAKOTA SANDSTONE
CONTOUR INTERVAL 500 FEET

Note:
Map is compilation from several sources including author's field reconnaissance.

C. L. Severy

GEOLOGY OF THE WILLIAMS PARK - FISH CREEK ANTICLINES
Routt County, Colorado

By C. L. SEVERY
Hiawatha Oil & Gas Company, Denver, Colorado

LOCATION

The Williams Park and Fish Creek structures are located in southwestern Routt County, Colorado in Townships 3 and 4 North, Ranges 87 and 88 West.

HISTORY AND DEVELOPMENT

The first well on the Williams Park Anticline was drilled prior to 1920 by the Twenty Mile Oil Company. It was spudded in the Mancos shale with the total depth reported as 1725 feet in the Dakota formation, and was followed by three more tests by the same operators. The Producers and Refiners Corporation also drilled a well on the north end of the structure in 1920. No complete records of the wells have been located and all information is of questionable reliability; but from that available, gas shows were reported in these wells from depths ranging from 42 feet to 1725 feet with several shows being reported in each well. Only the No. 1 well reported oil shows. All shows presumably were in stray sands or in the shales of the Mancos formation and in the Dakota formation. Stratigraphically, the deepest penetration was in the No. 4 well which reportedly drilled 380 feet of the Chinle formation at a total depth of 1518 feet. No information concerning casing of the wells or production from them is available. The largest estimated gas flow has been reported as 500,000 cubic feet per day from the No. 2 well. The field has never actually been in production.

On the Fish Creek structure, three wells have been drilled. The first, by the Mid-Colorado Petroleum Company was spudded in 1930 and finally reported as abandoned in 1941. The two other wells were drilled by O. D. Robinson in 1953 and the No. 1 Kagie was a granite test to 4794 feet. This well encountered oil shows in the Dakota, Lakota, Shinarump and Morgan. However, drill stem tests recovered only mud and water, and no commercial production was indicated.

GEOLOGY

General

Geologically these structures are folds plunging northward from the White River Uplift. The outline of the features are well delineated by the outcrop of massive sandstones in the basal Mesaverde formation, but the interior portions of the folds are more obscure, being expressed in the soft shale beds of the Mancos formation and lying partially in well wooded, rugged terrain. Tertiary basalt flows cap the highlands to the south, forming flat topped mountains, and in part obscure the southern portion of the Williams Park feature.

Structure

The Williams Park Anticline is a portion of a somewhat sinuous line of folding that includes the Little Poose Creek, Williams Park and Sage Creek Anticlines. The Williams Park feature trends generally north-south, but is slightly arcuate to the east. The fold is assymetrical with dips on the eastern flank ranging from 30° to 56° and those on the western flank from 12° to 20°. The structure shows approximately 1250 feet of closure covering an area some four miles long by a mile and a half wide. The northern plunge of the feature becomes the Sage Creek fold

The Fish Creek Anticline is a plunging nose trending northeasterly from the Williams Park closure. It is also an assymetrical anticline averaging 25° dip on the eastern limb and 10° on the western limb. No demonstratable closure is present.

The only faulting mapped on these structures is an axial fault exposed on the plunge of the Fish Creek nose in the Mesaverde sandstone beds. This appears to be a nearly vertical fault, downthrown on the west, with a displacement of some 100 feet.

STRATIGRAPHY

An abbreviated stratigraphic column is incorporated with the illustration; however, since shows have been encountered in several formations, a brief lithologic description of the more favorable of the potentially productive horizons is included here.

DAKOTA — sandstone, light gray to gray, fine-grained, quartzitic to tight, occasionally shaley, pyritic, grains sub-rounded to sub-angular, exhibits some porosity. Estimated 50 feet overall thickness approximately half being effective reservoir sandstone.

LAKOTA — sandstone, light gray to light tan, medium grained, well cemented, tight to quartzitic, fair sorting, abundant pyrite and chert. Chert is blue to milky. Sandstone conglomeratic in part. Estimated 30-35 feet thick.

SHINARUMP — sandstone, conglomeratic, white to gray, coarse grained to conglomeratic with pebbles up to ½" in size. Good porosity; estimated 40 feet thick.

Permo-Pennsylvanian strata penetrated in the No. 1 Kagie well include coarsely clastic arkose which may represent the time equivalent of several formations. Principally because of a lack of detailed correlation through the region, the formations present are tentatively identified as Morgan, Maroon and granite wash. Potential reservoir beds are absent in this part of the section.

BIBLIOGRAPHY

Barb, Clark F. (1946), "Selected Well Logs of Colorado," Quarterly of the Colorado School of Mines, vol. 41, no. 1, pp. 388-90.

Coffin, R. W., Perini, V. C., and Collins, M. J. (1924), "Some Anticlines of Western Colorado," Colo. Geol. Survey Bull. 24, pp. 34-35.

Crawford, R. D., Willson, K. M., and Perini, V. C. (1920), "Some Anticlines of Routt County, Colorado," Colo. Geol. Survey Bull. 23, pp. 17-19.

Wilson Creek Oil Field.

Wiggins Studio, Craig

SELECTED OIL AND GAS FIELDS OF NORTHWESTERN COLORADO AND SOUTHWESTERN WYOMING

By DANIEL S. TURNER
Geological Consultant, Denver, Colorado

As this summary of oil and gas fields is being composed for the Sixth Annual Field Trip Guidebook, the start of excavation for the Pacific Northwest Pipeline has been announced to the public.

Herein lies the key to the rapid development of oil and gas exploration on the western slopes of Wyoming and Colorado. Since the first announcement of plans for the new line, some three years ago, western area exploration has witnessed a boom of new locations and deeper drilling. In the historical vein, it has been interesting to note the U.S.G.S. press releases and bulletin discussions, dating back to the early 1900's. Many of our recent discoveries have been made on structures recommended for an oil or gas test some 50 years ago. Many of these were drilled to the limit of the existing mechanization with varying degrees of success. Others were reported to have potentials at too great a depth, (5000-6000 ft.) to warrant exploration.

By far the greatest majority of "western slope" fields were discovered 30 to 60 years ago on the basis of prominent surface expression of anticlinal closure. Cable tools and somewhat primitive drilling techniques necessarily limited the early tests to zones less than 600 feet in depth. Many of these early tests produced encouraging shows of oil and/or gas, but remote locations and the lack of transportation and market discouraged extensive development. Through the years, a characteristic pattern of periodic redrilling to greater depths has accompanied the improvement of exploration methods, drilling tools, techniques, markets and transportation. Comparatively recent subsurface exploration has revealed new pay horizons, and with their entry into the economic picture, many of the old records of shallow tests have been lost. Some fragments of information, regarding old tests and developments, now fall into the category of the lore of the oil fields, repeated by the "old timers" and garnished with fantasy. Much of this information is therefore questionable and inadequate to justify an attempt to map horizons in the subsurface. However, we cannot lose sight of the fact that economic factors of our day control the future development of untapped and undiscovered reserves just as they did in the days of the "bull wheel".

Western Colorado and Wyoming, with their many partially tested structures, shut-in gas fields, and plugged gas bearing horizons, cannot be overlooked when the market to the Pacific Northwest is made available through the new pipe lines.

In many respects the transportation and market problem has been the control of deeper exploration and the penetration into remote areas. The old geological reports referred to stage roads, tank wagons, and horseback exploration. Today's reports evolve from the seismograph, aerial photography, pipelines and helicopters.

The following resumes reflect the new technology, response of an industry to economic controls, and point out the great potential of a vast relatively unexplored oil and gas province.

BAGGS AND SOUTH BAGGS ANTICLINES
T. 12 N., Rs. 92 and 93 W.
Carbon County, Wyoming

The Baggs anticlinal structures lie principally along the east-west trend of the Cherokee Ridge zone of faulting and folding which separates the Washakie Basin from the Piceance Basin. Rocks along this zone have been folded into several en-echelon anticlines and synclines, the axes of which are generally parallel to the east-west striking faults. The folds are gentle, with low dips ranging from 3° to as high as 16° in the vicinity of Baggs, on the east end of the Baggs anticlinal structures.

The Baggs and South Baggs anticlines lie in a region of semiaridity at an average elevation of more than 6200 feet. The area is treeless, except near the major streams or in areas of irrigation. A high bluff, trending northeastward, lies about 2½ miles west of Baggs facing the townsite. The rocks exposed in the bluff are soft red, yellow, gray, and brownish shales and clays of the Tertiary Wasatch formation which has been carved into pinnacles, channels, and sink holes typical of badland topography. The bluff, which provides about 670 feet of relief, separates a sage-covered lowland from an upland that slopes gently to the northwest. The Baggs anticline, lying principally along the bottom tier of sections in T. 13 N., R. 92 W., shows exposures of the Tipton Tongue of the Green River formation and the Cathedral Bluffs tongue of the Wasatch formation. The South Baggs anticline is overlain on the north flank by the Browns Park formation and on the south flank by exposures of the Wasatch

formation. The South Baggs anticlinal axis may be plotted from the southwest corner of section 6, T. 12 N., R. 92 W., east-southeastward to the approximate center of section 14 of the same township. Seven wells have been drilled in the immediate vicinity of the Baggs structures, three of which have been dry holes testing the Baggs anticline. The earliest drilling in the area was the E. Bivens and others, No. 1 W. W. Grant drilled in 1929 to a total depth of 2300 feet. This well on the Baggs anticline was then followed by the second W. W. Grant well drilled in 193— to a total depth of 930 ft. About 50,000 CFGPD were reportedly produced from this second test. The third test of the anticline was the No. 1 Penlow Bear Bros. drilled by Marvin James in 1952. This test was drilled to 500 ft., shut down for the winter, and apparently has now been abandoned. The Superior Oil Company drilled a failure in the NW NW NW section 7, T. 12 N., R. 92 W., in November 1937 to a total depth of 2138 feet in "post Laramie" beds on the west end of South Baggs anticline.

The South Baggs structure was unitized, and approved by the U.S.G.S., on August 8, 1946 with a total unit acreage of 10,654.71. The first well in the unit was drilled by Kerr-McGee and Phillips in the SE section 10, T. 12 N., R. 92 W., to a total depth of 9,516 ft. in the Mancos shale. The well was plugged back and perforated in three intervals, presumably in sands of the lower Wasatch formation on March 29, 1947. The reported initial potential for this test was 3,750,000 CFGPD with no water and no oil in the upper two zones of perforations. The lower zone flowed water in addition to gas at the rate of 5 barrels of water per hour. This well is now shut in. Kerr-McGee and Phillips drilled a dry hole in June 1949 in SW NW NE section 15, T. 12 N., R. 92 W., to a total depth of 3940 ft. A third test in C NE NW section 9 drilled to 8125 feet, was plugged back to 4900 ft. and perforated from 4,692 to 4,720 and flowed at the rate of 734,000 CFGPD following acidization with 1,000 gallons of mud acid. The well produced 2,060,000 cu. ft. of gas in 21 hours thru ¾ inch choke. After sandfract, the well flowed 2,800,000 CFGPD from a stray sand in the Lewis formation. At this time, the fourth unit well is preparing to spud.

DANFORTH HILLS (Price Creek)
T. 5 N., R. 95 W.
Moffat County, Colorado

The Danforth Hills structure lies between Juniper Mountain and the syncline of the Citadel Plateau where the south flank of the Axial Basin is warped into a subsidiary fold. The Cretaceous Mancos shale is exposed on the northwest end of the anticline where it merges with the southerly dips on the south flank of the Axial Basin uplift. The anticline plunges to the southeast for several miles, beyond which it flattens into a broad low relief arch expressed in the Iles and William's Fork members of the Mesaverde formation. The axis shows a saddle just south of Coal Mountain and a low dome several miles to the southeast. Beyond this the axis is crossed by a deep saddle formed by the southwest fork of the Collum syncline. Extended on southward, the arch rises to the Wilson Creek dome. The Maudlin Gulch and Temple Canyon fields are located on this structural trend.

Oil and gas were discovered in lenticular sands of the Morrison formation in the Temple Canyon structure T. 4 N., R. 95 W., in July, 1953, by the Anderson Prichard Oil Company in section 17, on a Stanolind Oil and Gas Company farm-out. The Morrison formation produced 173 barrels of 32.5 degree API gravity oil per day flowing through perforations at 6,000 to 6,008 feet. Discovery of oil on the Danforth Hills anticline was announced in August, 1954. The Texas Company No. 1 Government Treleaven in section 32, T. 5 N., R. 95 W. pumped 323 barrels of oil a day from perforations at a depth of 6,625 to 6,647 and 6,652 to 6,658 in the Morrison formation. This test was drilled to a total depth of 6,819 feet in the Jurassic Entrada formation and plugged back to 6,743 feet. After sand-oil-squeeze with 4,000 gallons of oil and 4,000 pounds of sand, followed by acidization with 1500 gallons, the well swabbed approximately 2½ barrels of oil per hour. Four drill stem tests have been reported in the Morapos formation, Cretaceous Dakota formation, and Jurassic Morrison. The only show of oil, in other than the Morrison formation, was a rainbow of oil and 40 feet of water cut mud recovered in a 2 hour 45 minute open period test of the Dakota formation. To the date of this report, three wells have been drilled on the structure. The Texas Company No. 1 Government Scott, Center NE NW section 30, T. 5 N., R. 95 W., spudded on October 29, 1954, was drilled to a total depth of 6,022 feet in the Sundance formation, plugged back to 5,850 feet and perforated from 5,011 to 5,038 feet with six shots per foot. After acidization and sandfract, the well flowed 57 barrels of oil on a three hour 20 minute test and swabbed 60 barrels of oil in 14 hours.

The Texas Company No. 2, Government Scott, C NE SW section 30, T. 5 N., R. 95 W., spudded on March 11, 1955, drilled to a total depth of 6,934 feet and was plugged and abandoned on May 14, 1955 with drill stem tests showing only slightly gas cut fresh water.

The Texas Company No. 2, Government Treleaven Center, SW, NW, section 32, T. 5 N., R. 95 W., was last reported as waiting on rotary tools.

LAY CREEK
T. 8 N., Rs. 92 and 93 W.
Moffat County, Colorado

The Lay Creek structure is located on the south side of the Sand Wash Basin in the northeast corner of Moffat County, Colorado. The dome type structure, with about 300 feet to 350 feet of mapped surface closure, is expressed in the Eocene Wasatch formation. Apparent anticlinal closure is enhanced by a normal fault in the south end of the structure, which is one of the several faults associated with the Sparks Fault zone, extending southeastward toward the town of Craig. Traces of faults have been found at the surface. The Wasatch formation, which is found at the surface in the Lay Creek area, is in part the same section from which production is obtained in the Hiawatha and Powder Wash pools to the northwest.

This structure, lying about 20 miles northwest of Craig, Colorado, was the site of a discovery during January 1955. Moore and Gilmore, Halbert and Jennings, and the U. S. Smelting and Refining and Mining Company drilled the No. 1 Kern, C, SE, NE, section 13, T. 8 N., R. 93 W., as a Cretaceous Mesaverde test. This location, predicated on recent discoveries in the Mesaverde formation at Middle Mountain, Sugar Loaf, Slater Dome and others, was drilled along the predominate zone of folding and faulting on the north flank of the Uinta Mountain uplift. Typical of many of these tests, the No. 1 Kern was drilled to a total depth of 7,381 feet and plugged back to 1,904 feet to test productive potential in the lower part of the Wasatch formation. Several drill stem tests in the Mesaverde formation recovered slightly gas cut water and mud but no productive potential. A drill stem test attempted at approximately 700 feet blew out with an estimated 3,000,000 CFGPD. The well was completed for production between 622 ft. and 1,940 feet for an estimated production potential of 1,500,000 cubic feet of gas per day.

OAK CREEK (YAMPA ANTICLINE)
T. 3 N., R. 86 W.
Routt County, Colorado

The discovery of the Oak Creek field about 19 miles southwest of Steamboat Springs, Colorado is somewhat of a historical feature in the drilling of north central Colorado. Following the discovery of oil in a stratigraphic trap in the Triassic Shinarump conglomerate on the southeasterly plunging nose of this closed anticline, a flurry of drilling was set off throughout this part of the state. The Oak Creek anticline is cut on the south by an east-west fault. A high angle thrust fault flanks the northeast side with the upthrown block on the southwest. The discovery well (drilled in 1948 and 1949), the Havenstrite Oil Company of Los Angeles, No. 2, Ryles, NW, SW, NE section 2, T. 3 N., R. 86 W., was drilled sporadically under rigorous weather conditions during that winter at an altitude of about 9000 feet. Complete data on this test has not been released to the public; however, scattered reports indicate the well had an initial potential of 250 to 300 barrels of oil per day from a depth of 6670 feet to 6699 feet. Casing was landed at 6644 feet and the well was completed barefoot with an initial pressure of 2100 p.s.i.

Original reports indicated that the production was from a Morrison sand. However, more recent correlations indicate the Shinarump conglomerate as the producing horizon. A confirmation well drilled to the north of the discovery was a failure in the Shinarump conglomerate. A third test, the deepest on the structure, penetrated to the Triassic Moenkopi formation at a total depth of 7,135 feet. Intermittent production from the discovery well has totaled 39,023 barrels of 37.9° API gravity paraffin base oil.

Following this discovery, Gardner Brothers Drilling Company tested the Chimney Creek structure in section 4, T. 7 N., R. 86 W., on a farm-out agreement from the Carter Oil Company. The well bottomed in granite at a total depth of 2,064 feet. A North Trull failure was drilled by W. E. Atkinson with the No. 1 Dorr, SW SE NW, section 4, T. 7 N., R. 85 W., to a total depth of 2507 feet on a farm-out from Stanolind Oil and Gas Company. The flurry of drilling quieted down after failures at Buford Anticline, Rio Blanco County and a deep test in the Tow Creek field drilled by the Texas Company.

SHELL CREEK
T. 11 N., R. 100 W.
Moffat County, Colorado

The Shell Creek anticline, located in the northwest corner of Moffat County, Colorado, lies about 15 miles west of the Powder Wash field, six miles south of the Hiawatha gas field and one township east of the new discovery at Sugar Loaf.

The anticline is expressed in the Tertiary Wasatch formation by very low dips, in the magnitude of 1 to 3 degrees, bounding the axis which extends from section 1 to section 18, T. 11 N., R. 100 W.

Shell Creek, which is the northeast tributary to Vermillion Creek, roughly parallels the axis of the Shell Creek structure.

Mountain Fuel Supply Company opened the Shell Creek field with a discovery at the No. 1, Government, C, SW SW section 1, T. 11 N., R. 100 W., in March 1955. The well was spudded on December 16, 1954 with a rotary bushing elevation of 6844 feet and drilled to a total depth of 7948 feet in the Mesaverde formation. The hole was plugged back to 5615 feet in the Fort Union formation and perforated from 5349 feet to 5395 feet, with 4 holes per foot. Open flow test gave 7,450,000 cubic feet of gas per day under 1840 pounds pressure.

At the time of this report, Mountain Fuel Supply Company was drilling the No. 2, Government, Paul E. Cooley, SE SE NW section 11, T. 11 N., R. 100 W., and a locaion was announced for the No. 1, State 38, 450' N/S., 723' W/E of tract 38, section 1, T. 11 N., R. 100 W.

It is notable that the current trend of drilling with gas was applied in the discovery well in the Shell Creek structure. Mountain Fuel Supply Company reports drilling with natural gas from 7811 feet to 7865 feet and from 7872 feet to 7948 feet before the well was plugged back for production. The confirmation well, the No. 2 Government Cooley, is reported to be a projected Mesaverde test to a depth of about 9300 feet. Two old wells located on the Shell Creek structure, for which little information is available, were reported to have been drilled to depths of 3460 and 3761 feet prior to 1928.

SLATER DOME (LITTLE SNAKE RIVER DOME)
T. 12 N., R. 89 W.
Moffat County, Colorado

Slater Dome lies principally in sections 13 and 14 of T. 12 N., R. 89 W. The Little Snake River flows westward across the dome, north of and generally parallel to its axis. Slater Dome is located 10 miles southeast of Dixon, Wyoming, and 18 miles southeast of Baggs, Wyoming. The maximum topographic relief in the vicinity of Slater Dome is about 254 feet. The kelly bushing elevation of the discovery well on the structure is 6,792 ft. There are no prominent topographic features in the vicintiy of the dome. The broad flood plain covering the crest and north flank of the structure obscures much of the structural relief. Several basaltic plugs in evidence both north and south of the dome rise to elevations of over 7,000 feet. One of the most notable of these plugs is the 300 foot peak, one mile south of Slater Dome, through which Slater Creek has cut a picturesque canyon-like gorge.

Surface structure is expressed in the upper part of the Mesaverde formation by gentle dips both northwest and southeast. The dome is roughly oval with its axis trending northeast. Dips on the northwest side of the dome, range from 4 to 13 degrees. Structural contours drawn on a white sandstone marker, which outcrops around the dome, indicate surface closure in the magnitude of 225 feet. Mesaverde coals have been mined just south of the axis of the Slater Dome trend, and the earliest test for oil on record was drilled by the Ohio Oil Company late in 1924.

Gas was discovered on this structure in the fall of 1954 by Quintana Production Company at the No. 1 C.F.I. Corporation, NW SE SW, section 13, T. 12 N., R. 89 W. The discovery well was drilled to a depth of 8525 feet in the Jurassic Entrada, plugged back to 3,183 feet and perforated between 3,098 and 3,110 feet with 4 holes per foot in a sand of the Mancos formation. A zone from 2,524 feet to 2,542 feet was also perforated in the Mesaverde formation. Initial flow tests indicate production from the latter perforated interval of an estimated 1,500,000 to 2,000,000 cubic feet of gas per day through a quarter inch choke. The confirmation well drilled by Quintana Production Company, No. 1 Government, NE SE SE, section 14, T. 12 N., R. 89 W., drilled in October 1954 to a total depth of 3,045 feet, was plugged and abandoned after penetrating igneous intrusives at 2975 feet. A drill stem test between 2543 feet and 2669 feet recovered 426 feet of gas cut water. It therefore appears that the productive area on the structure is quite small.

ACKNOWLEDGEMENTS AND REFERENCES

Many of the fields discussed in the preceding paragraphs are new and still under intensive development. Operators are therefore justifiably rticent to release detailed maps, logs, and completion data on structures and wells which remain as the key to complete development.

The foregoing discussions have, therefore, drawn heavily upon the available literature and are in part, paraphrased and condensed from the following listed sources:

Ball, N. Y. and Stebinger, Eugene (1920), "Eastern Part of the Little Snake River Coal Field," U. S. Geol. Survey Bull. 24.

Bradley, W. H. (1945), "Geology of the Washakie Basin, Sweetwater and Carbon Counties, Wyoming, and Moffat County, Colorado." U. S. Geol. Survey Preliminary Map 32.

Campbell, M. R. (1923), "Twenty Mile Park District of the Yampa Coal Field," U. S. Geol. Survey Bull. 748.

Coffin, R. C., Perini, V. C., and Collins, M. J. (1920), "Some Anticlines of Western Colorado," Colo. Geol. Survey Bull. 24.

Fenneman, N. M., and Gale, H. S. (1906), "Yampa Coal Field, Routt County, Colorado," U. S. Geol. Survey Bull. 297.

Petroleum Infirmation (1954), "Resume Rocky Mountain Oil and Gas Operations."

.... (1949), "Resume Rocky Mountain Oil and Gas Operations."

U. S. Geological Survey (1924), "Prospects for Oil and Gas in the Slater Dome, in Northwestern Colorado," Press Release No. 17987.

.... "Prospects for Oil or Gas in an Anticline Near Baggs, Wyoming," Press Release No. 18092.

Sears, J. D. (1925), "Geology and Oil and Gas Prospects of Part of Moffat County, Colorado, and Southern Sweetwater County, Wyoming," U. S. Geol. Survey Bull. 751.

Unpublished summaries and authors field work.

THE URANIUM DEPOSITS OF THE FISH CREEK DISTRICT, COLORADO

By E. P. BERONI and R. C. DERZAY
U. S. Atomic Energy Commission, Denver, Colorado

The Fish Creek district is in the east-central part of Routt County, in the southern Rocky Mountain physiographic province. The area can be reached by taking U. S. Highway 40 east from Steamboat Springs, thence north on county highway 32 to Fish Creek Falls. The initial uranium-bearing discovery in the area in section 12, T. 6 N., R. 84 W., which was examined by Beroni and McKeown (1952) is approximately 1 mile northwest of the falls. The rocks in the vicinity of the claims consist principally of hornblende and biotite schists interlayered with and crosscut by granitic rocks. In general, the contacts between the schists and granitic rocks are gradational. The granitic rocks are composed of potassium—and sodium—feldspars with a minor amount of quartz and biotite, and are medium grained to pegmatitic in texture. These granitic bodies are commonly in the shape of elongated pods as much as 100 feet in length, and are conformable with the foliation of the schist.

The foliation of the schists and alinement of the granitic bodies trend north to northeasterly, and the foliation planes dip between 50 and 75 degrees to the northwest. Drag folds, fracture and flow cleavage, and slickensides are common, but the uranium mineralization does not seem to be genetically related to any of these structural features.

The secondary uranium minerals identified from the district are autunite, and uranophane. These uranium-bearing elongated zones are from 1 to 20 feet in length. Generally, the secondary uranium minerals are concentrated along the gradational contacts between the granitic rocks and the hornblende and biotite schist. Uranium-bearing material of the above type can be noted in the present workings of the original uranium discovery. Pitchblende was found microscopically disseminated in the biotite folia, especially with the biotite books that showed the greatest concentration of secondary uranium minerals.

Select samples taken from some of the uranium-bearing localities showed them to contain up to 0.3 percent uranium. Chip channel samples taken across the radioactive zones showed the uranium content to range from 0.002 to 0.054 percent.

Several dozen other exposures of abnormally radioactive rock were found within half a mile of the initial uranium discovery. All of them are limited to a few feet in areal extent and many are lithologically similar to the radioactive rock on the original uranium-bearing zone. Radioactivity in excess of 0.01 percent equivalent uranium was measured at these other localities.

One abnormally radioactive locality is on the west bank of the north fork of Fish Creek. The highest radioactivity is associated with granite that is stained with hematite and coated with an efflorescent incrustation. A chip channel sample collected from this locality assayed 0.026 percent equivalent uranium, and 0.036 percent uranium. The efflorescent incrustation was probably a combination of uranium and radium sulfates.

Of the numerous small uranium-bearing localities in the Fish Creek district, only a very small number have been examined in detail. Therefore, prospecting in the area might uncover additional uranium-bearing zones.

REFERENCE

Beroni, E. J., and McKeown, F. A. (1952), "Reconnaissance For Uraniferous Rocks in Northwestern Colorado, Southwestern Wyoming, and Northeastern Utah," U. S. Geol. Survey Trace Elements Investigation Report 308A.

URANIUM DEPOSITS IN THE SKULL CREEK AND URANIUM PEAK DISTRICTS, NORTHWEST COLORADO

By Y. WILLIAM ISACHSEN
U. S. Atomic Energy Commission, Grand Junction, Colorado

INTRODUCTION

Production of radioactive minerals in northwestern Colorado reportedly began in 1903 when a property near Skull Creek, Colorado yielded several tons of ore which were shipped to France for radium extractions. The most significant production of radioactive minerals from northwestern Colorado, however, has come from the Uranium Peak area near Meeker. The economic geology of deposits in these two areas is herein summarized.

SKULL CREEK DISTRICT

Location and History of Deposits

Uranium ore has been mined from two areas in the Skull Creek District. The oldest and most extensive workings are on the southern flank of Skull Creek anticline in the N½NE¼, sec. 35, T. 4 N., R. 101 W., where copper-uranium-vanadium ore occurs in a basal sandstone of the Curtis (?) formation. A second mining property resulted when a small uranium orebody was discovered in the Weber sandstone immediately north of the Skull Creek anticline in 1954. This deposit was largely mined out shortly after discovery.

The Curtis deposit is reached by travelling west from Skull Creek, Colorado, for one mile; thence north on an unimproved dirt road for 1½ miles.

This deposit was mined for radium in 1903 and thereafter lay idle until the 1920's when a small amount of ore was mined for vanadium content. The property again became inactive until 1953, when claims were relocated as the Blue Mountain Group, and development drifting for uranium was initiated (McDougald, in press).

Structure

Skull Creek anticline trends east-west, and is located north of Skull Creek, Colorado. Dips are greatest along the southern flank, site of the Curtis deposit, where they measure up to 45 degrees. Flanks of the upwarp are characterized by broad warps plunging down dip, and widespread fracturing which includes longitudinal, transverse, and diagonal joints. Locally displacements of a few inches occur parallel to or along longitudinal joints.

Stratigraphy

Rocks exposed in the area range from the Pennsylvanian Weber sandstone to the Mancos shale of Cretaceous age. Uranium deposits have only been found in the Curtis and Weber formations in the area, and the Weber deposit has not been studied. The Chinle (Shinarump ?) and Morrison formations, which are major uranium producing units on the Colorado Plateau, have not yet proven to be economically important in the Skull Creek area.

Stratigraphic relationships relating to the Blue Mountain Group Mine are described by McDougald (in press). The Curtis host rock overlies massive Entrada sandstone, and is a water-laid, ripple marked, thinly bedded, sandstone ranging in thickness from six inches to 3 feet. This unit grades upward into marine sandstone containing pelecypods, thus marking the transition from terrestrial to marine environment.

Economic Geology

Ore minerals in the Blue Mountain Group mine are all oxidized, and include azurite, malachite(carnotite, calcio-volborthite, and black vanadium minerals. These occur disseminated in the host sandstone, with concentrations along zones containing carbonized plant fragments. Ore is confined to the thin-bedded sandstone unit between the eolian Entrada sandstone and the overlying fossiliferous sandstone. Immediately below the ore-bearing unit, the Entrada is limonite-stained, thus facilitating recognition of the contact during drilling.

Known orebodies are elongated parallel to strike of host beds but no ore trends are known. The production potential of the basal Curtis is regarded as poor (Isachsen, et al., 1955).

URANIUM PEAK DISTRICT

Location and History of Deposits

Mines in the Uranium Peak District are located northeast of Meeker, Colorado, in T. 2 N., R. 91 W. and R. 92 W. The area may be reached by traveling 3 miles northeast from Meeker on State Highway 13 to County Highway 320, thence 11 miles to the Forest Service road which leads east from Yellowjacket Pass into the mining area.

The earliest discovery of uranium ore in the area was made in 1905 (Gale, 1906) at the localities now known as the Midnight Mines on the west flank of the Coal Creek anticline. A limited amount of mining for radium values was done until 1921 when work ceased. From 1940 to 1944, ore in the area was mined for vanadium content. After another period of dormancy, operations were resumed in 1948, this time for uranium. Production for the entire area aggregates somewhat less than 10,000 tons.

Structure

Uranium Peak, at an elevation of somewhat more than 9000 feet above sea level, forms the highest point on the Coal Creek anticline, which is a northwest plunging extension of the White River uplift. Uranium-vanadium deposits occur on both the steeply dipping (40-50°) west flank and the moderately tilted (5-15°) east flank.

Brittle rocks on the structure are affected by an intense conjugate joint system. Longitudinal joints are the more pronounced except on the anticlinal crest (Boyer, in preparation). High angle faults, with displacements ranging from a few inches to 20 feet are widespread in the area and are frequently reflected as stream courses.

Stratigraphy

Rocks exposed on the anticline in the vicinity of the deposits are the Triassic Chinle formation, the Jurassic Entrada, Curtis, and Morrison formations, and the Cretaceous Dakota sandstone and Mancos shale. Of these, only sandstones in the lower portion (Salt Wash member) of the Morrison formation are known to be mineralized.

The Salt Wash member ranges from 90 to 220 feet thick and consists of two sequences of light gray to buff sandstones with subsidiary shale, separated by a zone of red to green sandy shale containing minor thin sandstones (Boyer, in preparation). Overlying the Salt Wash are 230 feet of shale and mudstone constituting the Brushy Basin member of the Morrison formation. The Salt Wash thins on the Coal Creek anticline from about 200 feet on the flanks to 90 feet on the crest (Anthony, 1953).

Economic Geology

Oxidized uranium-vanadium deposits in the area occur in the portions of sandstone lenses exceeding 10 feet in thickness within the stratigraphic interval extending from 8 to 50 feet above the base of the formation according to Boyer (in preparation).

Major ore minerals are tyuyamunite and green to greenish-brown vanadium clay. The ratio of U_3O_8 to V_2O_5 is reported as 1 to 5 (Boyer, in preparation), and the ore is generally in radioactive equilibrium. Ore minerals are disseminated in gray sandstone containing carbonaceous material and lenses of green mudstone, to form localized groups of tabular orebodies and elongate pods averaging 2 feet in thickness.

Known mineralization is most intense near the crest of the anticline and decreases down the flanks; furthermore, no uranium deposits have been found off the anticline, despite persistence of Salt Wash sandstones (Boyer, in preparation).

Fracturing also seems to be influential in ore localization; thus it appears that fractured areas are more often mineralized. In addition, orebodies frequently terminate sharply against joint surfaces. The fact that uranium is held in a relatively insoluble vanadate, tyuyamunite, suggests that the orebodies were oxidized in place (assuming initial simultaneous introduction of low-valent uranium and vanadium). If this is the case such joint surfaces served to delimit the avenues of primary ore solutions. In some areas, displacement of ore layers along minor faults testifies to post-ore movement.

No definite ore trends are known in the area, but the localization of ore near the anticlinal crest coupled with a probable initial fracture control, makes the Uranium Peak District an important example of structural control of uranium localization.

BIBLIOGRAPHY

Anthony, M. V. (1953), "Reconnaissance and Preliminary Drilling in the Coal Creek Area, Rio Blanco County, Colorado," U. S. Atomic Energy Comm., unpublished report.

Boyer, W. H., "Geologic Reconnaissance of the Northwestern Section of the White River Plateau, Rio Blanco County, Colorado," U. S. Atomic Energy Comm., in preparation.

Gale, H. S. (1906), "Carnotite in Rio Blanco County, Colorado," U. S. Geol. Survey Bull. 315, pp. 110-117.

Isachsen, Y. W., Mitcham, T. W., and Wood, H. B. (1955), "Age and Sedimentary Environments of Uranium Host Rocks, Colorado Plateau," Econ. Geol., vol. 50, no. 2, p. 127-134.

McDougald, W. D. (in press), "Wagon Drilling in the Skull Creek Area, Moffat County, Colorado," U. S. Atomic Energy Comm.

URANIUM IN NORTHERN COLORADO AND SOUTHERN WYOMING

By EUGENE W. GRUTT, JR.
U. S. Atomic Energy Commission, Casper, Wyoming
and
JERRY F. WHALEN
U. S. Atomic Energy Commission, Rawlins, Wyoming

INTRODUCTION

Uranium, a relative newcomer to this region, was discovered in the Miller Hill area, 20 miles south of Rawlins, Wyoming, in October 1952. J. D. Love and J. D. Vine of the U. S. Geological Survey found samples containing uranium at localities in T. 17 N., R. 88 W., while checking ground anomalies in the Browns Park formation (Miocene), detected by airborne surveying.

The first significant discoveries were not made until a year later, when A.E.C. and U.S.G.S. reconnaissance found occurrences in T. 12 N., R. 92 W., in Poison Basin. The location of the anomalies, 16 in number, was released to the public November 15, 1953, on an A.E.C. airborne anomaly map. As a result, most of the Browns Park formation in this vicinity was staked in the following months and the wave of prospecting soon spread into Colorado resulting in new finds in the Browns Park formation near Maybell and Lay during March 1954.

These discoveries helped extend a pattern of "Off the Plateau" occurrences, which had developed in Wyoming. At present, a potentially important uranium province defined to a large extent by deposits in Tertiary clastics appears to exist north of the Colorado Plateau.

The index map shows a number of anomalies representative of the pattern recorded by A.E.C. airborne prospecting but of those shown, only some in the Browns Park formation have proven significant. As airborne coverage has not been complete in the area, many anomalies that are not recorded undoubtedly exist.

Browns Park Formation

The Browns Park formation, which unconformably overlies all older formations ranging in age from pre-Cambrian to Tertiary, covers approximately 1500 square miles. In Colorado it is found along the axis of the Uinta Range at its east end and extends eastward north of the Axial Basin Anticline as far as Craig, Colorado. Remnants are also present along the south side of the Washakie basin as far east as Baggs, Wyoming. East of Baggs, it occupies a large, high plateau-like area extending from the Colorado state-line north towards U. S. Highway 30.

This formation of continental origin is comprised of a basal conglomerate overlain by a series of gray and white sandstones, tuffaceous sandstones, tuffs and thin limestone beds. The basal conglomerate varies from less than 10' to over 200' in thickness and the entire formation is between 1000' and 1200' thick. In certain areas, sand grains are rounded and have a frosted appearance, which may at least in part, suggest eolian origin. There is a noticeable lack of mudstones throughout the section and typically the formation is comprised of beds containing uniform, very well sorted, fine-grained sandstones.

OCCURRENCE OF URANIUM

Miller Hill Area

This district is indicated on the index map by anomalies in T. 17 N., Rs. 87 and 88 W. Nearly horizontal Browns Park formation forms the top of dissected, rolling, plateau-lands which are above 8000' in elevation.

Most of the deposits, none productive, are in a freshwater limestone bed, which is up to 12' thick. Uranophane and a little meta-autunite occur as films coating small vugs or fractures in irregular masses of brown or gray chalcedony, which is common as a replacement mineral along fracture zones in the limestone.

Browns Hill

Anomalies in T. 15 N., Rs. 88, 89 and 90 W. were discovered by A.E.C. airborne reconnaissance in October 1953. The topographic and geologic setting is similar to that at Miller Hill; however, most of the anomalies are in fine-grained sandstone beds of the lower or middle Browns Park formation. Yellow and yellow-green uranophane-type mineralization with some meta-autunite disseminated interstitially in sandstone, form irregular deposits along bedding. The importance of these deposits has not been determined, but some are known to be low grade.

Poison Basin

Uranium occurs on a number of knolls and ridges in this small topographic basin, six miles west of Baggs, Wyoming.

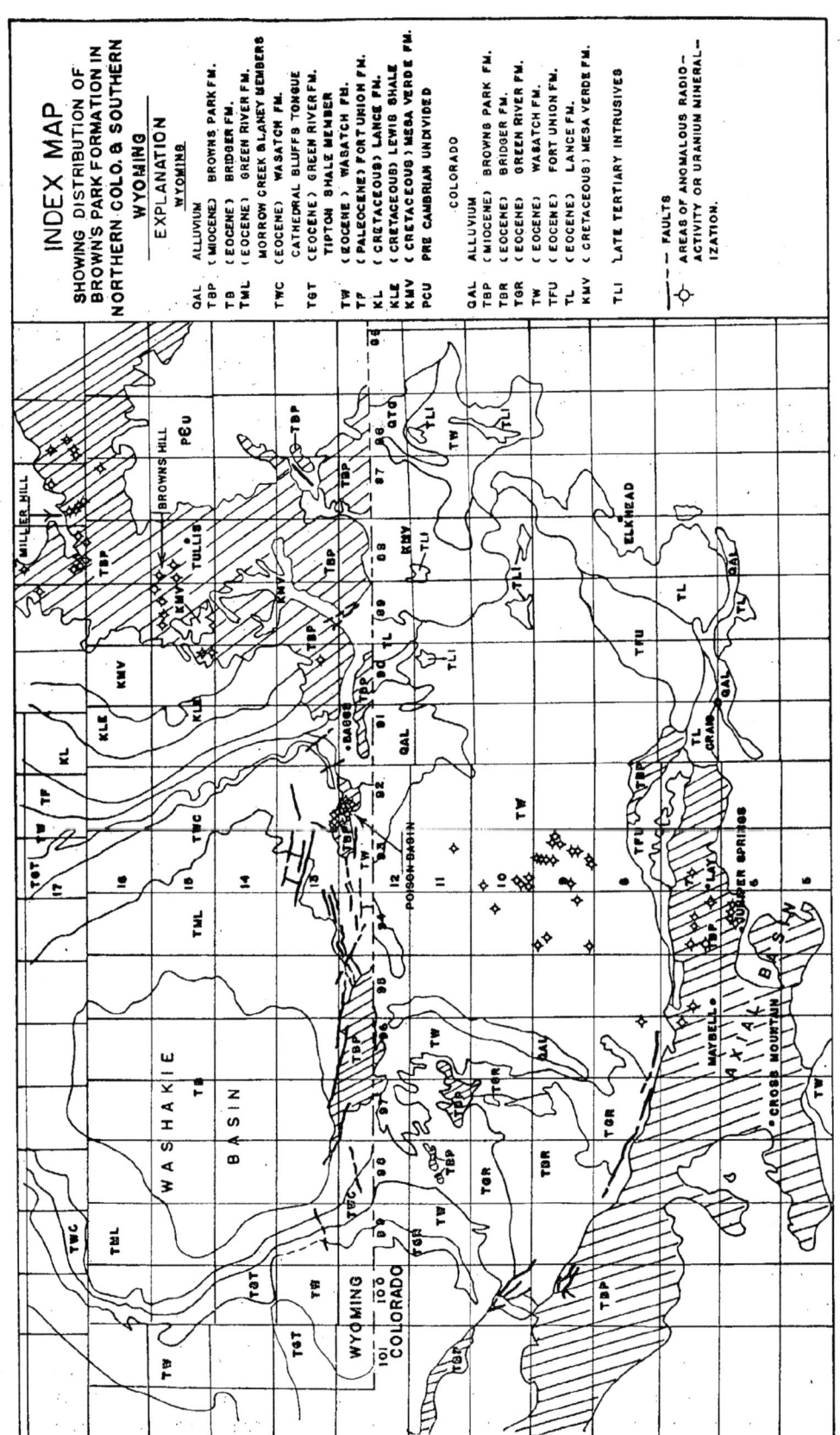

The Browns Park formation, host for all occurrences, is an erosional remnant about six miles long and up to two miles wide. Considerable faulting is present, but is difficult to detect due to the incompetent nature of friable strata and uniform lithology of the Browns Park formation.

A typical deposit contains pale yellow or light green uranium minerals disseminated as sand grain coatings or as interstitial cement in limonite and jarosite stained sandstone. This ferruginous staining often is so pronounced that uranium mineralization is obscured and difficult to detect. The deposits, irregular in shape and usually small in size, often follow along bedding planes in sandstone or along contacts between brown, limonite stained sandstone and white sandstones with bleached appearance. In some deposits, mineralization transects all sedimentary features, but in such cases, fracturing is believed to be the dominant ore control. Uranophane is the most common uranium mineral, but meta-autunite and schroeckingerite are locally important.

Oxidized ores often are lower in uranium than would be indicated by a radiometric assay. This disequilibrium situation is thought to be due to alteration of uranium ores with resultant leaching of uranium away from association with its more radioactive, more insoluble decay products. Another unusual feature of the ore is high selenium content, often over 100 parts per million. The poisonous nature of this element in soluble state and its effect on livestock grazing in the area was responsible for the naming of Poison Basin.

The secondary uranium minerals such as uranophane and meta-autunite, are probably derived from alteration of a primary mineral tentatively identified as uraninite. It has been found in subsurface excavations finely disseminated in dark ferruginous shells surrounding ellipsoidal shaped calcareous concretions up to 2′ long, which occur in sandstone strata with long axes parallel to bedding. Gypsum in seams, veinlets and as interstitial sandstone cement is a widespread gangue mineral.

Although no large deposits have yet been discovered, drilling may well develop commercial orebodies as uranium assays above .2% U_3O_8 have been secured from several prospects.

Maybell-Lay-Juniper Springs Area

The Browns Park formation surrounding these three small hamlets contains a number of radioactivity anomalies and uranium occurrences.

In one type of deposit uranium occurs in small irregular concretions which are sporadically distributed in friable, medium to coarse grained sandstones. No identifiable uranium mineral can be recognized, but uranium is thought to be present in phosphatic sandstones, which sometimes contain in excess of 4% P_2O_5. These deposits are not productive to date.

Deposits which appear to be controlled by a combination of faulting with attendant fracturing and by sedimentary features are the most important types and offer the most promise of being commercial. A prominent set of mineralized faults, as measured on several properties, strike between north, and west and dip in excess of 50 degrees. Uranium is by no means uniformly distributed along these, but fault breccias and fractures sometimes contain concentrations of uranophane, meta-autunite and small amounts of a uraniferous asphaltite. Locally, some faults are silicified or show replacement by fine grained gray calcium carbonate containing finely divided pyrite.

Other surface deposits contain important concentrations of uranium in light colored, friable sandstone beds as sand grain coatings and as interstitial filling. The prominent ore mineral in oxidized parts of such deposits is yellowish-green uranophane, and limonite and gypsum are the common gangue minerals. No definite uranium mineral has been identified in the corresponding unoxidized subsurface parts of these deposits, but the host rock is gray, medium-grained sandstone, which is sometimes calcareous. Some finely divided pyrite is present, but no fossil plant carbon can be discerned megascopically. An interesting feature, the tendency of unoxidized ore to contain more uranium than is indicated by radiometric assay, may indicate youthful formation of the deposits.

Although the present production from this area is small, the outlook is encouraging for development of significant deposits.

SUMMARY

The Browns Park formation, covering large areas in both Colorado and Wyoming is the most favorable formation to prospect. Detectable surface radioactivity and prominent limonite staining have been guides to most known deposits, but it is almost certain that unoxidized subsurface deposits with no such manifestations exist. The almost total lack of recognizable quantities of fossil plant carbon in the ores is a feature at variance with deposits in sedimentary strata in most

other uranium districts, particularly those on the Colorado Plateau.

Radioactivity and uranium are widespread in this area, especially in the Browns Park formation, and although much of the mineralization is low grade, the outlook for substantial tonnages appears good. As other formations also contain strata with favorable lithology, prospecting in them may disclose new uranium deposits.

REFERENCES

Bradley, W. H. (1936), "Geomorphology of the North Flank of the Uinta Mountains," U. S. Geol. Survey Prof. Paper 185(i), pp. 163-204.

Sears, Julian D. (1924), "Geology and Oil and Gas Prospects of Part of Moffat County, Colorado, and Southern Sweetwater County, Wyoming," U. S. Geol. Survey Bull. 751-g, pp. 269-319.

Stone, Leo, and Ove, Warren (in preparation), "Airborne Scintillometer Survey, South-central Wyoming," U. S. Atomic Energy Comm.

Vine, J. D., and Pritchard, George (1954), "Uranium in the Poison Basin Area, Carbon County, Wyoming," U. S. Geol. Survey Circ. 344, p. 8.

Wiggins Studio, Craig

Sand Mountain in the Elkheads, Late Tertiary Volcanics.

ROAD LOGS

Master Road Map

First Day (Log includes additional features)
> The Uinta Fault From Clay Basin Eastward
> Geology in Vicinity of First Night's Camp

Second Day (Log includes additional features)
> Irish Canyon Section
> Skull Creek Section
> Piceance Creek Gas Field

Third Day (Log includes additional features)
> Oil Fields Along Danforth Hills Anticline
> Pagoda Gas Field

Additional Logs

SIDE TRIPS

ROAD LOGS – SIDE TRIPS

Twentymile sandstone along U.S. 40 west of Mt. Harris.

H. R. Ritzma

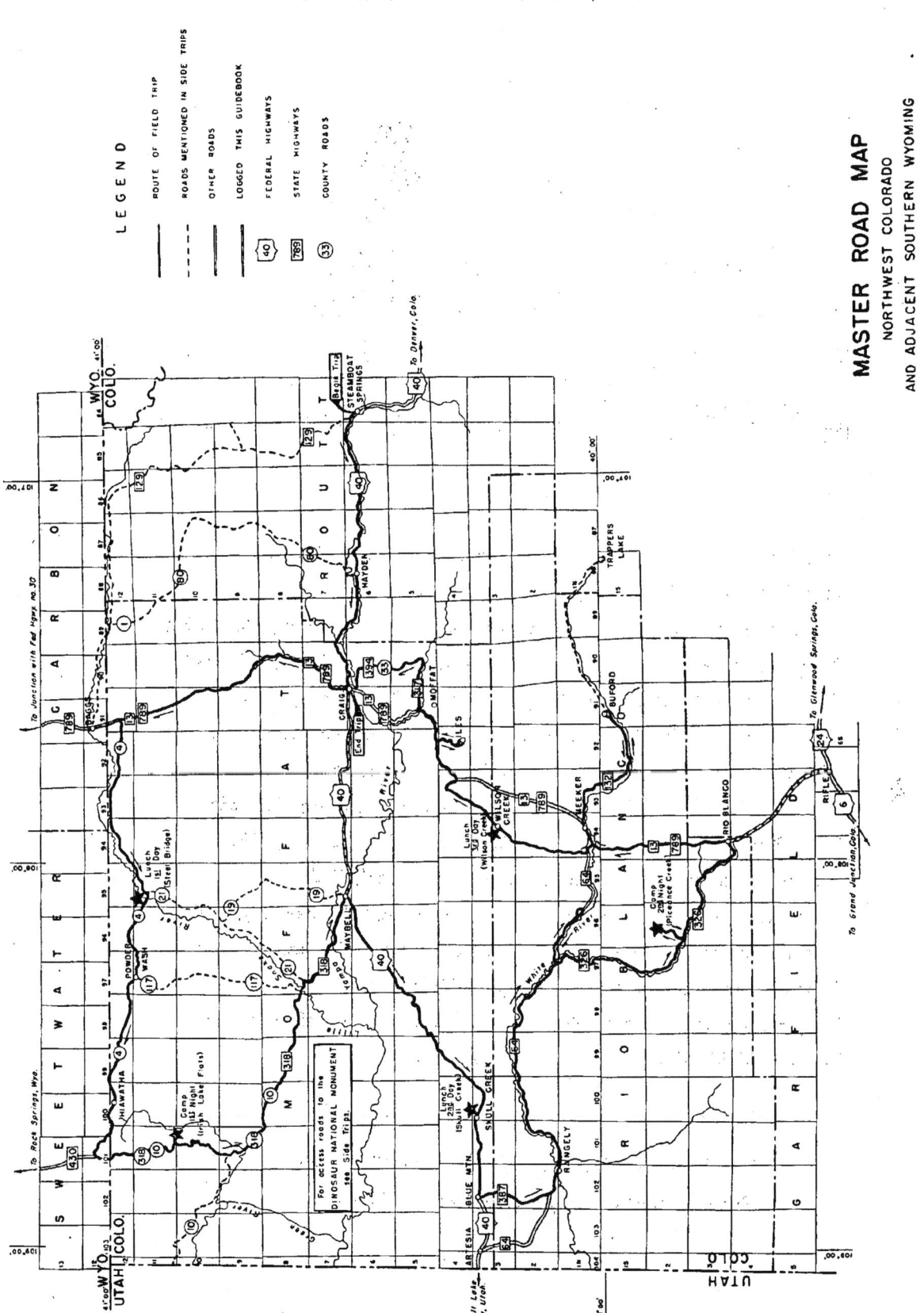

ROAD LOGS

LOGGING BY

John Dahm — El Paso Natural Gas Company, Salt Lake City, Utah
John Donnell — U. S. Geological Survey, Denver, Colorado
Victor B. Gras — Mountain Fuel Supply Company, Rock Springs, Wyoming
Ralph Heins — Phillips Petroleum Company, Steamboat Springs, Colorado
E. R. Keller — Mountain Fuel Supply Company, Rock Springs, Wyoming
Ted G. Larson — Phillips Petroleum Company, Steamboat Springs, Colorado
William H. Lee — Intex Oil Company, Ventura, California
Robert O. Linscott — General Petroleum Corporation, Craig, Colorado
H. R. Ritzma — General Petroleum Corporation, Salt Lake City, Utah
William W. Sloan, Jr. — Carter Oil Company, Vernal, Utah
Robert R. Smart — General Petroleum Corporation, Salt Lake City, Utah
John N. Terpening — General Petroleum Corporation, Durango, Colorado
Robert Thompson — Carter Oil Company, Vernal, Utah
L. Austin Weeks — General Petroleum Corporation, Salt Lake City, Utah
William W. Whitley — Intex Oil Company, Tyler, Texas

LOGGING COORDINATED BY

M. Dane Picard — Shell Oil Company, Salt Lake City, Utah
Robert L. Pott — Forest Oil Corporation, Denver, Colorado

FIRST DAY

STEAMBOAT SPRINGS TO EAST END OF UINTA MOUNTAINS VIA CRAIG, FOUR MILE CREEK AREA, POWDER WASH, HIAWATHA AND SPARKS.

Driving distance: 162.7 miles (For route see Master Road Map)

Dis-tance	Cumulative Mileage	
		Note: Log begins with side trip east from Steamboat Springs.
		LOG BY LARSON AND HEINS
0.0	0.0	Routt County Courthouse. (CARAVAN WILL LINE UP ALONG U.S. 40 HEADING *EAST*.)
0.2	0.2	Bathhouse.
0.1	0.3	Road left to Fish Creek Falls. Stay on U.S. 40. (See paper on uranium by Beroni and Derzay, this guidebook.)
0.3	0.6	Blocks of travertine at 3 o'clock across river.
0.7	1.3	Junction, road to Bear Lake. Stay on U.S. 40.
0.1	1.4	Cross Fish Creek.
0.1	1.5	Junction, old road. Stay on U.S. 40.
1.0	2.5	Junction (TURN LEFT ON GRAVEL ROAD.)
0.8	3.3	(STOP. DISCUSSION OF EMERALD MOUNTAIN EXPOSURES.) Stop beside large granite boulder on right side of road. Looking west toward top of Emerald Mountain, brown sandstone on skyline at 1 o'clock is Dakota. Sandstone outcrop at 11 o'clock is Entrada. Shale exposure above Entrada is Morrison. Chinle is not exposed. Lower hill at 10 o'clock and bench dipping north from it are held up by bed of angular quartz sandstone. This bed is in right position to be the Shinarump, but its appearance suggests something else, possibly Maroon. (See paper by Larson, this guidebook.)
0.4	3.7	(REJOIN U.S. 40, TURN RIGHT BACK TOWARD STEAMBOAT SPRINGS.)
1.7	5.4	Passing courthouse. End of side trip.
0.0	0.0	Begin road log of main trip. Routt County courthouse.
0.6	0.6	(TURN LEFT INTO PARK.)
0.05	0.65	(TURN RIGHT. DO NOT CROSS BRIDGE.)
0.05	0.7	(STOP. DISCUSSION OF STRATIGRAPHY AND STRUCTURE.) Sulphur Spring. Dakota outcrops visible north and south of river. Faulting is visible in outcrop on south. Sulphur Springs on both sides of river probably related to faulting. On west side of Dakota hogback south of river, the thin bedded siliceous Mowry is visible. (A spring near railroad cut to the east emitted a whistling noise like a boat whistle; hence the name Steamboat Springs. Railroad cut in 1905 ruined prenomena.) (CONTINUE ON GRAVEL ROAD.)
0.1	0.8	(REJOIN U.S. 40. TURN LEFT TO WEST.)
1.1	1.9	Junction, Colorado 129 north to Clark, Columbine, Hahns Peak and Little Snake River. Stay on U.S. 40. (See papers by Hunter and Barnwell, this guidebook.) (See Side Trips section, Elkhead Loop Trip.)
0.7	2.6	Lower Mancos shale along road. Dip is to west.
0.5	3.1	Contact between Benton and Niobrara equivalents visible in upper part of shale bank.
0.1	3.2	White outcrop ½ mile at 3 o'clock, Niobrara equivalents. (See section by Larson, this guidebook.)
0.5	3.7	Rifle range. Niobrara outcrops.
0.2	3.9	Road cuts in Niobrara next 1.4 miles.
1.8	5.7	East dips in Niobrara on flank of Trull Anticline. Trull is a long narrow fold paralleling Park Range. It has had numerous tests without success.
0.4	6.1	West dips on west flank of Trull Anticline at right.
2.3	8.4	Cross Elk River. Elk Mountain (Sleeping Giant) to north is large acidic intrusive of late Tertiary age.
1.2	9.6	Sandstone on point of hill at 9 o'clock is about 600 feet below top of Mancos.
1.0	10.6	Sandstone at 2:30 o'clock is Trout Creek sandstone of Mesaverde.
0.7	11.3	Passing through town of Milner.
0.7	12.0	Road left to Osage coal strip mine. Stay on U.S. 40.
0.8	12.8	Steep dips on Lower Mesaverde sands on east flank of Tow Creek Anticline.
0.5	13.3	Passing Mancos - Mesaverde contact.
0.1	13.4	Tow Creek sandstone on top of escarpment at 1:30 o'clock. Two upper Mancos sandstones visible below Tow Creek.
0.7	14.1	(TURN RIGHT INTO TOW CREEK FIELD. STOP. DISCUSSION OF FIELD AND STRATIGRAPHY FROM THIS POINT WEST TO CRAIG.) Turnout into field not included in log. (See paper by Saterdal, this guidebook.)
0.0	14.1	(REJOIN U.S. 40.) Continue log.
0.6	14.7	Outcrops on north side of road are somewhat slumped.
1.1	15.8	Tow Creek sandstone on west flank of anticline.
0.3	16.1	Sandstone toward top of hill ahead is Trout Creek.

Dis-tance	Cumulative Mileage	
1.0	17.1	Road skirts along Trout Creek sandstone. Twenty-mile sandstone on skyline at 12 o'clock.
0.9	18.0	Town of Mt. Harris, once a very active coal mining center.
0.3	18.3	Mt. Harris school right. Twenty-mile sandstone escarpment straight ahead.
1.1	19.4	Crossing D. & R. G. W. R. R. tracks. A few hundred feet of beds intervene between top of Twenty-mile sandstone and base of Lewis shale. (For Cretaceous stratigraphy, see O'Boyle, this guidebook.)
0.1	19.5	Mesaverde - Lewis contact.
0.1	19.6	Cross Yampa River.
0.6	20.2	Hooker Mountain intrusive at 2:30 o'clock, a basalt sill in Lewis shale. Wolf Mountain intrusives at 3:30 o'clock.
1.4	21.6	Valley in Lewis shale. Hills ahead and to right are mainly Lance.
0.4	22.0	Double peak on skyline to north is Bears Ears, a landmark in the volcanic Elkhead Mountains. Flat Tops, also volcanics, at 9 o'clock distant skyline. (See paper by Carey, this guidebook.)
1.6	23.6	Lewis - Lance contact in lower bluff at 3 o'clock across river.
1.3	24.9	East edge of Hayden. Town named for F. V. Hayden, noted pioneer geologist. (Elkhead Loop Trip in Side Trips section ends here.)
0.7	25.7	West edge of Hayden.
0.9	26.6	Lance formation exposed on both sides of valley.
0.7	27.3	Crossing Yampa River. Lance exposed at right.
2.3	29.6	Hills at right are Lance.
2.5	32.1	Cross D. & R. G. R. R. tracks.
2.8	34.9	Passing sandstones in Lance.
0.5	35.4	Routt - Moffat County line.
0.5	35.9	Cross Elkhead Creek.
0.5	36.4	Road right to Elkhead. Stay on U.S. 40. Breeze Mountain intrusives across river at 9 o'clock. Lance outcrop in road cut ahead with small fault exposed.
5.2	41.6	East edge of Craig. Craig, western terminus of the D. & R. G. R. R., is county seat of Moffat County and center of Northwest Colorado's wool, wheat, and cattle industries. Coal, oil, gas, and gilsonite also contribute to its economy.
0.4	42.0	Cross Fortification Creek.
0.1	42.1	Cosgriff Hotel left.
0.2	42.3	Stop light, junction, U.S. 40 with Colorado 789-13. (TURN RIGHT ON COLORADO 789-13.)

LOG BY LINSCOTT

CRAIG'S GEOLOGIC SETTING

Craig rests on north dipping outcrops of Lewis shale on the north flank of Breeze Anticline (Craig Dome) (See paper by Wyeth, this guidebook.) The formation consists of gray marine shale and a few thin beds of soft sandstone, reaching a thickness of 1600 feet in the vicinity of Craig. It is believed that the absence of the Lewis shale and Lance formations south of the Axial Basin are the result of regressive offlap produced by north retreating late Cretaceous seas, and that in reality the Lewis shale pinches out to the southwest and the Lance formation merges with the Williams Fork member of the Mesaverde formation.

0.5	42.8	Craig High School. For the next 11 miles the trip will follow the valley cut by Fortification Creek. All beds dip northerly into the Sand Wash Basin. As the trip progresses north, the dips become more westerly.
0.2	43.0	Massive sandstone cliff at 10 o'clock represents the basal sand of the Lance formation. Originally mapped by Gale as "Laramie", the Lance consists of a succession of shales, thin coal beds, and soft sandstones. Plant remains from the upper part indicate a Laramie age; however, marine invertebrates from the lower 250 feet are of Lewis age. The prominent bluffs formed by the Lance sands are conspicuously unique, for elsewhere they have little topographic definition and are almost indistinguishable from the overlying Fort Union formation. At Craig the Lance formation is 1020 feet thick.
0.1	43.1	Road crosses basal Lance. The prominent massive sandstone at 12 o'clock on the horizon represents the top of the Lance formation. The medial beds of the Lance are much less resistant and form strike-valleys.
0.9	44.0	Road swings sharply to right.
0.6	44.6	Top of Lance formation just off road to left.
0.4	45.0	Basal Fort Union conglomerate can be seen about 150 yards ahead just to the right of the road. It is a very inconspicuous and subtle surface exposure occurring as a zone of residual pebbles.
0.1	45.1	STOP. DISCUSSION OF TERTIARY STRATIGRAPHY. This conglomerate marks a notable unconformity between the Lance and Fort Union formations, although there is no discordance of dip above or below. The thickness of the conglomerate and the size of the contained pebbles are strong evidence favoring rapid uplift in the vicinity of the Park Range. The conglomerate contains pebbles of granite, two or more porphyries, vein quartz, quartzite, ultrabasic rocks, and slabs of agatized wood. The last is very diagnostic. Look west where the Lance crosses the highway. Note its sudden loss of topographic definition.
0.4	45.5	Cross bridge. For the next few miles the road follows fairly close to the strike of the Fort Union formation. The Fort Union consists of 800 feet of drab shales, soft shaley sands, and coal beds.

Dis-	Cumulative	
tance	Mileage	
0.7	46.2	Black Mountain at 11 o'clock, a flat-topped wooded mountain on horizon capped by late Miocene volcanics. The high point is at an elevation of 10,300 feet. Bears Ears at 12 o'clock.
0.6	46.8	The typically drab colors of the Fort Union are exposed on the hillside at 2 o'clock.
0.3	47.1	Cross Sand Gulch.
1.4	48.5	Pilot Knob in Elkhead Mountains on horizon at 2 o'clock.
0.6	49.1	Road bears to north, still in north-dipping Fort Union sediments. Featureless topography very characteristic of Fort Union exposures.
2.0	51.1	Cross Cottonwood Gulch.
1.9	53.0	Road crosses Fortification Creek. In 1947, Carter Oil Co. drilled its North Craig well three miles west of this point. It reached total depth at 9000 feet in the Mancos shale after having shows of oil in the Mesaverde.
0.6	53.6	Little Buck Mountain straight ahead.
2.5	56.1	Passing through transition zone between Fort Union and Wasatch formations. No objective physical criteria delineates the transition zone. Sears describes the Wasatch-Fort Union contact as an inconspicuous quartz pebble conglomerate. Further complicating this boundary is the fact that the basal Fort Union conglomerate is overlain by Wasatch south of the Axial Basin suggesting that the Fort Union may be lowland facies of the Wasatch formation. Regional dip is now west.
2.2	58.3	Coarse gritty sand and conglomerate in road-cut to right is Miocene Browns Park formation. The Browns Park overlies unconformably and usually at great angularity all older rocks from the pre-Cambrian to late Eocene. Nearly all exposures on the right for the next few miles are believed to be isolated remnants of the Browns Park formation preserved by the overlying volcanic flows and rubble.
2.2	60.5	Fortification dike at 11 o'clock. (See paper by Carey, this guidebook.)
1.5	62.0	Fortification dike crosses highway on next turn. Small ramifications of main dike can be seen as it is traced west. There is no evidence that contact metamorphism accompanied intrusion. The period of vulcanism in this area has been dated as late Miocene. In most occurrences it forms a resistant cap rock on the Browns Park formation. However, some authors have reported Browns Park above the "flows." (See paper by Carey, this guidebook.)
0.7	62.7	First typical Wasatch exposures a few miles to the north at 10 o'clock, characterized by brightly banded benonite shales.
0.3	63.0	(STOP NORTH OF FORTIFICATION DIKE OPPOSITE EXPOSURES OF SMALLER SUBSIDIARY DIKES OF THE FORTIFICATION DIKE SYSTEM. DISCUSSION OF ELKHEAD VOLCANICS.) Specimens of the Fortification Dike have been described as ". . . because the rock contains about 20% nepheline; a nepheline minette. The mineralogy is essentially like minettes described in the literature. Microphenocrysts ($\frac{1}{2}$ to 2 mm) of idiomorphic calcitized olivine, diopsidic augite, and brown biotite are set in a ground mass of idiomorphic diopsidic augite, biotite, and nepheline, and xenomorphic orthoclase (probably sodic). There is a conspicuous amount of what appears to be analcite, secondary after nepheline. Minor accessories are idiomorphic apatite and pyrite. Lamprophyres, of which minette is the most common type, are widely distributed throughout the Rocky Mountain region and the lower Mississippi Valley but they are extremely rare in other parts of the country. Their spatial association with granitic basement rocks suggests a genetic relationship. Although the origin of these unique potassic basic rocks is still a matter of much debate, the most widely accepted hypothesis suggests that they have been derived from olivine or nepheline basalts which have selectively assimilated the potassic granitic basement." (J. H. Halsey, Magnolia Field Research Laboratories, Dallas, Texas.)
0.4	63.4	Note old flume crossing Fortification Creek at 11 o'clock. It was designed to divert water from Black Mountain to the gold placers in the Great Divide district. Placers later proved to be too fine for commercial extraction. An early geologic report on the Four Mile placer district was by Herbert Hoover in 1897.
2.0	65.4	Cross Fortification Creek. The remainder of trip on Colorado 789 will be over gentle westerly dipping Wasatch beds.
2.7	68.1	Baker Peak at 3 o'clock.
0.4	68.5	Brightly colored banding in Wasatch at 3 o'clock. Thick cover of volcanic rubble obscures exposures higher up the mountain.
5.2	73.7	Baggs anticline is defined in Wasatch exposures at 11 o'clock approximately 15 miles to the north. It is the eastern complement of the structurally high Cherokee Ridge which effectively separates the Sand Wash and Washakie Basins. About 7 miles west of Baggs, Kerr McGee, Phillips, and Ray McDermott drilled an 8500 foot Mesaverde test. Blowouts from shallow (700') Wasatch sands burned the rig down twice. The well was completed as a Lewis discovery for 4000 MCF/D. (See SOUTH BAGGS, paper by Turner, this guidebook.)
2.4	76.1	Cross Sand Spring Draw. Still on Hiawatha Wasatch. Nearing locale of vertebrate fossil quarry of University of California Museum. (See paper by McKenna, this guidebook.)
2.0	78.1	(TURN LEFT ON MOFFAT COUNTY ROAD 4.)

LOG BY GRAS AND KELLER

| 0.1 | 78.2 | On Tertiary for next 70 or so miles. |

Distance	Cumulative Mileage	
1.2	79.4	(STOP. DISCUSSION OF GENERAL GEOLOGY AND VERTEBRATE PALEONTOLOGY.)
3.2	82.6	Keep right. Left hand fork to Great Divide.
1.0	83.6	Ridge to right across Little Snake River is Baggs anticline, and is capped by Browns Park formation. At four o'clock over the ridge is North Baggs Unit Well No. 1 discovery of Lewis gas production. The well was drilled by J. Ray McDermott Company on farmout from Kerr McGee - Phillips.
4.8	88.4	Off to right is visible arching in Wasatch formation on Baggs anticline.
10.9	99.3	At one o'clock on horizon is Sunny Point, sharp peak in Laney shale of Green River formation.
5.9	105.2	Keep right. Left fork to Craig via Great Divide.
0.4	105.6	Cross Little Snake River (LUNCH STOP AHEAD. CARAVAN TURNS RIGHT UNDER ONLY TREES IN COUNTY, OR SO IT SEEMS.)
0.2	105.8	(REJOIN ROAD.) Ascending Nipple Rim, still on Wasatch formation.
4.4	110.2	Juniper covered ridge to right is Cherokee Ridge. Flat topped butte on Cherokee Ridge at two o'clock is Powder Mountain. Cherokee Ridge is capped by Browns Park formation.
1.4	111.6	Keep left. Right fork to Cherokee Ridge.
.5	112.1	To left is contact of Cathedral Bluffs member of Wasatch formation and Laney shale member of Green River formation. Laney shale caps ridge.
1.3	113.4	Contact Cathedral Bluffs and Laney shale.
1.1	114.5	Leave Laney shale drive on to Cathedral Bluffs.
0.5	115.0	Leave Cathedral Bluffs, drive on to Laney shale.
1.0	116.0	At right is location of Ace Unit Well No. 4. Dry hole drilled by Mountain Fuel Supply Company.
0.4	116.4	Leave Laney shale and drive on to Cathedral Bluffs. Powder Wash field ahead.
2.2	118.6	Straight ahead after (STOP. DISCUSSION OF POWDER WASH FIELD.) Road to left to Powder Wash field camp. Compressor plant on left. (See paper by Folsom, this guidebook.)
3.1	121.7	Drive from Cathedral Bluffs on to Laney shale.
2.8	124.5	(CAUTION) Sharp curves and steep hill ahead. Laney Rim. (STOP. DISCUSSION OF VERMILION BASIN.) (See paper by Gras, this guidebook.)
0.4	124.9	Contact of the Laney shale member of the Green River formation and the red Cathedral Bluffs member of the Wasatch formation.
0.6	125.5	At eleven o'clock on the horizon are the Uinta Mountains.
2.1	127.6	(CAUTION) Sharp corners and steep hills.
0.2	127.8	Guiterrez Ranch located on Shell Creek. Decation courtesy Mountain Fuel Supply Company Rig No. 1.
0.4	128.2	Contact of the Cathedral Bluffs member of the Wasatch formation and the Tipton tongue of the Green River formation.
4.8	133.0	Contact of the Tipton tongue and the Hiawatha member of Wasatch formation. Hiawatha member on surface to end of road log.
1.6	134.6	On Hiawatha anticline. (See paper by Gras, this guidebook.)
0.4	135.0	Hiawatha Unit No. 3 gas well on left.
1.0	136.0	Hiawatha Unit No. 1 gas well on left. Hiawatha compressor plant on right.
0.7	136.7	(Keep right.) Road to left to Hiawatha Camp. Ahead, Cold Spring, Diamond, Middle and Pine Mountains are visible from left to right on horizon.
1.6	138.3	(CAUTION) Curves and steep hill. Note lacustrine character of Hiawatha formation. This sequence has been called "Lower Tipton".
1.0	139.3	Vermilion Creek.
3.0	142.3	Sugar Loaf field at ten o'clock, approximately five miles.
5.3	147.6	Canyon Creek field. Unit No. 4 at three o'clock and Unit No. 6 at eight o'clock. (See paper by Gras.)
0.3	147.9	(TURN LEFT ON CUTOFF ACROSS ROAD JUNCTION.) Road right continues to Rock Springs, Wyoming (Wyoming 430).
0.3	148.2	(TURN LEFT ON COUNTY ROAD TO IRISH CANYON.) End of portion of road log.

LOG BY RITZMA AND WHITLEY

Distance	Cumulative Mileage	
0.1	148.3	(STOP FOR DISCUSSION OF CANYON CREEK GAS FIELD.) (See paper by Gras, this guidebook.)
0.1	148.4	View of south plunge of Canyon Creek Anticline ahead. Many small wrinkles are observable on crest of fold. Well on left at 10:30 o'clock.
1.3	149.7	Road right to 4—J Basin, a topographic hollow of Hiawatha (Wasatch) surrounded by Tipton (Green River). Basin partially reflects structure of Middle Mountain Anticline (gas field) which culminates under Pine Mountain. Canyon Creek and tributaries have cut impressive canyons through brownish orange Tipton sandstone. This sandstone was built into the Tipton phase of the Green River Lake from streams discharging off the highlands of the Uinta overthrust belt. Clastics totally replace the lacustrine facies of the Tipton.
0.4	150.1	Bridge over Canyon Creek.
0.3	150.4	Canyon Creek coal mine left (abandoned). Coal is a local development in the uppermost Hiawatha. Axis of Canyon Creek anticline ahead and to left.

Dis-tance	Cumulative Mileage	
0.5	150.9	Bridge. Entering Colorful Colorado. Boundary not well marked.
0.6	151.5	View to left to east and south. Derricks of Hiawatha Dome on skyline at 10 o'clock. Lookout Mountain straight ahead is a high point of the Laney shale rim around the Vermilion Basin Uplift. The mountain is capped by a conglomerate (Basal Browns Park (?) or Bishop (?)) which contains fragments of Uinta quartzite and Madison limestone of considerable size. This indicates two things: (1) that the Uintas once stood much higher in relation to the basin than at present (2) that the Uinta overthrusting may have carried some of this large material a considerable distance to the north and east.
0.6	152.1	Surface expression of Canyon Creek anticline fades into small, low Haymower Dome in drab, flat-lying beds at 2 o'clock (right). Note Tipton sandstones in scarp (3 o'clock) disappearing into side of Diamond Peak. Traced to south Tipton becomes a loosely compacted coarse sand and conglomerate, loses its topographic prominence and weathers into soft, juniper covered bouldery, sandy hills along base of Diamond Peak. Above Tipton on Diamond Peak is Cathedral Bluffs, reddish, sandy and conglomeratic. Diamond Peak (9975') is capped by Bishop conglomerate (Miocene?).
1.1	153.2	Canyon Lake playa on right (9 o'clock).
0.9	154.1	Descend badlands cut into Hiawatha Wasatch. As the Uinta fault is approached Hiawatha becomes continental, colorful and conglomeratic, indicating a sharp change in environment from lake-swamp-mud flat deposition seen near Hiawatha and Canyon Creek. Road has crossed syncline between Canyon Creek Anticline and the Hiawatha Sugarloaf axis, and beds are rising to Sugarloaf Anticline. Note coaly beds and red clays in close proximity indicating widely contrasting depositional environments close to the rising Uinta Arch and overthrust belt.
1.7	155.8	Road right to well site. (TURN RIGHT TO WELL SITE, TURN AROUND AND PARK FACING OUT TO MAIN ROAD.) (ASSEMBLE ON TOP OF SMALL KNOB FOR DISCUSSION OF GENERAL GEOLOGY.) Side trip not logged. See map "The Uinta Fault from Clay Basin Eastward," we are in vicinity of cross section "a-b", just north of Sparks Fault.
0.0	155.8	(RETURN TO MAIN ROAD.)
0.6	156.4	Fault in road cut left ahead has unknown displacement down to south. It is probably part of the graben faulting which in this area totals 1500 to 2000 feet of movement down to the south. Main graben faults are probably concealed in complex zone between Sparks and Uinta faults.
0.3	156.7	Approaching crest of Sugarloaf Anticline. Sugarloaf Butte ahead.
0.3	157.0	Road right to Sparks Ranches. Good exposures of Sparks and Uinta faults along this road.
0.8	157.8	(STOP FOR DISCUSSION OF GEOLOGY.) Talamantes Creek Valley. Ahead is seen Sparks Fault with Mesaverde slice faulted against Tipton. Sugarloaf Butte is the locus of a large fan built into upper Hiawatha (?) and Tipton by a stream discharging from the rising Uinta Arch and thrust belt. Sugarloaf Butte conglomerate contains fragments of Paleozoics but no Uinta quartzite (pre-Cambrian). This is the case with the Tipton conglomerate on Diamond Peak and presumably the Sugarloaf conglomerate is a Tipton equivalent. All younger (post-Tipton) sediments carry distinctive cobbles of Uinta quartzite. Looking from left to right up Talamantes Creek, gray Mancos shale is seen capped by pinkish gray to white unconformable Browns Park beds which cover most of the important details of this complex area. Paleozoics and pre-Cambrian of Cold Spring Mountain on skyline are the Uinta Overthrust sheet which has placed pre-Cambrian over Mancos. Mancos is in an intermediate fault slice between the Uinta and Sparks faults. Sparks Fault is seen almost due west above corrals and pens with Mancos faulted against Hiawatha Wasatch which has been dragged up and is vertical to overturned along the fault zone.
0.8	158.6	Cross Talamantes Creek. Crushed Mesaverde to right (1 o'clock) in fault slice with Browns Park draped over it. Tipton conglomerate on the left in Sugarloaf butte.
0.2	158.8	Continuation of Sparks Fault in canyon wall at left at 10:30 o'clock. Tipton conglomerate is faulted against crushed Mesaverde which has been almost removed by pre-Browns Park erosion. Browns Park "conveniently" obscures most of the details.
1.0	159.8	Side road left affords cross section views of Sugarloaf Butte. (OPTIONAL SIDE TRIP.) Sugarloaf conglomerate beds may be seen rapidly fingering out to east into lacustrine Tipton and "lower Tipton".
0.2	160.0	Road on "pavement" of basal Browns Park conglomerate. Cold Spring Mountain right with high hogback of Madison dipping east. Low ridges at base of hogback are Morgan and Weber. At Talamantes Creek, pre-Cambrian and Madison were thrust over Mancos. Southward the complete section intervenes between Cold Spring Mountain and the Sparks fault which lies east of the road. This indicates decreasing throw on the Uinta Overthrust which presumably dies out in steep to vertical beds north of Irish Lake Flat.
0.4	160.4	Road ahead to left leads to Sparks Fault with spectacular exposures. (CARAVAN TURNS LEFT AHEAD.)

SIDE LOG

0.0		(LEAVE MAIN ROAD.)
0.3		(TURN LEFT.)
0.6		At 9 o'clock at break of trees and slope across gully is trace of Sparks Fault. Fault

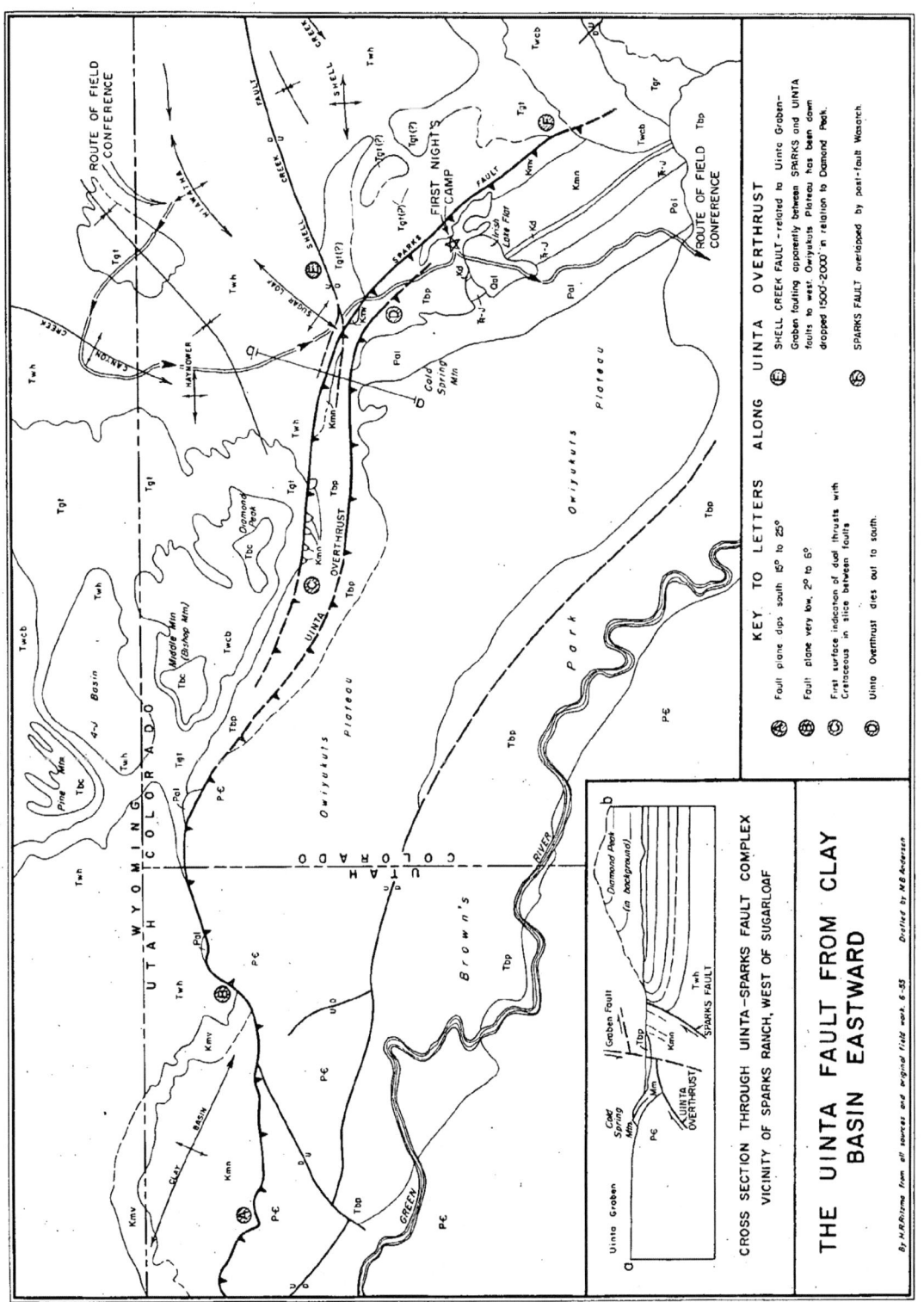

Dis-tance	Cumulative Mileage	
		passes under road and to right with roughly similar break of slope and topographic expression. Exact trace is difficult to determine because of width of crushed and jumbled zone.
1.1		Spectacular exposures in gullies to right of road. Caravan will stop for discussion of faulting and general geology. Turn around and return to main road.

Dis-tance	Cumulative Mileage	
0.0	160.4	(REJOIN MAIN ROAD.)
0.7	161.1	Jeep trail to right leads up Cold Spring Mountain and on to Owiyukuts Plateau. Stay on main road.
0.1	161.2	Note forest fire devastation. (PLEASE BE CAREFUL WITH MATCHES AND CIGARETTES.)
1.5	162.7	(TURN LEFT TO FIRST NIGHTS CAMP.)

GEOLOGY IN VICINITY OF 1ST. NIGHTS CAMP

On second morning cars will proceed to overlook of Sparks Fault where conference will begin walking trip of section. Drivers will take cars to parking area north of Irish Lake where they will wait for walking tour to return. Some drivers will want to drive trail to north to observe Moenkopi to Navajo section

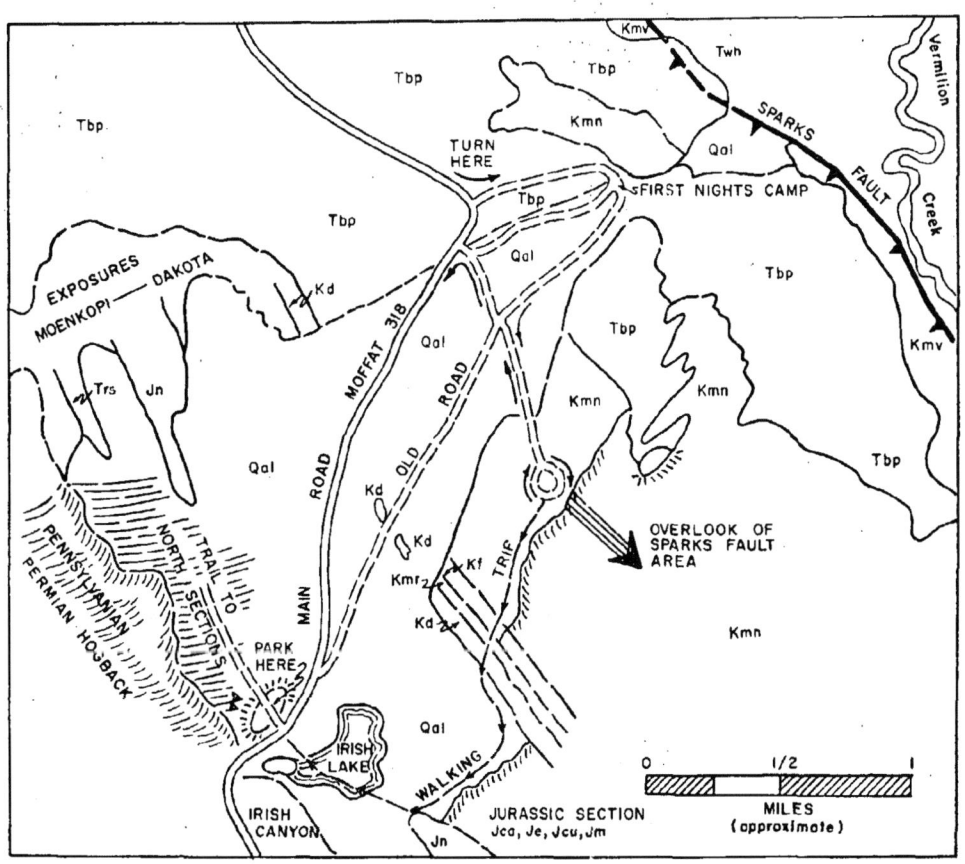

DIAMONDS!

Diamond Gulch, Diamond Creek, Diamond Peak, Ruby Gulch — strange names to crop up in an area where occurrences of gem stones are unknown. And stranger is the story of one of the nation's greatest mining swindles commemorated by these place names in northeastern Utah and northwestern Colorado.

In 1871 two prospectors, Arnold and Slack, deposited several sacks of rough diamonds in the Bank of California in San Francisco, confiding to William Ralston, the bank's president, that they had discovered diamonds 1000 miles east of California. A railroad official and a mining engineer were dispatched to the "field" and their glowing reports of "acres of gems" and "a million dollars a month" fanned the glowing fires of excitement to a conflagration. A $10,000,000 corporation was founded with stock-holders prominent among the jewelry houses of New York and reaching to the House of Rothschild in Europe.

Clarence King, of 40th Parallel Survey fame, investigated. His findings quenched the fires — gems strewn on the ground but none in place — diamonds in the cracks in rocks and forks of trees — some with partially cut faces. The "gems" proved to be South African "niggerheads" purchased as rejects from the Amsterdam markets. Banker Ralston lost $660,000 and ultimately his life. The perpetrators were never brought to justice, and the Uinta diamond fields faded into dusty history.

SECOND DAY

EAST END OF UINTA MOUNTAINS TO PICEANCE CREEK DOME VIA IRISH CANYON, MAYBELL, SKULL CREEK, RANGELY AND WHITE RIVER.

Driving distance: 194.2 miles   (For route see Master Road Map)

Dis-tance	Cumulative Mileage	
		LOG BY RITZMA AND WHITLEY
0.0	0.0	Road left to first night's camp. (CARAVAN WILL NOT FOLLOW THIS LOG. SEE SEPARATE MAP "GEOLOGY IN VICINITY OF FIRST NIGHT'S CAMP" FOR CARAVAN ROUTE.)
0.2	0.2	Road left to spectacular overlook of Sparks Fault and overlap of fault by Cathedral Bluffs (Wasatch) beds.
0.1	0.3	Irish Lake Flats, former course of Vermilion Creek before diversion by combination of stream piracy and drainage changes due to Uinta Graben faulting. Hills on skyline left are near vertical Mancos capped by Browns Park. From left to right at 10 o'clock is white Frontier sandstone escarpment and tree covered Dakota ridge with Mowry shale in swale between. Dakota ridge extends well out into alluvium in Irish Lake Flat.
0.5	0.8	Begin complete section visible on right with purple basal Browns Park conglomerate truncating nearby vertical Mesozoic beds.
0.4	1.2	Sharp tree covered ridge at 11 o'clock between Dakota hogback and large Weber hogback is basal Navajo sandstone with a small amount of Chinle visible immediately below.
0.8	2.0	(CARAVAN PARKS HERE. WALKING TRIP REASSEMBLES. LOG AND CARAVAN NOW TOGETHER.) Trail to right leads to top of knoll for view of Irish Lake Flats and surrounding area. Looking north from top of knoll from left to right is gray-green Moenkopi; low escarpment of Shinarump; deep red Chinle; massive, cross-bedded Navajo; thin streak of red Carmel; thin Entrada; gray-green Curtis; red and gray-green Morrison; and thin Dakota in rounded, tree covered hill. To south, same section can be seen from right to left from Shinarump to Dakota.
0.3	2.3	Irish Lake playa to left. Fremont party of 1844 stopped for lunch here. Entering Irish Canyon. Large hogback is Weber sandstone. Phosphoria (Park City) is mostly covered by talus and vegetation.
0.3	2.6	Weber-Morgan contact in notch on skyline ahead. (CARAVAN WILL STOP FOR BRIEF SCENIC AND PHOTO STOPS.)
0.5	3.1	Road crosses Weber-Morgan contact. Morgan red shales, sandstones and dolomites in dip slope ahead right. Road proceeds roughly along Weber-Morgan contact.
0.8	3.9	Pictographs on Weber block to left.
0.3	4.2	Morgan exposures right and left of road.
0.2	4.4	Note Weber-Morgan contact in canyon wall ahead; light colored Weber contrasts with reddish Morgan sandstones, dolomites and cherty dolomites.
1.0	5.4	Road veers right into gap cut into Madison limestone. Between Morgan and Madison is 80' of

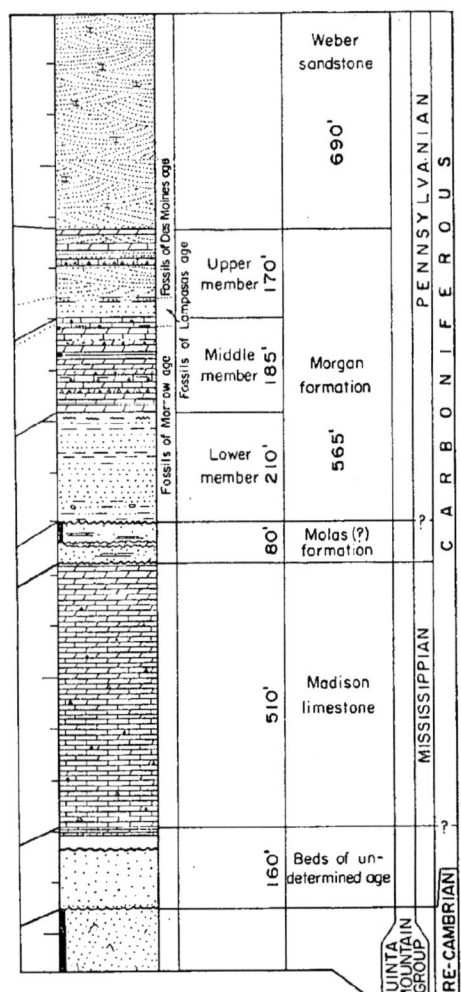

IRISH CANYON SECTION
After U.S.G.S. Oil and Gas Preliminary Chart No. 16

Drafted by M.B. Andersen, 6-55

Distance	Cumulative Mileage	
		sandstone and dolomite attributed to the Molas? formation.
0.4	5.8	Madison and pre-Cambrian Uinta Mountain Group contact in break of slope on skyline to left and right of road. Between Madison and deep red Uinta quartzite is 150' of dolomite, sandstone and limestone which may be basal Madison or may be thin remnants of Devonian or Cambrian. This interval of "beds of unknown age" can be seen in the tree covered interval between the cliff forming Madison and the somber Uinta quartzite.
0.1	5.9	Madison basal? breccia in block at right of road. Road descends grade through pre-Cambrian.
0.2	6.1	Madison-Uinta quartzite contact in cliff above road ahead.
0.9	7.0	East-west graben fault (down to south) of unknown magnitude cuts through notch at right of road at 2:30 o'clock and continues across road to east through low saddle where rubbly Madison is truncated and in contact with Browns Park and gravel (right of saddle).
0.4	7.4	To right, Browns Park laps on to Uinta quartzite. Probably there is fault contact in places in this complex area.
0.9	8.3	Road to left leads to ranch and then by jeep and foot trail to classic Vermilion Creek section where pre-Cambrian to Mesaverde is continuously exposed. Bull Canyon, lower part of abandoned course of Vermilion Creek, to right.

LOG BY RITZMA, DAHM AND WEEKS

Distance	Cumulative Mileage	
1.4	9.7	North point of triangle in east portion of Browns Park. Ahead on skyline from right to left are the Uinta Mountains, at 3 o'clock Diamond Mountain; and at 2:30, the "Gates of Lodore", north opening of Lodore Canyon. The Green River here has cut a canyon across the Uintas over 2000 feet deep. To left of Lodore Canyon is Douglas Mountain (Escalante Hills of Hayden Survey). Highest knob visible left of Gates of Lodore is Zenobia Peak, elevation 9006', highest in Dinosaur National Monument. (See Additional Logs for trip to Clay Basin.)
0.2	9.9	Road right leads to Browns Park area proper. Continue on Moffat 318-10.
0.5	10.4	Descend slight grade. Browns Park exposures lapping onto pre-Cambrian of Douglas Mountain ahead.
1.1	11.5	Begin descent into canyon cut by Vermilion Creek into Browns Park formation. Excellent exposures of Browns Parks in canyon and in road cuts to left. Note darker beds (see paper by Carey, this guidebook).
0.6	12.1	Waterfall in Vermilion Creek at head of canyon. Highway equipment shed.
0.6	12.7	Bridge over Vermilion Creek.
0.4	13.1	Road right to Greystone, Douglas Mountain and "Cottonwoods", a picnic site and camp ground at "Gates of Lodore" in Dinosaur National Monument. Continue ahead on Moffat 318-10. (See Side Trips.)
1.6	14.7	Cross Douglas Draw.
2.5	17.2	Road swings left up West Boone Draw.
1.9	19.1	Road left leads to Marshall Spring and exposures of a highly complex window of Mesozoic overthrust material protruding through blanket of Browns Park. See Sears, U.S.G.S. Bulletin 751-g. Abandoned cabin left. Continue ahead on Moffat 318-10.
4.2	23.3	Junction with Moffat 12 which leads to Greystone and Douglas Mountain. Continue on Moffat 318-10.
0.4	23.7	Cow camp left.
1.5	25.2	Road left leads (6 miles) to site of abandoned British-American No. 1 Govt.-Raeder (Dry Mountain Anticline) test. Well site affords excellent view of center of Sand Wash Basin and structural complications along Sparks Overthrust. Well bottomed at 6346' in steep dipping beds, Fort Union (?) or Mesaverde (?). Continue on 318-10.
0.6	25.8	View of peaks of eastern Uintas on skyline right. To right of road at 1:30 is Cross Mountain, a detached segment of the Uinta Arch (See paper by Kanizay, this guidebook).
4.2	30.0	Ranch.
1.3	31.3	Note increasing dip in Browns Park formation. Collapse of east end of Uinta Arch into a large complex graben took place in middle of Browns Park deposition. Pre-graben Browns Park has strong dip into various segments of the graben complex. Post-graben Browns Park assumes confusing attitudes — dips into graben, dips draped over older topography and graben fault scarps— all of which conceal details of the complicated graben fault system and the older overthrust belt.
1.4	32.7	Cross Mountain at 1:30.
0.7	33.4	Steep dip in Browns Park, south into graben.
1.9	35.3	Continued string dipping Browns Park.
0.8	36.1	Bridge over Sand Wash. Red beds on skyline ahead and to left are basal conglomerate of Browns Park formation here incorporating much material from pre-Cambrian Uinta quartzite. Browns Park has strong south-southwest dip into graben.
1.2	37.3	Continue along Browns Park dip slope. Cross Mountain to right. Morgan (Penn.) and Madison (Miss.) cap the mountain and dip steeply off north end (right). Uinta quartzite forms core of mountain. East facing front of mountain is likely overthrust.
0.8	38.1	Two Bar Ranch.

Dis-tance	Cumulative Mileage	
0.05	38.15	Road to left just before crossing bridge (Moffat 117) leads along north bank of Little Snake River and affords close up views of fault complex and axis of Two Bar anticline that cross river a short distance upstream from the bridge. (See Side Trips.)
0.05	38.2	Bridge over Little Snake River.
0.4	38.6	Browns Park formation still dipping southwest into graben ahead. Two Bar Ranch on all sides.
0.6	39.2	Road to left leads up south side of Little Snake River as far as Baggs. (See Side Trips.)
1.1	40.3	(STOP) Look back to left across Little Snake River for cross section view of faulted southwest flank of Two Bar Anticline. Red bed is basal Browns Park conglomerate here resting unconformably on Green River (Laney). A graben fault cuts the Green River and Wasatch (Cathedral Bluffs) in this same vicinity. Cathedral Bluffs is exposed in crest of fold. For a detailed map of this complex area see Sears, U.S.G.S. Bulletin 751-g. At 10 o'clock, Green River (Laney) and Wasatch (Cathedral Bluffs) on west end of Godiva Ridge (10 o'clock) are on up side the same as on the north side of the river.
3.2	43.5	Cross Mountain right at 3 o'clock.
0.2	43.7	View to right (south). At 2 o'clock are the Gray Hills, Green River on the south flank of the Uinta-Axial Basin Arch. At one o'clock are the Danforth Hills. Ahead is Juniper Mountain.
2.4	46.1	At 3 o'clock, Cross Mountain Canyon, and low notch to left of Juniper Mountain ahead. Course of the Yampa River is across both of these uplifts, through canyons at these points.
0.2	46.3	Descend hill into Yampa River valley at Sunbeam, Colorado; a presently inactive station on the Rangely to Wamsutter pipeline.
2.2	48.5	Bridge over Yampa River.
1.0	49.5	Glimpse into Sand Wash Basin through saddle on skyline left.
0.3	49.8	Bald Mountain on skyline through gap left is faulted outlier of Godiva Ridge. The ridge is a cuesta of Green River (Laney) which dips north into the Sand Wash Basin. The lower escarpment in foreground through the gap is Tipton as extended by Ritzma (see paper this guidebook).
0.5	50.3	Through gap left is east end of Godiva Ridge with suggestion of faulting that separates Bald Mountain from ridge proper. Note Tipton escarpment below Godiva Ridge. Careful observation reveals two identical Tipton ridges. Unit is repeated by faulting (down to south) associated with Uinta Graben.
2.2	52.5	Beds on skyline to left with steep north basinward dip are Mesaverde, Lance and Fort Union on the nearby vertical north flank of the Axial Basin Arch. These beds protrude through a thin skiff of Browns Park. The Uinta-Sparks Overthrust System presumably passes into these nearly vertical beds and dies out. The flank of the arch becomes much less steep east of Maybell.
0.6	53.1	Juniper Mountain ahead. Light colored Madison limestone makes slope from top of mountain dipping to left. Small knob to left is Little Juniper Mountain. Yampa River flows in canyon between the two mountains. Juniper Mountain is easternmost outlier of Uinta Arch, apparently the "prow" of the arch that was shoved from west to east. It is half encircled by high angle reverse faults on the northeast, east and south and has been involved in the easternmost phases of Uinta Graben faulting. (For details of geology see paper by Abrassart and Clough, this guidebook.)
1.5	54.6	Junction with U.S. 40. (CARAVAN WILL DISASSEMBLE FOR GAS IN MAYBELL AND WILL REASSEMBLE FACING WEST. NEXT LAP OF TRIP TO LUNCH STOP.)

LOG BY RITZMA

0.0	54.6	(BEGIN AGAIN) Junction of Colorado 318 with U.S. 40, Maybell, Colorado and proceed west along U.S. 40.
1.6	56.2	View into Sand Wash Basin right. Tipton (Green River) escarpment repeated by faulting can be seen with red Hiawatha (Wasatch) below. Above Tipton are occasional glimpses of red Cathedral Bluffs (Wasatch), Godiva Ridge and Bald Mountain on skyline are Laney shale (Green River). Highway traversing Browns Park outcrops.
0.6	56.8	Gravel cap on Browns Park in road cut. View into Sand Wash Basin in distance right.
1.7	58.5	Left at 9 o'clock in broad arch of north end of Danforth Hills Anticline with Mesaverde sands arched over crest. Lower part of cliffs are Mancos shale. Hills on skyline to right of Danforth Hills are the north end of Citadel Plateau, topographically high Wasatch and Green River occupying the trough of Coyote Basin, a north segment of the Piceance Basin.
2.2	60.7	Right at 1:30 o'clock is Cross Mountain with Madison and Morgan capping pre-Cambrian Uinta quartzite, reddish brown beds visible in draws along east face of mountain and in canyon through which Yampa River flows. Madison can be seen dipping toward viewer on east face of mountain and also dips strong to left (south) on south plunging end of uplift.
0.9	61.6	Right. Paleozoics can be seen plunging off north end of Cross Mountain.
0.5	62.1	Note Pleistocene (?) sand and gravels capping Browns Park in road cut ahead.
1.9	64.0	Rounding curve. Road cuts in Browns Park. Low mountain ahead is Wapiti Peak composed of Wasatch which lies close to trough of syncline separating Danforth Hills Anticline from

Dis-tance	Cumulative Mileage	
		Cross Mountain segment of Uinta-Axial Basin uplift. West from Wapiti Peak are hogbacks of steep dipping (45 to 70 degrees) Mesaverde and Wasatch. Recognizable Lance, Lewis and Fort Union are missing in the Wapiti Peak area as one proceeds west toward the late Cretaceous-early Tertiary Douglas Creek Arch positive area.
1.2	65.2	Pleistocene (?) sand and gravel caps Browns Park in road cuts.
1.6	66.8	Right. Madison in rim of Cross Mountain Canyon.
1.5	68.3	Bridge over Mud Spring Draw. Tree covered hills at 1:30 are Morgan dolomites and red shales, southernmost exposures of Cross Mountain before complete overlap by Browns Park. Uplift is presumably cut off by a concealed high angle reverse fault which bounds the south flank of the Uinta-Axial Basin Uplift from Cross Mountain to Juniper Mountain.
1.0	69.3	Cross Mountain store and post office (abandoned). Hills at 9 o'clock are south dipping Mesaverde (50 to 60 degrees).
0.3	69.6	Right at 1:30 o'clock. View into Lily Park. Yampa River flows through pre-Cambrian core of Cross Mountain, across faulted west flank (pre-Cambrian thrust over Mancos) and into Lily Park, an open area of Mancos shale. West of Lily Park river enters spectacular Yampa Canyon. Mountain on skyline at 1:30 o'clock is Douglas Mountain in Dinosaur National Monument.
1.6	71.2	Road right leads to Lily Park and exposures along faulted west flank of Cross Mountain. Grass covered knobs to left, 9 to 10:30 o'clock, are steep Mesaverde hogbacks dipping south and southeast.
2.3	73.5	Road curves through Browns Park. Capping of Pleistocene (?) gravels in road cuts.
1.1	74.5	Top of hill. Sweeping view into east end of Uintas ahead and to right. At 2 o'clock, upper

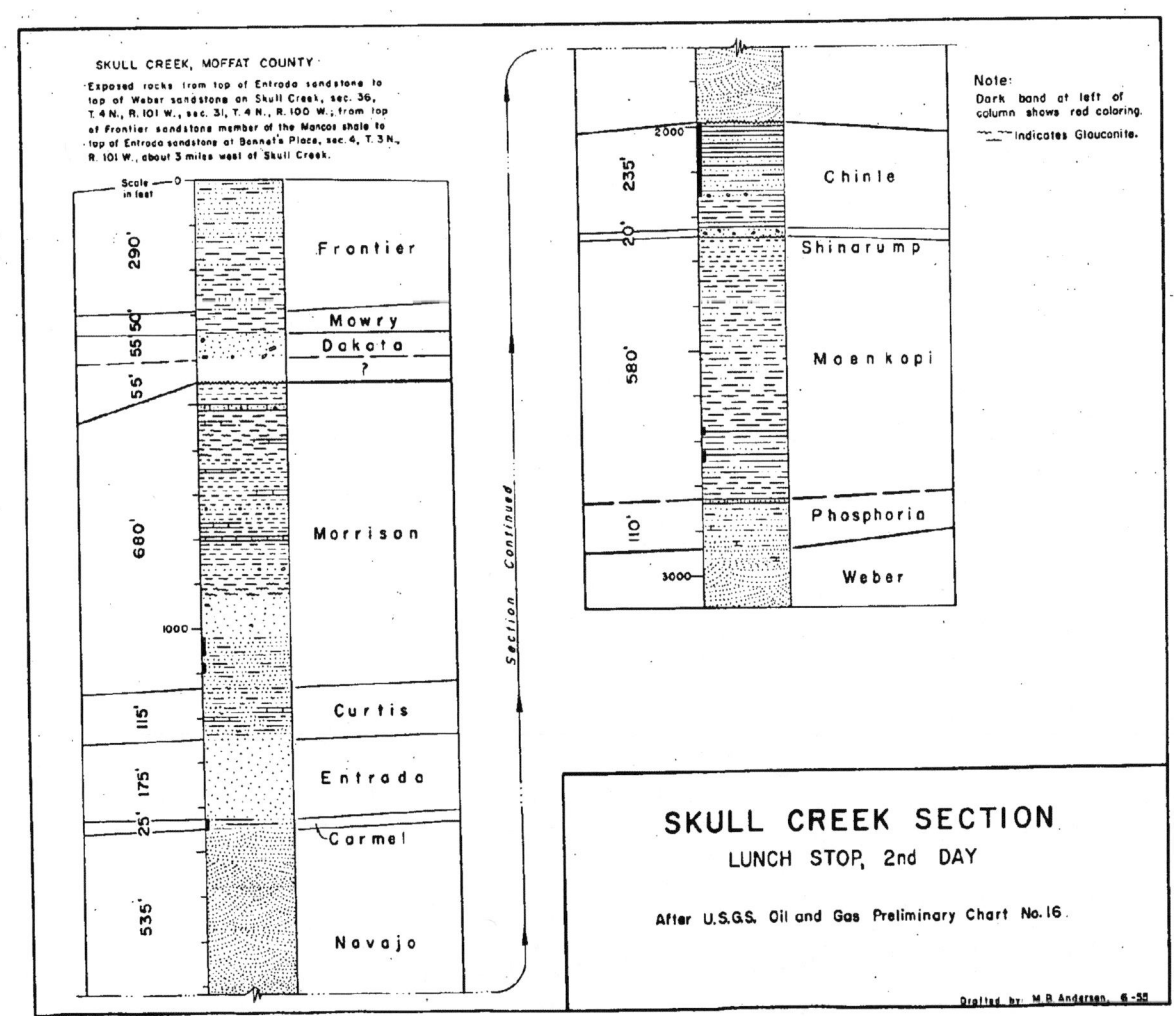

SKULL CREEK SECTION
LUNCH STOP, 2nd DAY

After U.S.G.S. Oil and Gas Preliminary Chart No. 16.

Distance	Cumulative Mileage	
		rims of Yampa Canyons can be seen. The Yampa River is incised 1000' into narrow goosenecks cut into Weber sandstone. Flat mountain on skyline at 1 o'clock is Red Rock Peak (8,800') on north flank of Skull Creek Anticline.
1.3	75.8	Gash across country left is Rangely to Wamsutter pipeline. Note uranium prospect high on hill at 10 o'clock.
1.2	77.0	Prospect pits left. Area has been extensively trenched, pitted and core-drilled for uranium.
1.5	78.5	Elk Springs. Highway Department buildings right. From Elk Springs, Moffat County Road 14 leads west into Dinosaur National Monument. (See Side Trips.)
0.8	79.3	Elk Springs Oil Field left. Basal Browns Park conglomerate-Mancos contact in bank to right. Gravel sluffs over contact. (STOP FOR DISCUSSION OF ELK SPRINGS OIL FIELD.) (See paper by Fenex, this guidebook.)
0.4	79.7	Road left into field.
3.3	83.0	Cross draw. Road on Mancos shale.
0.5	83.5	Cross draw.
1.2	84.7	Pipeline gash right through spur of hill of Mancos shale. Mancos shale hills ahead. Road traverses Mancos shale on south flank of Skull Creek Anticline. Skull Creek is east-west oval closed anticline 8 miles long and 5 miles wide within the closing contours. Weber sandstone is exposed in the core and Triassic and Jurassic rocks form high colorful rims. Two noses plunge southeast from bulges on the southeast end of the structure. A northerly nose is called Coyote Basin (Pinyon Ridge) anticline. The southerly and most prominent bulge develops into the strong Massadonna anticline which continues east into White River and Powell Park anticlines.
1.7	86.4	Note dips in Mancos shale hill at 10 o'clock. On skyline south is Mesaverde escarpment. The broad arch of Coyote Basin (Pinyon Ridge) anticline and the sharper Massadonna anticline can be seen back to the left.
0.8	87.2	Cross Wolf Creek.
0.1	87.3	Road right (Moffat 16) leads into Dinosaur National Monument.
0.2	87.5	Bridge over Little Wolf Creek.
3.0	90.5	Mesaverde on south flank of Skull Creek Anticline makes hogback left foreground. Hills on skyline at 9 o'clock are Wasatch and Green River on the north end of the Cathedral Bluffs. These hills are the extreme northwest corner of the Piceance Basin and are in the trough between Massadonna and Rangely anticlines.
0.9	91.4	Mesaverde in syncline at 10:30 o'clock with beds on skyline rising to south to Rangely Anticline.
0.3	91.7	Mowry and Dakota visible in notches through near vertical Frontier (Ferron) hogback to right for next 0.7 miles.
1.1	92.8	Cross Bridge. Entering Massadonna.
0.9	93.7	Panoramic view ahead of Skull Creek Anticline with Chinle visible in eroded core and Navajo, Carmel (?), Entrada, Curtis, Morrison, Dakota, Mowry and Frontier (Ferron) dipping off flanks.
1.1	94.8	Skull Creek. Store, filling station and highway department buildings. Road beyond highway department buildings, beyond fence (not through gate) leads up Skull Creek.
0.1	94.9	(TURN RIGHT TO SKULL CREEK SECTION. LUNCH STOP. WALKING TOUR OF SKULL CREEK SECTION. DISCUSSION OF GEOLOGY.) Mileage to lunch stop not logged.

LOG BY WHITLEY AND LEE

Distance	Cumulative Mileage	
0.0	94.9	Start mileage at bridge over Skull Creek. Proceed west on U.S. 40. Driving on Mancos shale. First scarp on left is formed by Iles sandstone capping Mancos shale. On right first hogback is Frontier sandstone. White sandstone ridge behind Frontier is Entrada. Dakota not visible at this point.
1.1	96.0	Road cuts through terrace developed on Mancos shale.
0.5	96.5	Bridge across Miller Creek.
0.7	97.2	On right are Frontier hogbacks paralleling highway. Dakota sandstone forming ridge behind Frontier.
1.7	98.9	On right note silver-grey weathering Mowry shale in saddle between Frontier and Dakota hogbacks.
0.5	99.4	On right note steepened south dip in Dakota. Frontier absent due to erosion. Quarry is in a series of grey-green clay shales and thin-bedded sandstones underlying the Dakota and designated as Upper Cretaceous "beds of undetermined age". Entrada-Navajo sandstones form rounded hogback behind Dakota.
1.4	100.8	Bridge across Red Wash Gulch. On right Dakota barely visible in front of higher Entrada ridge. Weber sandstone on crest of Skull Creek Anticline in distance. On left hogback is Iles sandstone overlying Mancos. Gravel road at left (County No. 61) joins Colorado Highway No. 64 at White River, 8 miles northeast of Rangely.
1.4	102.2	To right on horizon is good view of Navajo, Chinle, Shinarump, and Moenkopi formations on north flank and west plunge of Skull Creek Anticline.
1.5	103.7	On right first hogback is Frontier succeeded by underlying Dakota and Morrison.
1.6	105.3	Blue Mountain. (TURN LEFT (SOUTH) ON COLORADO HIGHWAY 387 TO RANGELY.)

Distance	Cumulative Mileage	
0.8	106.1	Road cuts through east-west ridge at Iles-Mancos contact. Note massive red-orange sandstone of Iles on south side of ridge.
0.4	106.5	On left in Willow Creek are exposures of south-dipping interbedded buff sandstone and grey shale of Iles. Low, juniper-covered hills on right are Iles sandstone.
0.4	106.9	Crossing axis of syncline separating Skull Creek Anticline from Rangely Anticline. Ahead on right is gentle northeast dipping Iles sandstone.
0.1	107.0	Bridge across Willow Creek.
0.2	107.2	Leave Moffat County. Enter Rio Blanco County.
0.6	107.8	Buff sandstones and interbedded shale of Iles. At left note thin coal beds in shale member.
2.3	110.1	Road cuts across strike of uniformly north-dipping Iles cavernous sandstones and grey-brown shales.
0.1	110.2	Ridges on both sides of road are formed by basal buff-orange sandstone of Iles capping Mancos. Note thinly bedded sandstones at top of Mancos. These sandstones form rim which closes around Rangely Anticline (Rimrock sandstone).
0.6	110.8	Descending into Raven Park Basin formed in non-resistant Mancos shale.
2.2	113.0	On right are outpost wells of Rangely Field. Ridge on horizon is Mancos capped by Iles.
1.1	114.1	Road follows approximately along axis of anticline.
2.1	116.2	At right note Mancos shale outcrops in Stinking Creek. Ahead is Iles-Mancos ridge denoting southwest closures.
0.4	116.6	(TURN LEFT AT INTERSECTION OF COLORADO 387 WITH COLORADO 64.) (See papers by Campbell and Peterson, this guidebook.)

LOG BY SLOAN AND THOMPSON

Distance	Cumulative Mileage	
0.0	116.6	INTERSECTION COLO. 387 AND COLO. 64 (CARAVAN WILL TAKE CONDUCTED TOUR OF RANGELY FIELD). Tour not logged. Proceed east.
2.1	118.7	Bridge over White River.
0.6	119.3	Rangely city limits.
1.2	120.5	Historical marker on left. Road to Douglas Creek Pass on right. (See historical sketch, Father Escalante, this guidebook.)
0.5	121.0	Bridge over Douglas Creek.
1.7	122.7	Bridge over Gillam Draw.
2.0	124.7	Covered contact Mancos - Mesaverde (Iles) on right of road. Exposed contact may be observed across river on left.
2.6	127.3	Sign "Jesus is Coming" painted on sandstone right of road. Lignitic beds immediately below.
0.9	128.2	Note lenticular nature of sandstone in Mesaverde (Iles) in cliff across river to left.
0.5	128.7	Road left to coal mine. Contact between Iles and Williams Fork can be seen just beyond point where side road crosses river. Burned out carbonaceous beds of Mesaverde (Williams Fork) are exposed by old excavation.
0.7	129.4	Bridge over Spring Creek.
2.6	132.0	Bridge over Fletcher Gulch.
1.3	133.3	Contact Mesaverde (Williams Fork) and Wasatch in north bank of White River across meadow to left.
2.3	135.6	School house on hill to right.
1.1	136.7	Bridge over Boise Creek. Steep dip of Wasatch and Mesaverde beds into the Piceance Creek Basin can be observed across valley to north. Green River beds lie up valley to right. This point is near the northwest end of the synclinal axis of the basin.
1.6	138.3	Bridge and sharp turn.
2.6	140.9	Small road to left. On skyline to left Mesaverde - Mancos contact can be seen arching over Midland (Coyote) Basin anticline. Wasatch - Green River contact in slope ahead at one o'clock.
5.3	146.2	Bridge over Yellow Creek. Ahead and to the right, is exposed a nearly complete section of Wasatch and Green River.
8.3	154.5	Abandoned ranch on left. Wasatch - Green River contact exposed on cliff face across valley to the north. Basal Green River sands exposed in upper third of cliff.
1.8	156.3	Bridge over White River.
1.9	158.2	Piceance Creek turnoff right. Continue ahead on Highway 64.
0.7	158.2	Blacks Gulch. Good exposure of Wasatch on left.
2.5	161.4	Kellog Gulch — observe Wasatch - Green River contact, and pillared (loess-like) weathering of upper Wasatch. (CARAVAN WILL MAKE U-TURN NEAR BRIDGE AND RETURN TO PICEANCE CREEK ROAD.)
2.2	163.6	(STOP FOR TALK ON WHITE RIVER DOME.) (See paper by Helmke, this guidebook.)
1.0	164.6	Piceance Creek road. (TURN LEFT.)
0.2	164.8	Bridge over White River.
1.5	166.3	Bridge over Piceance Creek. Wasatch - Green River (Douglas Creek member) contact midway up cliff on right. Basal Green River sands cap hill.
0.5	166.8	Break in slop right marks approximate contact Douglas Creek - Garden Gulch members of Green River formation.
0.8	167.6	Note paper shales of Garden Gulch on right. The top of these shales may possibly mark the Garden Gulch - Parachute Creek contact.

Dis-tance	Cumulative Mileage	
0.4	168.0	Oil shales of Parachute Creek exposed in cliff across Piceance Creek to left.
0.9	168.9	Slope above road on right contains Parachute Creek - Evacuation Creek contact. The Evacuation Creek is typified by a darker shade of buff than the lighter gray buff beds of the Parachute Creek. The term Evacuation Creek supplants use of term "Bridger formation" in Piceance Basin.
0.6	169.5	Road to Yellow Creek right. Note channel sands in cliff above where Little Hills road can be seen turning right on far side of valley.
2.8	172.3	Road to left to Little Hills Game Station. Excellent exposure of channel sands in cliff above where Little Hills road can be seen turning right on far side of valley.
2.4	174.7	Bridge over Piceance Creek.
0.2	174.9	Fenced off road, right into abandoned location of The Carter Oil Company, Square S-Herring No. 1 (Yellow Creek Unit).
1.1	176.0	Square-S ranch headquarters.
2.1	178.1	Abandoned location General Petroleum 51-28P in field to right. This is also on Yellow Creek Unit.
5.3	183.4	School house left.
2.0	185.4	Location Equity Oil No. 1 Johnson is ½ mile to right up Hunter Gulch.
1.5	186.9	Road right to Willow Creek and PL Cattle Company Ranch.
2.4	189.3	Road left to Piceance Creek Gas Field (TURN LEFT).
4.9	194.2	Piceance Creek Gas Field Camp — End log for day. (Talk on Piceance Creek Gas Field will be given on morning of third day prior to leaving field.)

PICEANCE CREEK DOME
Rio Blanco County, Colorado

Piceance Creek Dome is located in the "Little Hills country", about 20 miles due southwest of Meeker, Colorado, in the heart of the Piceance Creek Basin. The surface structure is expressed in the Eocene "Bridger" formation (about 1000' thick) more recently and accurately described as the upper Evacuation Creek member of the Green River formation. These beds consist of massive sandstone, shaly sandstone and siltstone and thin bedded whitish marlstone. These Bridger (?) or upper Evacuation Creek beds are underlain by approximately 2300 feet of older Green River rocks representing, in descending order, the Parachute Creek, Garden Gulch and Douglas Creek members. Basal sandstones of the Douglas Creek member are the gas-bearing zones in the field. Beneath the Green River is 3500 to 5000 feet of Wasatch (lower Eocene) which rests on the Mesaverde (?) formation of late Cretaceous age. Formation boundaries are difficult to pick in the one deep well in the field. Abnormally thick Fort Union, Lance and Lewis equivalents are possibly present, and the Mesaverde may be considerably deeper than estimated tops indicate.

The northwest-southeast trending anticline is in a basin area of gently dipping beds. It is flanked on the east by the steeply dipping Grand Hogback and on the west by shallow synclines separating it from the Yellow Creek and East Douglas Creek noses that plunge east from Rangely and the Douglas Creek Arch. The anticline is cut on the south flank by a set of parallel faults and/or fractures aligned with the axis. These faults are slightly displaced down to the south.

Structural movement commenced in Eocene during early Green River deposition when near-shore sediments were deposited on the east side of the Piceance Creek Basin. Along the Grand Hogback steep tilting and faulting occurred after deposition of the Green River oil shales. The surface structure of the Piceance Creek dome has some 250' of closure, but at producing depth the structure is expressed as an eastward plunging nose without closure. The trap appears to be a sand pinchout along the north flank of this nose.

Discovery well in the field was White Eagle Oil - Magnolia Petroleum No. 1 Fordham, SW SW Section 9, T. 2 S., R. 96 W., drilled in 1929-30 to 5130'. Its initial open-flow potential was 2,350 mcf gas per day with 764 psi pressure at 2655-57 and 2860-85. To date 14 wells have been drilled in the area. Four wells have been abandoned and ten with potentials from 2 to 13 mmcf gas per day await pipeline outlet. Some 19,000 acres are considered productive in the field.

The deep test drilled by General Petroleum in 1945-46, successor-operator to Magnolia, No. 84-15 in Section 15, T. 2 S., R. 96 W., bottomed in Mesaverde (?) formation at T.D. 12,019'. No important shows were reported below the Green River formation.

SOUTH PICEANCE CREEK

As of April 1955, Equity Oil Company had a 2,500 mcf gas per day discovery in their No. 1 Clubine, Section 18, T. 3 S., R. 95 W., Rio Blanco County, Colorado. Their gas discovery from the Douglas Creek member of the Green River formation indicated considerable extension of the Piceance Creek gas producing area to the southeast. Several other wells by Equity in this same area have extended this farther in a scattered pattern along nearly 15 miles of the south flank of Piceance Creek Dome. The new field has been tentatively designated South Piceance Creek.

See map in pocket, back of this guidebook.

REFERENCES

Duncan, D. C., and Belsor, Carl (1950), "Geology and Oil Shale Resources of the Eastern Part of the Piceance Creek Basin, Rio Blanco and Garfield Counties, Colorado." Oil and Gas Investigations Map OM 119.

Kramer, William B. (1939), "Geologic Map and Section of Piceance Creek Dome, Rio Blanco County, Colorado," U. S. Geol. Survey (open file).

Kramer, William B. (1945), "Geologic Map and Section of Piceance Creek Dome, Rio Blanco County, Colorado (Revised)," U. S. Geol. Survey (open file).

Turner, Daniel S. (1954), "The Oil and Gas Fields of Colorado, a Symposium," Rocky Mountain Association of Geologists, Denver, Colorado, p. 66.

G. H. Fentress
Lion Oil Company, Denver

THIRD DAY

PICEANCE CREEK DOME TO CRAIG VIA RIO BLANCO, BUFORD, MEEKER, WILSON CREEK, ILES AND MOFFAT.

Driving distance: 153.3 miles   (For route see Master Road Map)

Distance	Cumulative Mileage	
		LOG BY SLOAN AND THOMPSON
0.0	0.0	(START AT INTERSECTION PICEANCE CREEK GAS FIELD ROAD AND COLORADO HIGHWAY 326 — GO SOUTHEAST.)
1.1	1.1	Old school house on left.
2.0	3.1	Gerald Oldham Ranch.
1.1	4.2	Equity Oil No. 1 Oldland Brothers.
1.5	5.7	Equity Oil No. 2 Clubine gas well.
0.7	6.4	Clubine Ranch.
0.5	6.9	Equity No. 1 Clubine right.
5.2	12.1	Gulch on left. Contact Evacuation Creek - Parachute Creek in spur on far side of gulch.
2.2	14.3	Road right.
0.8	15.1	Prospect holes in oil shale on right across creek. Beyond this point oil shales of Parachute Creek line both sides of road.
1.9	17.0	Contact Parachute Creek - Sandy member in left bank of road.
0.7	17.7	Small road left. Contact Green River (Basal sandy member)-Wasatch at slope break approximately 100 feet left of road.
0.6	18.3	Rio Blanco Store. Junction Colorado Highway 326 and Colorado Highway 13 (GO NORTH ON HIGHWAY 13.)
		LOG BY DONNELL, R.M.A.G. Guidebook, 1953
0.95	19.25	Brown shale to east in basal part of Wasatch formation. White sandstone below shale is Ohio Creek conglomerate.
3.65	23.0	Fourteen Mile Creek. Mesaverde-Wasatch contact crosses highway. Road on Mesaverde formation.
0.1	23.1	Road down Fourteen Mile Creek. Good exposures of Green River formation about 1 mile west. Lenticular sandstones at base of Green River formation are asphaltic.
1.5	24.6	Thirteen Mile Creek.
0.45	25.05	Road cut to east exposes deeply weathered Ohio Creek conglomerate. Wasatch-Mesaverde boundary crosses highway again. Road on Wasatch formation.
0.55	25.6	Twelve Mile Creek.
2.45	28.05	Road leading over bluff on skyline to west to asphaltic sandstone in base of Green River formation. Prospect hole in asphaltic sandstone about 3 miles northwest.
1.0	29.05	Mesaverde formation crosses highway. Road on Mesaverde formation.
0.35	29.4	Vertical sandstone west side of road has Ohio Creek conglomerate at its base. The brown sandstone above conglomerate is basal sandstone of Wasatch formation. Road is on Mesaverde formation.
1.1	30.5	Beds of Mesaverde sandstone ahead are overturned slightly.
0.4	30.9	Sandstones and shales in Mesaverde formation. Red color of rock is due to baking by burning of adjacent coal beds. Crest of hills 1 mile west in Green River.
6.1	37.0	On horizon to north is valley of Strawberry Creek. West of hogbacks ahead on highest point to northeast on east side of Strawberry Creek is Wilson Creek oil field.
0.9	37.9	Gravel terraces of White River on west.
0.7	38.6	White River.
0.3	38.9	Junction of Colorado Highways 13 and 64. Monument to commemorate the Meeker Massacre and Thornburgh battle that took place in September 1879. The Indian Agency was in the valley about a mile below the monument. Meeker was the Indian agent. He had offended the White River Utes by plowing up their pony pasture and attempting to plow up their race track. The immediate cause of the massacre was the breaking by Major Thornburgh of a promise to Ute Jack, an Indian leader, to send 5 men to meet with 5 Indians at the Agency before he sent his soldiers onto the reservation. Maaker, and many others were killed. Mrs. Meeker (68) and her daughter (20), Mrs. Price and 2 small children were held captive 23 days. The Utes held Major Thornburgh's troops under siege at Milk Creek, northeast of Meeker, for five days.
0.45	39.35	Coal beds in Mesaverde formation on north side of road.
0.85	40.2	Mesaverde sandstone and shale form crest of ridge to north. Base of slope formed by Mancos shale. Cultivated fields to south are on Mancos shale.
0.8	41.0	Meeker city limits. Continue straight through town. High dip slopes several miles south formed by Dakota sandstone.
0.1	41.1	Trout Creek sandstone in white ledge north in hills below red beds (burned rocks).

Dis-tance	Cumulative Mileage	
1.0	42.1	East city limits. Scarp formed by persistent sandstone member of Mancos, about 1600 ft. below base of Mesaverde, ¼ mile north of highway at 10 o'clock.
1.2	43.3	Junction, State Highways 789-13 and 132. (TURN RIGHT ON HIGHWAY 132.)
		LOG BY SLOAN AND THOMPSON
1.2	44.5	(STOP FOR MEEKER DOME. CONTINUE ON UP COLORADO 132.) This will be logged from Buford back therefore only mileage checks are given on route to Buford.
3.1	47.6	K-T Ranch (For lower Paleozoic stratigraphy, see Bass, this guidebook).
10.5	58.1	El Rio Rancho.
1.7	59.8	Sleepy Cat Ranch.
2.2	62.0	Turn right on road with sign to State Trout Rearing Station.
0.1	62.1	Bridge across White River.
1.6	63.7	Intersection with Rio Blanco 17 — (TURN LEFT).
0.0	63.7	Rolling "R" Ranch — View up South Fork of White River — Hills on skyline capped by basalt.
0.05	63.75	Bridge over South Fork of White River.
0.05	63.8	Road left into trout rearing station.
1.0	64.8	Bridge over North Fork of White River.
0.05	64.85	Junction with Colorado 132 — Buford on right — (TURN LEFT).
0.15	65.0	Maroon formation outcrop right and continuing along roadside. Observe gypsiferous nature of this formation.
0.5	65.5	White River terrace gravels.
0.5	66.0	Maroon red beds exposed in road cut. (See Landon and Thurman, this guidebook.)
1.0	67.0	(RIO BLANCO 10 LEFT. LOOP COMPLETED — CONTINUE ON RIO BLANCO 132.)
2.2	69.2	Sleepy Cat Ranch.
1.0	70.2	Small slump block in road cut. Maroon formation.
0.7	70.9	El Rio Rancho.
2.9	73.8	Gravel pit on right — Gravel from White River terrace.
0.6	74.4	White River gravels on irregular surface of Maroon.
0.5	74.9	Small fault in Maroon in road cut.
0.5	75.4	Quarry right — White River terrace.
0.2	75.6	Contact Maroon - Weber right. Marked by change of beds to buff sandstone.
0.4	76.0	Chinle, Entrade-Navajo and Morrison exposed on hills to right. Ridge on skyline capped with Dakota. Maroon formation makes up hills across river on left.
0.2	76.2	Lower Moenkopi exposed in road cut right. Contact with Weber obscured in slope behind.
0.2	76.4	At roadside right is a color change separating grey-green lower Moenkopi from red upper Moenkopi. Up slope to right approximately 100 feet from road is Moenkopi - Shinarump contact. Approximately 10' of Shinarump is exposed with Chinle capping the hill above.
0.4	76.8	Culvert and small road plowed out up hill to right. Chinle - Navajo, Entrada contact at base of cliff to right.
0.9	77.7	White River terrace slumping over Entrada - Navajo.
0.2	77.9	Entrada - Navajo - Morrison contact.
0.7	78.6	Contact Morrison - Dakota.
0.2	78.8	Contact Dakota - Mancos in slope right confused by slump.
0.4	79.2	Mancos outcropping from under slump.
0.3	79.5	Thin-bedded sandstones (Mowry ?).
1.8	81.3	K-T Ranch.
1.9	83.2	Old road right. Contact Mancos - Dakota at two o'clock.
0.1	83.3	Contact Dakota - Morrison obscured by rubble in slope to right.
0.3	83.6	Morrison outcrop in road cut on right.
0.3	83.9	Mineral spring on right. Near crest of Meeker Dome.
0.2	84.1	Contact Morrison - Dakota.
0.2	84.3	White River terrace.
1.2	85.5	Junction with Colorado 789-13. (RETURN THROUGH MEEKER TO JUNCTION COLORADO 789-13 AND 64.
4.5	90.0	Junction Colorado 789-13 and 64. (TURN RIGHT ON HIGHWAY 64). After turn, beds exposed ahead are Williams Fork formation of Mesaverde group.
0.8	90.8	Contact Williams Fork - Wasatch (Ohio Creek age) to right of road. Wasatch is characterized by darker brown color than underlying buff Williams Fork sandstones.
0.45	91.25	Junction Rio Blanco No. 7 right. (TURN RIGHT). Road goes north on Wasatch.
1.15	92.4	Hilltop, gates right and left. Derrick on skyline at 1 o'clock marks Wilson Creek pool.
3.2	95.6	Bridge over Strawberry Creek.
1.3	96.9	Road right to gravel pile. Approximate axis of east-west trending Sulfur Creek syncline which forms south closure for Wilson Creek structure. Derrick on skyline. Wilson Creek straight ahead. (For information on Wilson Creek field see penetration chart, this guidebook.)
1.9	98.8	Junction Wilson Creek road (Rio Blanco 9). (TURN RIGHT.)
0.8	99.6	Sharp turn left. Williams Fork in bottom of draw approximately hundred yards left. Ohio Creek conglomerate capping small ridge left.

Dis-tance	Cumulative Mileage	
0.4	100.0	Sharp turn right. Williams Fork dip slope ahead on southeast flank of Wilson Creek structure. Beds can be followed up valley left to a point several miles away where they swing north to form the west side of the structure.
0.4	100.4	Double cattle guard. Network of roads for Wilson Creek pool can be observed on skyline through notch at 1 o'clock.
1.6	102.0	Cattle guard.
0.05	102.05	Trout Creek sandstone left of road marking top of Iles formation. At 11 o'clock on skyline the Williams Fork - Iles contact can be observed just above excavation. Red burned carbonaceous beds of Williams Fork overlie Trout Creek sandstone with dark grey carbonaceous beds of Iles below.
0.55	102.6	Cattle guard.

OIL FIELDS ALONG THE DANFORTH HILLS ANTICLINE

The Danforth Hills, a northwest-southeast trending range of hills, is a structural high as well as a topographic high with the upfolded sandstones of the Mesaverde formation forming the backbone of the rugged range. The relief of the area is the result of a youthful stage of dissection and a wide variance in resistance of different Mesaverde members. (See photo map by Oburn, pocket at back of this guidebook.)

WILSON CREEK is a domal closure on the long Danforth Hills axis. Irregularly exposed on the crest of the dome in canyon bottoms is the Iles formation. Most of the canyon sides and hills are in the younger Williams Park formation.

50° gravity paraffin base oil is produced from the Jurassic basal Morrison and Entrada sandstones. The Entrada is a uniform sandstone across the field and the trap is purely structural with a water drive. Basal Morrison sandstones are erratic in character, and this reservoir has many aspects of a stratigraphic trap as well as simple domal closure. It has combination gas-water drive.

Wilson Creek, discovered in 1938, was Colorado's most prolific oil field until development of the Rangely Weber pool. It then held second place in Colorado's daily production figures until the advent of newer fields in eastern Colorado. Daily production averages 7500 barrels. Cumulative production exceeds 30,000,000 bbls.

MAUDLIN GULCH, 8 miles northwest of Wilson Creek is a smaller and more elongate structure which also produces from the basal Morrison and Entrada. Daily production is about 375 barrels per day, 36° gravity oil (July 1955).

Farther north along the Danforth Hills anticline are TEMPLE CANYON and DANFORTH HILLS oil fields. (See papers by Clough and Turner, this guidebook.)

H. R. Ritzma

Dis-tance	Cumulative Mileage	
2.7	105.3	Good turnout for panoramic view to southeast of Strawberry Valley, White River Valley and Piceance Creek Basin.
0.4	105.7	Intersection — Road to Wilson Creek camp. (TURN LEFT.)
0.8	106.5	Recycling plant.
0.4	106.9	Field office — Road to camp left. (GO RIGHT.)
1.0	107.9	Small road left.
1.5	109.4	Small road left.
1.0	110.4	Pumping station.
0.6	111.0	Contact Williams Fork - Iles.
0.1	111.1	Road left.
2.8	113.9	Approximate axis of northwest-southeast trending Collum syncline which forms north closure for Wilson Creek structure. Beyond the point beds climb onto south flank of Axial Basin anticline.
0.1	114.0	Small road right.
1.6	115.6	Exceptionally good contact Williams Fork - Iles.
0.7	116.3	Contact Iles - Mancos obscured by rubble in slope left.
0.6	116.9	Junction Moffat Co. 32 continue ahead. (For information on Maudlin Gulch field, see penetration chart this guidebook.)
1.0	117.9	Junction Moffat County 17. (TURN RIGHT.) This point is near crest of Axial Basin anticline.

LOG BY DONNELL, R.M.A.G. Guidebook, 1953.

1.8	119.7	Junction with Colorado Highway 789-13. (TURN LEFT.) Iles Mountain is Mesaverde scarp on north side of Axial Basin, skyline at 12 to 1 o'clock.
1.1	120.8	Approximate axis of Axial Basin anticline.
1.8	122.6	Milk Creek bridge.
2.4	125.0	Road left to Iles Ranch. Low scarp at 12 o'clock formed by Morapos sandstone member of Mancos, about 800 ft. below the base of the Mesaverde group.
1.0	126.0	Loyd (Iles Grove, monument to Thomas Iles, pioneer settler of area). Road right to Iles oil field. (LUNCH STOP IN ILES GROVE. IF TIME PERMITS CARAVAN WILL TOUR ILES FIELD.) Tour not logged.
0.4	126.4	Iles field at 4 o'clock. This field produces mainly from the Jurassic Entrada sandstone but has some Morrison production. Some oil has been produced from fractured shale near the base of the Mancos. (See paper by Nelson, this guidebook.)
0.4	126.8	The highway is about to pass behind the scarp formed by the Morapos sandstone.
1.1	127.9	Road right. Moffat (Hamilton) dome and oil field at 12 o'clock. This field produces from the Dakota and Entrada; has also produced a small amount of oil from the Morrison.
1.4	129.3	Road right up Morapos Creek. Moffat dome at 3 o'clock outline by Morapos sandstone scarp. Note steep dip slope on Morapos sandstone on southwest flank of field. Highway now crossing axis of Round Bottom syncline, which plunges northwest. (CARAVAN WILL STOP IN THIS VICINITY FOR DISCUSSION OF MOFFAT FIELD.)
0.5	129.8	Note pinnacle on Morapos sandstone at 3 o'clock. Derrick stands over dry hole, G. L. Reasor No. 1 Carpenter, test to Weber on northwest plunge of Moffat Dome.

Dis-tance	Cumulative Mileage	
1.5	131.3	Hamilton. Road right to Moffat field. Mancos-Mesaverde contact in cliff at 12 o'clock.

LOG BY RITZMA AND TERPENING.
R.M.A.G. Guidebook, 1953.

0.3	131.6	Junction Colorado Highways 789-13 and 317 at town of Hamilton. (TURN LEFT. FOLLOW COLORADO 317 EAST ALONG NARROW VALLEY OF WILLIAMS FORK RIVER.) Cliffs formed by basal Mesaverde sandstones on left. Beds are dipping north with structure rising to south on Moffat Dome.
0.6	132.2	Hay shed to right.
0.6	132.8	Note gentle northwest dipping Mesaverde beds across river ahead. Flattening structure occurs at juncture of Pagoda-Beaver Creek anticline, which fades into north flank of Moffat dome in this vicinity.
2.2	135.0	Riding along crest of Pagoda-Beaver Creek anticline. North flank of this structure can be seen in cliffs along left side of valley ahead for many miles. Mesaverde-Mancos contact in base of cliff left. This contact is transitional. From 3 to 7 sands are found in the 800'. Shale interval between the Morapos sandstone of the Mesaverde. At places the Morapos is difficult to identify with certainty.
0.7	135.7	Southwest dip in beds across river to right. Pagoda-Beaver Creek anticline becoming more prominent.
0.6	136.3	Strong southwest dip across river right.
1.2	137.5	Road to south to Waddle Creek. Keep straight ahead on Colorado 317 along north flank of anticline just below Mesaverde-Mancos contact. At 2 o'clock note gently dipping Mancos beds across river on crest of structure. Browns Park (Miocene) and lava rubble caps brush-covered and wooded hills at higher elevations at 2 o'clock. At this point Pagoda-Beaver Creek anticline broadens and bifurcates. Mesaverde rim swings south at 4 o'clock and then southeast in distant wooded hills which form steep south flank.
0.9	138.4	At 3 o'clock in Harper Hill (elev. 7554), crest of Pagoda-Beaver Creek anticline. At 6 o'clock to rear in notch of valley is crest of same structure. At point of broadening the anticline bifurcates with south axis fading into Moffat dome. North axis persists and fades into north flank of Moffat dome. Pagoda-Beaver Creek anticline culminates in Pagoda Dome to the southeast.

PAGODA GAS FIELD
Routt County, Colorado

Pagoda gas field, T. 4 N., R. 89 W., Routt County, Colorado is located on the northwest plunging anticline designated as the Pagoda - Beaver Creek anticline. The structure is broadly expressed in the Mancos shale; but the details are mostly concealed by Browns Park, basalt flows, lava debris and dense vegetation. It is presumed that closure exists in the southeast part of the structure under this cover.

Gas was discovered in 1948 by General Petroleum Corporation No. 55-34-G, Section 34, T. 4 N., R. 89 W., in the Triassic Shinarump formation. The well had an initial potential on tests of 7,700 mcf gas per day from perforations at 3957-70, 3980-90, 4002-26 and 4035-37. A confirmation test in 1949, No. 33-34-P, was completed for 1930 mcf gas per day. The Shinarump reservoir is sandstone and conglomerate of extremely variable character intercalated with red and purple shales and siltstones. A third well drilled in 1954 (No. 12-27) on the northwest plunge of Pagoda anticline was not successful.

The gas wells are located on the heavily timbered north slope of Beaver Flat Top at elevations ranging from 7850 to 8100 feet. The lava flat top rises another 1000' above the field. The structure was mapped by surface methods with some seismic corroboration.

The deepest well, 55-34-G, tested the Weber formation topped at 4696. A questionable top of "Maroon" was picked at 4800. Total depth was 4902. The two wells are shut in pending pipeline outlet, and the available structural information is insufficient for estimation of the size of the field.

Adapted from:
Dan S. Turner, "Oil and Gas Developments of Western Colorado" in
R.M.A.G. Symposium Volume (1954).
Supplemented by editorial staff.

Dis-tance	Cumulative Mileage	
0.6	139.0	Sweeney Ranch.
0.1	139.1	Road left over mountains to Craig. (GO LEFT.) The Mesaverde hogback and dipslope to left of valley form the Williams Fork Mountains, divide between the Yampa and Williams Fork drainages.

LOG BY SLOAN AND THOMPSON

0.05	139.15	Beginning of transitional contact Mancos Iles.
0.65	139.8	Fence corner left. End of transition zone road continues on Iles.
1.2	141.0	Coal mine left.
0.2	141.2	Sharp turn left, small road right. Trout Creek sandstone right of road and across valley on left.
0.3	141.5	Cross drainage. Approximate Iles-Williams Fork contact.
1.8	143.3	Contact Williams Fork - Lewis marked by rich dark soil above sandstone ledge at right of road.
0.45	143.75	Small road right.
0.45	144.2	Beginning of S-turn. Panoramic view ahead. Craig at 12 o'clock, Cedar Mountain at 11 o'clock beyond Craig, Black Mountain at 1 o'clock, Bears Ears at 2 o'clock.
0.7	144.9	Road right. (GO RIGHT.)
0.3	145.2	Road left. (GO RIGHT.) Covered contact Lewis - Williams Fork.
3.1	148.3	Approximate axis of Breeze Basin anticline.
0.5	148.8	Road on Quaternary alluvium.
0.5	149.3	Junction Colorado 394 (TURN LEFT.)
1.0	150.3	Craig airport on right.
1.7	152.0	Junction Colorado Highway No. 13. (GO RIGHT.)
1.3	153.3	Junction Colorado 13 - U.S. 40 — town of Craig. End trip.

ADDITIONAL LOGS

LOG OF U.S. 40 — CRAIG TO MAYBELL

Cumulative
Mileage

LOG BY LINSCOTT

0.0 Stoplight Craig. Junction U.S. 40 and Colorado Highway 789. Proceed west on U.S. 40.

0.3 Basal sand of Lance formation at 3 o'clock. Highway crosses Lewis shale for the next few miles.

1.1 Gently sloping hills at 9 o'clock are Williams Fork Mountains formed by dip slope of Mesaverde formation.

2.0 General north dip here is off Craig dome. A 10,000 foot Weber test on Craig dome by General Petroleum found gas in the Frontier and Shinarump. The hole was plugged and abandoned in January, 1951.

2.5 Cedar Mountain at 3 o'clock. It is composed of Browns Park with a capping of basalt.

3.1 Road following strike-valley in Lewis shale.

3.6 Lance outcrops at 12 o'clock. Basal sand is poorly defined here, but upper sand which is seen through stream cut to the north is well developed forming prominent bluff.

5.2 Road cut in Lewis shale.

5.6 Dip slope of Williams Fork Mountains from 8-11 o'clock. White River flattops can be seen through gap in Williams Fork Mountains at 10 o'clock. Distant hills at 11 o'clock are along the Danforth Hills anticline.

6.7 North dip is off Bell Rock dome. Whitish-gray sand from 1 to 3 o'clock is Browns Park formation. Dip slope at 11 o'clock is Mesaverde veneered with Browns Park. Remainder of trip will cross Browns Park formation.

8.7 Juniper Mountain, pre-Cambrian Uinta quartzite, at 11 o'clock on horizon.

13.3 Road turns down Big Gulch.

17.3 Massive character of Browns Park sands demonstrated in exposures at 12 o'clock.

Cumulative
Mileage

18.1 Cross Big Gulch. Williams Fork member of Mesaverde formation crops out across valley to left from 8 to 11 o'clock.

19.4 Enter Lay. Leave Lay.

19.6 Small coal prospects in Williams Fork member of Mesaverde formation at 9 o'clock across Lay Creek. A few white sands are visible.
The next few miles we cross over Browns Park sandstone which has been the scene of a uranium rush. Large low grade deposits of uranium salts have been drilled out. As the ore has no vanadium (very desirable in the economy of milling operations) and as the metallurgical problems of extraction are great, development of the deposits has been greatly retarded. There are two conflicting theories of ore emplacement—concentration by ground waters and hydrothermal.

26.0 Juniper Mountain—pre-Cambrian Uinta quartzite at 11 o'clock. Stratification can be seen in the quartzite near the crest of the mountain. By and large the flanking sediments are masked by Browns Park formation.

27.0 Little Juniper Mountain composed of Weber sandstone, Morgan sands and limes and the Leadville limestone. Outcrop of Dakota and Navajo formations can be observed on the east flank near the hot springs. The area is greatly faulted.

28.2 Cross Yampa River.

28.7 Drab exposures on skyline at 2 o'clock are in the Green River formation.

29.5 Exposures across river from 1 to 3 o'clock are faulted Mesaverde sands.

31.1 Enter Maybell.

31.7 Junction U.S. 40 and Colorado county highway 318. Log joins log of conference tour.

ROAD LOG OF COLORADO 789-13 FROM JUNCTION WITH MOFFAT COUNTY ROAD 4 NORTH TO BAGGS, WYOMING

LOG BY RITZMA

0.0 Junction Colorado 789—13 and Moffat County Road 4. (Caravan turns west on Moffat County 4 at this junction.)

0.8 Cross Four Mile Creek.

2.0 Ascending "Yellow Ridge." Enter Wyoming ahead. Now on Wyoming 789 (formerly 330).

2.3 Top of grade. Note basal Browns Park conglomerate on Hiawatha (Wasatch) beds in road cuts right and left. Begin descent into Little Snake River Valley.

4.7 Junction of Wyoming 789 with Wyoming secondary road. Town of Baggs immediately ahead. Secondary road to right leads to Dixon, Savery, Slater and the upper Little Snake River Valley. (See Side Trips.) End of log.

LOG FROM EAST END OF BROWNS PARK, COLORADO, TO CLAY BASIN, UTAH, VIA JESSE EWING CANYON

LOG BY RITZMA AND SMART

Cumulative Mileage

0.0 Log begins at north point of "triangle" in east end of Browns Park, Colorado (mile 9.7 in second day's log). "Gates of Lodore," north entrance to Lodore Canyon at 11:30 Precambrian Uinta quartzite dips south on south flank of Uinta Mountain Arch. Divergent dips in Uinta quartzite and Browns Park formation (Miocene) in beds at 3:00 o'clock are involved in graben faults, down to south.

1.1 Road to left is Moffat 10. Log continues along Moffat 10.

1.3 Bull Canyon right is abandoned course of Vermilion Creek which once flowed through Irish Canyon, Bull Canyon and thence into the Green River prior to diversion by stream piracy and/or graben faulting (Miocene). Fremont expedition of 1844 traversed Bull Canyon and Irish Canyon leaving Browns Park en route east.

3.4 Ranch right.

4.2 School right. Vermilion Creek left.

5.0 Bridge.

5.4 Flatirons of Browns Park right at 1:30 dipping south (20 to 45 degrees) into graben. These are pre-graben Browns Park beds. Graben fault (down to south) is probably at base of flatirons concealed by post-graben Browns Park beds and alluvium.

9.6 First view of Green River. In 1836 or 1837, traders Thompson, Craig and Sinclair (St. Clair) built a fort in this vicinity and called it Fort Davy Crockett. They displaced earlier traders who had frequented "Browns Hole" since the early 1820's. The site of the fort is not known for certain, and it was in ruins and abandoned in 1844 when the Fremont party camped in the area.

11.8 Browns Park and Uinta quartzite in fault contact at base of escarpment at 1:30. This is a graben fault.

14.5 Pinkish gravels on Browns Park formation.

15.1 Canyon of Beaver Creek right at 3:00 o'clock. Large fan at canyon mouth.

16.8 Dip ahead. (CAUTION)

17.0 Road to Browns Park Bridge left. This leads to Vernal —eventually. Continue on Moffat 10.

18.8 Bridge over Beaver Creek.

19.0 Proterozoic Uinta quartzite in contact with Archeozoic Red Creek Complex in canyon at 2:00. Fault at base of escarpment right.

20.2 Colorado-Utah boundary.

22.1 Road forks. Continue left. Airstrip right.

23.8 Descend hill. Sweeping view of Green River Valley.

24.5 Road forks. Continue right. Complex relation of Uinta quartzite and Red Creek Complex ahead.

25.6 Cabins left.

25.8 Mailbox. Cross ditch.

26.9 Keep on main road. Loess, gravels and reworked Browns Park on Browns Park formation ahead.

28.8 Road curves to right to ascend Jesse Ewing Canyon. Fault (down south) contact of Browns Park and Red Creek Complex ahead. Begin steep ascent of Jesse Ewing Canyon through Red Creek Complex. Note the variety of lithologies. Also watch road.

31.6 Top of grade. (STOP FOR VIEW.) Ahead is Ericson sandstone formation of Mesaverde group (Cretaceous) 900 feet thick. Above is a 500 foot soft, drab clay interval, possibly Fort Union (Paleocene), probably the basal drab or coaly sequence of the Hiawatha (Wasatch). The top of Tepee Mountain ahead (salmon orange beds) are Hiawatha bouldery conglomerate. This is the northeast limb of Clay Basin Anticline. The Ericson when followed to the right is seen to be dragged up under the Uinta Overthrust. Exposures here indicate that the fault plane dips about 5 degrees to the south. Continue ahead down grade.

31.7 On skyline right, white quartzite of the Red Creek Complex (Archeozoic) is thrust for nearly a mile over the Ericson and Wasatch at a low angle. Fault contact of Red Creek Complex with Wasatch is north (left) of trail seen ascending to skyline at 2:00. Note dragged up Ericson at 3:00.

32.1 View of Clay Basin. Mancos shale forms topographic basin with Mesaverde rimrocks making a half circle around the anticline. Uinta thrust sheet forms south "flank" of structure. To west note Ericson dragged up along thrust. Thrust plane dips south at about 15°.

34.3 Ahead Ericson dragged up by thrust. Fault plane dips south, trace roughly at base of trees. Mancos shale below thrust sheet does not support juniper-pinyon forest. Masses of thrust material have slumped down over soft Mancos shale obscuring details of fault.

34.9 Road forks. Keep straight ahead.

35.7 Passing Mountain Fuel No. 1 Lauzer well left. Clay Basin has a long history of gas and distillate production from the Frontier and Dakota sandstones.

36.6 Road to left. Continue straight ahead. Ericson scarp ahead. Tan beds below Ericson are remnants of Rock Springs formation almost completely replaced by Mancos marine shale.

37.0 Bridge.

37.5 Clay Basin Camp.

37.6 School right. Turn left. Keep on main road.

39.4 Notch cut by Red Creek through Mesaverde rimrock.

39.6 Enter Wyoming. Road continues to Rock Springs. End log.

LOG OF COLORADO 789-13 — RIO BLANCO TO RIFLE, COLORADO

LOG BY WHITLEY AND LEE

Cumulative Mileage	
0.0	Rio Blanco Post Office, at intersection of Piceance Creek Road and Colorado No. 789-13.
0.4	Bridge across Piceance Creek. At left ahead is Williams Fork sandstone and shale; Iles sandstone on higher ridges. On right brush-covered slopes are on Wasatch.
1.9	Leave Rio Blanco County and enter Garfield County. At left are variegated sands and shales of Wasatch.
3.4	On high ridge at right are good exposures of Green River white-weathering shale. Steeply dipping Wasatch and Williams Fork at left.
4.5	On left are good exposures of Wasatch variegated beds. Midway up timbered slopes at right is Wasatch-Green River contact. Note predominance of evergreens and aspens on Green River shale.
5.5	Steeply dipping Wasatch on both sides of road. Good example of badland topography.
6.9	On left are steep dip slopes on Williams Fork and Iles beneath Wasatch.
7.7	Iles sandstone on high ridge at left.
8.6	On right note uniform, even-bedded Green River shale on ridge.
9.1	Spectacular exposure of Wasatch and Williams Fork in canyon to left of road.
10.8	On skyline at left note sharp ridges formed on Williams Fork and Iles sandstones. Road still on Wasatch.
12.0	Grand Hogback diverges southeastward from highway.
14.1	On left dip slopes are on Wasatch.
15.8	At left is intersection of Colorado No. 325. On right skyline is Green River shale. Ahead on horizon is Battlement Mesa.
17.5	Bridge over Government Creek. Ahead, across Colorado River, are terraces capping Wasatch.
17.8	City limits of Rifle (elevation 5322 feet).
18.5	Bridge over Rifle Creek.
19.1	Junction of Colorado No. 13 with Highways No. 6 and No. 24. Colorado River ¼ mile ahead. End of log.

BROWN'S PARK

Brown's Park is a beautiful mountainous valley situated on the banks of the Green River in extreme northeast Utah and northwest Colorado. The area is sometimes referred to as Brown's Hole, after a trapper named Baptiste Brown who was snowed in here during the fall and winter of 1835-36. Here was located historic Fort Davy Crockett; and here also was established "Robber's Roost," rendezvous of western bad men, including the celebrated Butch Cassidy and his notorious gang of cattle thieves. Because of its protected location and mild climate Brown's Park was a favorite winter encampment for both Indians and whites. Besides a plentiful supply of fish and game, the valley provided abundant pasturage for horses and mules. Writing in 1839, Thomas Farnham, famed western American traveller, says:

"Brown's Park is something more than eight thousand feet above the level of the sea. It appeared to be about six miles in diameter; shut in, in all directions, by dark frowning mountains, rising one thousand five hundred feet above the plain. The Sheetskadee, or Green River, runs through it, sweeping in a beautiful curve from the north-west to the south-west part of it, where it breaks its way through the encircling mountains, between cliffs, one thousand feet in height, broken and hanging as if poised on the air. The area of the plain is thickly set with the rich mountain grasses, and dotted with little copses of cotton wood and willow trees. The soil is alluvial, and capable of producing abundantly all kinds of small grains, vegetables, etc., that are raised in the northern states. Its climate is very remarkable. Although in all the country, within a hundred miles of it, the winter months bring snows, and the severe cold that we should expect in such a latitude, and at such an elevation above the level of the sea, yet in this little nook, the grass grows all the winter, so that, while the storm rages on the mountains in sight, the drifting snows mingle in the blasts of December, the old hunters heed it not. Their horses are cropping the green grass on the banks of the Skeetskadee, while they themselves are roasting the fat loins of the mountain sheep, and laughing at the merry tale and sing."

L. H. Creer

SIDE TRIPS

Dinosaur National Monument
Sand Wash Basin Roads
Baggs to Slater
Elkhead Loop Trip

Gates of Lodore, Dinosaur National Monument.

DINOSAUR NATIONAL MONUMENT

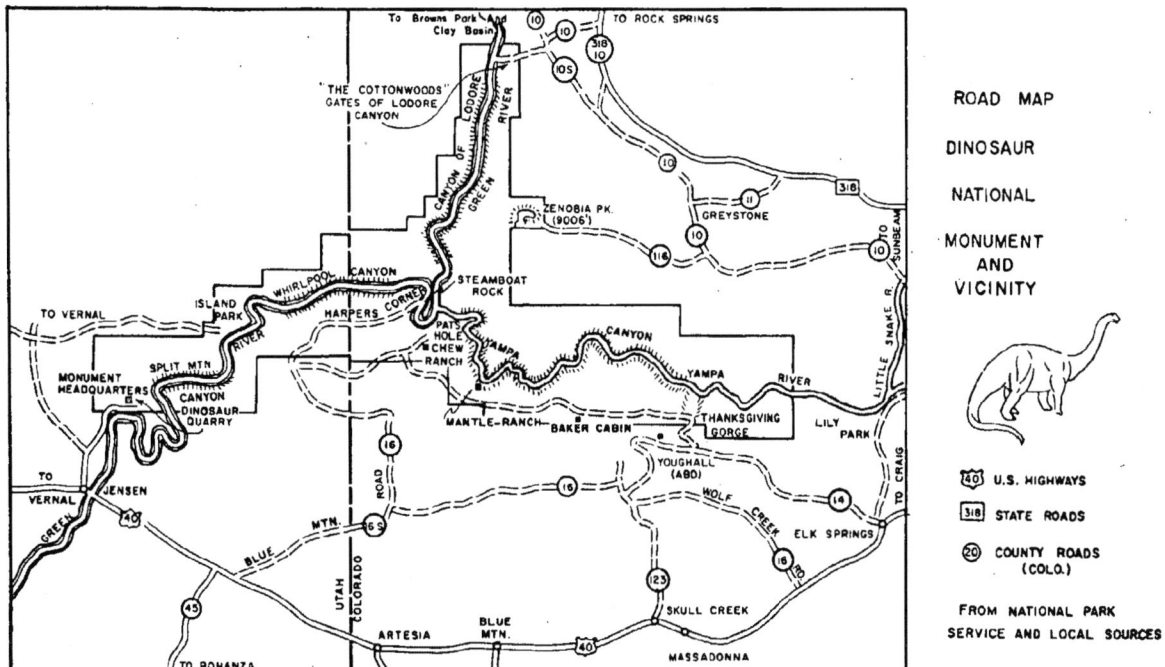

Access roads to Dinosaur National Monument are almost unmarked except for the paved road which leads to the Monument Headquarters and the Dinosaur Quarry north of Jensen, Utah. Some 95% of the Monument's area is not seen by the casual visitor, particularly that portion which lies in Colorado. So that visiting geologists may see the scenic grandeur of the Monument and its spectacular geologic exposures, this guidebook includes a road map and short explanation of trips into the more remote areas of the Monument.

(1) BLUE MOUNTAIN ROAD — Leaves U.S. 40 0.9 miles east of the junction of U 45 (Bonanza Road) with U.S. 40 east of Jensen, Utah; 5.2 miles west of Colorado-Utah boundary. Road is not marked but there is a stop sign at its junction with U.S. 40. It strikes off to the northeast to the left of a low ridge.

This road is by far the best way to Harpers Corner and the Echo Park and Pats Hole areas. It can be driven without difficulty in a passenger car. Within and adjacent to the Monument signs have been placed marking the way to points of interest. A parking area and turn-around have been constructed at the approach to Harpers Corner. A good trail leads to the overlook which affords a view comparable to the Grand Canyon of the Colorado.

(2) WOLF CREEK ROAD — (Moffat 16) — Leaves U.S. 40 at Wolf Creek Bridge about 8.5 miles west of Elk Springs.

This road skirts the south boundary of the Monument and joins the Blue Mountain Road. It can be negotiated in a passenger car.

(3) MOFFAT 14 — leaves U.S. 40 at Elk Springs and joins Moffat 16 south of Youghall (abandoned).

From Moffat 14, east of Youghall, a road leads north into the Monument by way of Thanksgiving Gorge and traverses the length of the Monument to Pats Hole. East to west across the Monument is a rough trip, not recommended for passenger cars. For the adventurous it is a memorable excursion usually requiring a day or more.

(4) MOFFAT 123 — Leaves U.S. 40 at Skull Creek and joins Moffat 16 south of Youghall (abandoned).

A rough road to be used with caution by passenger cars. Excellent exposures of the Skull Creek section and anticlinal structure can be visited by travelling this road.

(5) Between Sunbeam and Irish Canyon along Colorado 318 (Moffat 318) are several roads which lead west to Greystone and Douglas Mountain ranches and into the Monument. Most become difficult to traverse by passenger car as they ascend the Uinta foothills. It is possible to drive by jeep close to the summit of Zenobia Peak from which a spectacular view is obtained. By jeep, then by horse or on foot, one can reach Limestone Overlook on the west end of Limestone Ridge where there is a 2850' view straight down into Hells Half Mile in Lodore Canyon. This part of the Monument should not be visited without considerable preparation and consultation with detailed road and topographic maps.

The easiest trip into the Monument from the east or north is to "The Cottonwoods," a campsite and picnic ground at the north entrance to Lodore Canyon. Many Lodore Canyon river trips launch from this area. A log to this point follows:

From U.S. 40 at Maybell, Colorado (slightly west of town) turn north on Colorado 318 and follow for 41.5 miles to a point just south of where 318 crosses Vermilion Creek. Incidentally, you have run out of state highway and are on Moffat County 318. Turn left on Moffat 10 (signs point to Greystone and Douglas Mountain). Log begins here:

0.6 Y-intersection at "triangle". Turn right. If this road is in bad condition, continue 0.05 miles and turn hard right at old sign which dimly points to Lodore Canyon.
5.0 Y-fork. Go left. Watch for deep soft sand ahead.
7.5 Enter Dinosaur National Monument.
8.8 Cross Cottonwood Creek valley.
9.6 "The Cottonwoods". Dinosaur National Monument cabin.
10.0 Picnic area and campground.

In all cases the above described roads should be considered as "summer roads". Summer thunderstorms may temporarily cause the more primitive roads to become difficult to travel even with specialized vehicles. Travel in and around the Monument should not be undertaken without an adequate supply of gasoline, spare tires and tools and a check of general weather conditions.

RIVER TRIPS

Not to be overlooked in seeing the Monument are the never to be forgotten trips through the canyons by boat. Trips down the Green and Yampa rivers through spectacular Yampa, Lodore, Whirlpool and Split Mountain canyons are conducted by the Hatch River Expeditions. Trips of varying length can be arranged from one day to several weeks depending on the time at the disposal of the traveller. Guides are skillful river-runners experienced in the canyons of the Yampa, Green and Colorado.

Information can be obtained by addressing:
Hatch River Expeditions,
411 East Second North,
Vernal, Utah.

MAPS AND SOURCES OF INFORMATION

Recommended guides, maps and sources of information for the Monument and vicinity are:

"Road Map and Ranch Directory of Moffat County, Colorado," published by the Craig *Empire-Courier,* Craig, Colo. $1.00.

"Topographic Map of the Dinosaur National Monument, Colorado-Utah" by the U. S. Geological Survey and National Park Service. $0.75.

"Geology of Dinosaur National Monument and Vicinity, Utah-Colorado," Utah Geological and Mineralogical Survey Bull. 42 (1954) by G. E. and B. R. Untermann. Available at the Utah Field House of Natural History, Vernal, Utah, from the U. G. & M. S., and from other sources. This is undoubtedly the most complete volume available on the geology, scenery and lore of the Monument — a scientific labor of love by two persons intimately associated with the Dinosaur National Monument area. $2.50 — and well worth it.

"This is Dinosaur," edited by Wallace Stegner. Alfred A. Knopf, New York, 1955. A beautifully illustrated volume in eight chapters each on a different aspect of the Monument by authorities in each field — archaeology, history, exploration, wildlife, river-runners and geology. The geological section is by Eliot Blackwelder. $5.00.

Information can also be obtained by addressing:
Superintendent,
Dinosaur National Monument,
Jensen, Utah.

National Park Service

Steamboat Rock, a 1000' monolith of Weber sandstone.

THE "RIVER RUNNERS"

To most geologists, the story of Powell's epic expeditions down the Green and Colorado in 1869 and 1871 are quite familiar. But Powell was not first— he has his predecessors in river exploits along the Green.

Almost unknown to the outside world and unrecorded until 1918 when his journals were published, was the canyon voyage of General William H. Ashley. In 1825, Ashley's party ran the canyons of the Green in bull boats, a craft described as "nothing but a pole frame thirty feet long, with twelve feet of beam, covered with pitch-smeared buffalo skins." In Red Canyon he inscribed his name and the date 1825, and farther along the river he recorded laborious portages, most of them in Lodore Canyon. His boats were adaptable cargo carriers but hardly "white water" craft.

Second recorded canyon travellers were Manly and his group of impatient gold seekers of 1849. In the rehabilitated hulk of a ferry they reached Red Canyon where their clumsy craft foundered and stuck fast. The rest of their trip to Green River, Utah was completed in dugout canoes lashed together into rafts. Manly's journal records a cache of abandoned equipment and a message posted on a tree in Lodore Canyon. Obviously there were other canyon voyagers whose tales of exploits were silenced or forgotten before they became history.

By the time of Powell's first voyage down the Green in 1869, the Uinta canyons were reasonably well known —if poorly understood—and the "parks" and "holes" into which the canyon-rimmed river valley occasionally opened had been intermittently inhabited. Finding the inscription "Ashley 1825" in Red Canyon but not knowing of Ashley, Powell misread the date as 1855 and supposed that his predecessor was a trapper or prospector. He named the falls in the Green at this point Ashley Falls, a name that survives today to commemorate the first "river runner" of the West.

—H. R. Ritzma

SAND WASH BASIN ROADS

Three roads are shown on the master road map traversing the central portion of the Sand Wash Basin.

MOFFAT 21 — (Marked with signs.) Follows the south bank of the Little Snake River from Two Bar Ranch northward to the vicinity of "Steel Bridge", lunch stop first day.

Along the road at its southern end are good views of Two Bar Anticline and the associated fault complex. Farther northeast, the north-sloping dip slope of Godiva Ridge and Bald Mountain can be seen to the south of the road. Other scenery along this road is not too inspiring.

MOFFAT 117 — (Not marked by signs.) Begins a few feet north of the Little Snake River Bridge (Colorado 318) at Two Bar Ranch and follows north bank of river to the east.

This road affords access to and closeup views of Two Bar Anticline and associated fault complex. It climbs the bluffs of Laney shale north of the Little Snake River and takes off north toward Powder Wash. Sweeping views west into the deep of the Sand Wash Basin and to the Uinta Mountains and views east and south to the Sierra Madre of Wyoming and the Elkheads, Park Range and Flat Tops of Colorado are afforded along this road. The bigness and emptiness of this country are very evident. The road follows a gravel capped ridge, probably Bishop conglomerate. The size and angularity of the cobbles of Madison limestone and Uinta quartzite are surprising for a locale 25 miles east of the Uinta Mountains. The road then ascends the Powder Rim to a gravel-paved plateau of basal Browns Park conglomerate deposited on the Gilbert Peak erosion surface. Many of the wooded hollows south of Powder Rim and Powder Rim itself are good artifact hunting localities. Moffat 117 ends at junction with Moffat 4 at Powder Wash camp.

MOFFAT 19 — (Marked by signs.) Begins at Maybell, Colorado on U.S. 40 and proceeds north. This road crosses the belt of vertical beds called the Maybell Monocline (here partly covered by the Browns Park formation) that marks the north flank of the Uinta-Axial Basin line of folding. North of the Maybell Monocline is a sweeping view opening into the Sand Wash Basin with Hiawatha, Tipton, Cathedral Bluffs and Laney visible from the foreground to Godiva Ridge on the skyline. Farther north, the road crosses Tipton (as extended by Ritzma, this guidebook) and skirts to the east of Bald Mountain, a down-faulted outlier of Godiva Ridge. The road joins Moffat 21 about midway between "Steel Bridge" and Two Bar Ranch.

H. R. Ritzma

BAGGS TO SLATER

From Wyoming 789 (formerly 330) at Baggs, a paved secondary road leads up the valley of the Little Snake River. From Baggs to Dixon (7 miles) one gets an excellent view eastward into Elkhead Mountains and its volcanic outliers north of the Wyoming line. On the distant skyline are the Sierra Madre (Wyoming) and Park Range (Colorado), westernmost of the three northern prongs of the Colorado Rockies. The Continental Divide follows the crests of these ranges.

Just before crossing the Little Snake River west of Dixon, the road passes the soft hogback of the uppermost sandstone of the Lance. The basal conglomerate of the Fort Union makes the gravelly soils just west of the hogback. From Dixon to Savery (4 miles) outcrops are scarce; one occasionally sees Lewis shale with a cap of Browns Park in road cuts or stream banks. The sharp volcanic peak to the south is Baker Peak. The grave of Jim Baker is a few hundred yards south of the road about 2½ miles east of Dixon near the site of the Baker Cabin.

At Savery the Mesaverde formation is encountered, and the road continues to wind along the river in a narrowing and more verdant valley, skirting cliffs of Mesaverde sandstone, shale and coal. North of the road are the Horse Mountains, a group of basic intrusives and jumbled extrusives. The large flat-topped mountain glimpsed ahead is Battle Mountain (9100'), basalt capping 1000' or more of Browns Park formation. (Savery to Slater, 4 miles.)

Slater was a once important local trading point through which goods and supplies from the Union Pacific Railroad to the north reached the upper Little Snake River Valley and the Hahns Peak Mining District. Here we join the route of the Elkhead Loop Trip.

H. R. Ritzma

JIM BAKER

Jim Baker, bear fighter and fur trapper of the Forties, came to the west in 1837 with the American Fur Company as a young man of nineteen. He was reputedly present at the 1841 battle at Fraeb's Post when Sioux destroyed that fur company outpost on the upper Little Snake River. That same year, in company with Peg-Leg Smith he traversed Utah while en route to Arizona. On this expedition he reconnoitered the Bear, Weber and Provo Rivers, noting on the latter stream, that "the surrounding country was covered with sunflowers on which the Indians used to get seeds and eat them with grasshoppers." In 1847, Baker returned to Utah, and he was known to have been at Fort Bridger in 1848. In subsequent years he was a freighter, trapper, guide for Fremont and associate of Jim Bridger.

In 1873, Baker settled near Dixon in the Little Snake River Valley of southern Wyoming where he built a three-story log blockhouse. This structure was moved to Cheyenne in 1917 where it is preserved in Frontier Park. Baker died in 1898 at the age of 80 and was buried near Dixon, Wyoming. A prominent peak in the northern Elkhead Mountains nearby commemorates the name of Jim Baker, one of the last surviving "Mountain Men".

—L. H. Creer

ELKHEAD LOOP TRIP

Steamboat Springs to Hayden via Elk River Valley, Hahns Peak, Little Snake River, Slater Park and California Park.

It is suggested that those taking this trip provide themselves with lunch. Trip distance is 128 miles. This is especially recommended as a combination geologic trip and a photographic tour during the Fall. The Elkhead Mountains are renowned as the ultimate in autumn coloring.

LOG BY LARSON AND HEINS

Cumulative Mileage

- 0.0 Junction of U.S. 40 and Colorado 129 one mile west of Steamboat Springs. Proceed north on Colorado 129.
- 1.2 Passing through cut in Niobrara portion of Mancos shale.
- 1.4 Steamboat Springs Airport on left.
- 4.1 Elk Mountain at 11 o'clock is a large late Tertiary intrusive.
- 4.6 Highway turns parallel to Elk River.
- 5.1 Lower Mancos black shales along road.
- 5.8 Crossing Mad Creek.
- 6.1 Pre-Cambrian rocks along road with sediments on opposite side of river which follows contact at this point.
- 7.6 Permo(?)-Triassic redbeds on pre-Cambrian rocks.
- 7.8 Crossing Big Creek.
- 8.6 Road junction; continue right.
- 9.1 Pilot Knob on horizon ahead is a late Tertiary intrusive protruding through extensive area of flows and lava rubble.
- 9.2 Trull anticline visible at 9 o'clock. Hay meadows in this valley produce around four tons of timothy and clover hay per year from one cutting.
- 11.1 Sand Mountain intrusive at 11 o'clock; Pilot Knob at 9 o'clock.
- 14.1 Road junction at Moon Hill Bridge; continue right.
- 14.3 Moon Hill School.
- 14.9 Hahns Peak on horizon ahead.
- 17.1 Good view of Sand Mountain at 9 o'clock.
- 17.6 Farwell Mountain on horizon ahead east of Hahns Peak.
- 18.5 Clark store and post office.
- 19.4 Glen Eden Resort Ranch. Crossing Elk River.
- 19.5 Junction with road right to Seedhouse. Log continues left. Good exposures of Dakota and older sediments ahead. The Dakota and Morrison are essentially vertical with dip lessening to the east toward the granite.

EDITOR'S NOTE: Section here measured through heavy brush is:

Dakota formation	100'
Morrison formation	310' app.
Prominent basal sandstone	60'
Curtis (Sundance)	50' app.
Entrada	75'
Chinle	140'
Shinarump (?)	15'
Moenkopi	350'

Pre-Cambrian red-brown coarse-grained granite.

- 22.9 Good view of Hahns Peak. Fire lookout tops the peak.
- 23.4 Steep dipping Dakota and Mancos about ½ mile ahead.
- 23.7 Willow Creek.
- 24.3 Road on Arizona Bar named in the days of placer mining.
- 26.3 Sand Mountain at 9 o'clock.
- 27.0 Hahns Peak, once county seat of Routt County (included all of Moffat County also) and largest settlement on the western slope. Gold was discovered here by Joseph Hahn in 1865. The town flourished into the 1880's although it never became a bonanza mining town. Production probably did not exceed half a million dollars. For geology of this general area see papers and maps by Hunter and Barnwell, this guidebook.

Columnar jointing in basalt, Slater Creek (mile 70.6)

Cumulative Mileage

27.4 Old placer workings visible at right.

28.6 Road surfaced with Mowry shale.

29.6 Dakota (Cloverly) makes cuesta-like outcrops at left.

31.6 Low dipping Dakota outcrop left of road.

32.1 Columbine store and post office on divide between Yampa River watershed and the Little Snake River.

32.8 Quarry at right where county obtains Mowry shale for road metal.

33.0 Dakota is quarried here for building material. It makes an attractive veneer for house exteriors. Occasional poor outcrops of Morrison (?) and Triassic redbeds ahead next mile or so.

37.2 Junction for Whisky and Red Parks; continue left.

38.9 Road cut in granite.

39.1 Browns Park formation on granite.

40.0 Browns Park in roadside.

44.0 Browns Park formation at left of road.

45.1 Leaving Routt National Forest.

46.5 Little Snake River valley with Browns Park at valley level. Relief on base of Browns Park is several thousand feet in this area, and there is good evidence that the Little Snake River is exhuming a more ancient river course.

47.0 Three Forks Ranch. North of ranch, up valley of Roaring Fork is section of Dakota to Chugwater (Moenkopi) with thin limestone intervening between the Permo(?)-Triassic redbeds and the granite. This has been questionably identified as Madison, the farthest south Paleozoic occurrence on the wedge-out of Paleozoic on the north end of the Pennsylvanian "Ancestral Rockies".

50.4 Mesaverde outcrops on north side of valley. Battle Mountain ahead down valley is a large lava-capped mesa resting on 1000 or more feet of Browns Park. Along the north side of the valley and in road cuts for the next mile or so are good exposures of exceptionally thick basal Browns Park conglomerate.

52.9 Crossing Little Snake River.

53.0 Focus (Temple) Guest Ranch. "Where the Sons Raise Meat."

54.7 Mesaverde sandstone above road.

55.1 Coal bed in Mesaverde.

56.0 Cross Battle Creek, tributary from north, ahead. Saddle Pocket (Salisbury) Guest Ranch. In this area was the site of Fraebs Post, destroyed in a fierce battle with Sioux in 1841.

59.5 Mesaverde shale above road.

63.1 Cross Little Snake River again.

64.9 Slater Dome. The valley cuts through north side of the structure. Road left leads to Quintana gas well. (See paper by Turner, this guidebook).

Cumulative Mileage

66.8 Crossing Little Snake River.

69.1 Road junction at old Slater post office and store. Turn left on Moffat County No. 1.

69.4 Cross Little Snake River (last time).

70.3 Crossing Slater Creek. Volcanics ahead. Look for inclusions of Lewis shale in lava in road cuts.

70.6 Spectacular jointing in basalt across creek left. Junction ahead; continue on Moffat No. 1 to left. Road follows Slater Creek valley in Browns Park formation.

76.0 Columnar jointing in basalt on Long Mountain at 9 o'clock.

76.4 Mesaverde outcrops along creek at 3 o'clock.

80.6 Mesaverde outcrops at 10 o'clock.

83.1 Dike at 9 o'clock.

83.8 Junction; continue left.

88.2 Bears Ears Peaks at 3 o'clock.

90.0 Crossing dike.

92.0 Entering Routt National Forest. Slater Park.

93.7 Dike at left forming sharp ridge.

95.0 Adams Creek.

96.6 Divide between Slater Park and California Park.

98.6 Elkhead Creek.

99.2 Lewis shale in roadcut.

99.3 California Park Guard Station.

99.6 Torso Creek.

101.3 Lewis shale exposures at 3 o'clock.

102.5 Armstrong Creek. Dikes ahead left.

104.3 First Creek. Lance exposed in the west end of California Park reflects the broad syncline of the Sand Wash Basin which extends far to the east under the volcanics of the Elkhead Mountains.

106.7 Hayden Divide, watershed between California Park and Hayden Valley.

108.1 Leaving Routt National Forest.

111.5 Hooker Mountain ahead, a sill-like intrusive in the Lewis shale.

112.7 Lewis exposures along creek.

113.1 Junction; follow Routt County No. 80.

118.9 Hooker Mountain at 9 o'clock.

122.2 Crossing contact of Lewis shale and Lance formation. Beginning descent into Hayden Valley, the local name for this part of the Yampa Valley.

124.5 Junction; continue to southwest.

126.8 Junction; turn left.

127.0 Cross Yampa River.

127.7 Junction with U.S. 40 in Hayden. End of trip.

SELECTED BIBLIOGRAPHY

Compiled by JOHN CHRONIC
University of Colorado, Boulder, Colorado

Abrassart, C. P. (1952), "Stratigraphy and Sedimentation of the Juniper Mountain Area," unpublished Master's thesis, University of Colorado.

Anonymous (1924), "Exploration Near Craig and on Moffat Dome," Oil and Gas Jour., vol. 23, no. 8, pp. 56, 68.

Atwood, W. W. (1907), "The Glaciation of the Uinta Mountains," Jour. Geology, vol. 15, pp. 790-804.

.... (1915), "Glaciation of the Uinta and Wasatch Mountains," U. S. Geol. Survey Prof. Paper 61, pp. 7-93.

Baker, A. A., Dane, C. H., and Reeside, J. B., Jr. (1936), "Correlation of the Jurassic Formations of Parts of Utah, Arizona, New Mexico, and Colorado," U. S. Geol. Survey Prof. Paper 183, 66 pp.

.... (1947), "Revised Correlation of Jurassic Formations of Parts of Utah, Arizona, New Mexico and Colorado," Bull. Amer. Assoc. Petrol. Geol., vol. 31, pp. 1664-1668.

Baker, A. A., and Williams, J. S. (1940), "Permian in Parts of Rocky Mountain and Colorado Plateau Regions," Bull. Amer. Assoc. Petrol. Geol., vol. 24, pp. 617-635.

Ball, M. W. (1925), "Waters from the Green River Shale," Bull. Amer. Assoc. Petrol. Geol., vol. 9, no. 1, p. 171.

Barb, C. F. (1944), "Hydrocarbons of the Uinta Basin of Utah and Colorado; Review of Geology and Field Work by Clark F. Barb; Survey of Bitumen Analyses and Extraction Methods by James Ogden Ball," Colorado School of Mines Quart., vol. 39, no. 1, 115 pp.

.... (1945), "The Origin of the Hydrocarbons in the Uinta Basin (Colorado and Utah)," Mines Mag., vol. 35, no. 10, pp. 555-557.

.... (1946), "Selected Well Logs of Colorado," Colorado School of Mines Quart., vol. 41, no. 1, pp. 388-390.

Bass, N. W. (1946a), "Rangely Field," Oil and Gas Jour., vol. 45, no. 35, pp. 48-50, 52.

.... (1946b), "Geological Survey (U.S.) Completes New Structure Map for Rangely Field (Colorado)," Oil and Gas Jour., vol. 45, no. 30, pp. 86-87, 93.

.... (1946c), "Subsurface Maps of the Rangely Anticline, Rio Blanco County, Colorado," U. S. Geol. Survey Prelim. Map 67, Oil and Gas Inv. Ser.

.... (1950), "Pre-Pennsylvanian Paleozoic Rocks in Western Colorado and Southeastern Utah," New Mexico Geol. Soc. Guidebook, First Field Conf., pp. 56, 57.

.... (1954), "Geology and Mineral Fuels of Mount Harris, Pilot Knob, Elkhart Creek, and Daton Peak Quadrangles, Routt and Moffat Counties, Colorado," U. S. Geol. Survey Open File Rept.

Bass, N. W., and Northrop, S. A. (1950), "South Canyon Creek Dolomite Member, a Unit of Phosphoria Age in Maroon Formation Near Glenwood Springs, Colorado," Bull. Amer. Assoc. Petrol. Geol., vol. 34, no. 7, pp. 1540-1551.

Berkey, C. P. (1905), "Stratigraphy of the Uinta Mountains," Geol. Soc. America Bull. 16, pp. 517-530.

Beroni, E. P., and McKeown, F. A. (1952), "Reconnaissance for Uraniferous Rocks in Northwestern Colorado, Southwestern Wyoming, and Northeastern Utah," U. S. Geol. Survey Trace Elements Inv. Rept. 308-A.

Berry, E. W. (1915), "Paleobotanic Evidence of the Age of the Morrison Formation," Bull. Geol. Soc. America, vol. 26, pp. 340, 341.

Blair, R. W., and Heiny, L. W. (1952), "Piceance Creek Basin of Colorado Portion of Uinta Basin," Petrol. Eng., vol. 24, no. 3, pp. 48-60.

Bradley, W. H. (1925), "A Contribution to the Origin of the Green River Formation and Its Oil Shale," Bull. Amer. Assoc. Petrol. Geol., vol. 9, no. 2, pp. 247-262, 410.

.... 1929a), "The Varves and Climate of the Green River Epoch," U. S. Geol. Survey Prof. Paper 158, pp. 87-110.

.... (1929b), "The Occurrence and Origin of Analcite and Meerschaum Beds in the Green River Formation of Utah, Colorado, and Wyoming," U. S. Geol. Survey Prof. Paper 158, pp. 1-7.

.... (1931), "Origin and Microfossils of the Oil Shale of the Green River Formation of Colorado and Utah," U. S. Geol. Survey Prof. Paper 168.

.... (1935), "Petroleum Geology in Colorado and Southwestern Wyoming," Amer. Assoc. Petrol. Geol., vol. 22, pp. 1020-1047.

.... (1936), "Geomorphology of the North Flank of the Uinta Mountains," U. S. Geol. Survey Prof. Paper 185 i.

.... (1945), "Geology of the Washakie Basin, Sweetwater and Carbon Counties, Wyoming, and Moffat County, Colorado," U. S. Geol. Survey Prelim. Map 32.

.... (1952), "Jurassic and Pre-Mancos Cretaceous Stratigraphy of the Eastern Uinta Mountains, Colorado-Utah," unpublished Master's thesis, University of Colorado.

Brill, K. G., Jr. (1944), "Late Paleozoic Stratigraphy, West-Central and Northwestern Colorado," Bull. Geol. Soc. America, vol. 55, pp. 632, 633.

Campbell, M. (1923), "Twentymile Park District of Yampa Coal Field, Routt County, Colorado," U. S. Geol. Survey Bull. 748, 82 pp.

Coffin, R. C. (1933), "Colorado Symposium" (Preface), Bull. Amer. Assoc. Petrol. Geol., vol. 17, pp. 351, 352.

Coffin, R. C., and DeFord, R. K. (1934), "Waters of the Oil- and Gas-Bearing Formations, Rocky Mountains," Problems of Petroleum Geology (Amer. Assoc. Petrol. Geol. Sidney Powers Memorial Volume), pp. 927-952.

Coffin, R. C., Perini, V. C., and Collins, M. J. (1920), "Some Anticlines of Western Colorado," Colorado Geol. Survey Bull. 24, pp. 1-45, 56-67.

Crawford, R. D., Wilson, K. M., and Perini, V. C., Jr. (1920), "Some Anticlines of Routt County, Colorado," Colorado Geol. Survey Bull. 23, pp. 17-19.

Crittenden, Max D., Jr. (1950), "Pre-Carboniferous Stratigraphy and Structure of the Uinta Basin, Utah and Colorado," Petroleum Geology of the Uinta Basin (Guidebook to the Geology of Utah, no. 5, Intermountain Assoc. of Petrol. Geol.), pp. 61-69.

Cupps, C. Q., et al. (1951), "Petroleum in Northwest Colorado," World Oil, vol. 133, no. 7, pp. 192-194.

Deely, C. (1950), "Petroleum in Northwest Colorado," World Oil, vol. 130, no. 6, pp. 83-84, 86.

DeMohrenschildt, G. (1949), "Rangely Field," Oil and Gas Jour., vol. 48, no. 7, pp. 124, 126-128, 130.

Eardley, A. J. (1949), "Paleotectonic and Paleogeologic Maps of Central and Western North America," Bull. Amer. Assoc. Petrol. Geol., vol. 33, pp. 655-682.

Eardley, A. J., et al. (1950), "Petroleum Geology of the Uinta Basin," Guidebook to the Geology of Utah, no 5, Intermountain Assoc. Petrol. Geol., 151 pp.

Eby, J. B. (1924a), "Coal in Elkhead District of Yampa Coal Field, Northwestern Colorado," U. S. Geol. Survey Press Notice 16653.

.... (1924b), "Prospects for Oil and Gas in the Slater Dome in Northwestern Colorado," U. S. Geol. Survey Press Notice 17987.

Eby, J. B., and Reeside, J. B., Jr. (1925), "Prospects for Oil or Gas in an Anticline Near Baggs, Wyoming," U. S. Geol. Survey Press Notice 18092.

Emmons, S. F. (1897), "The Origin of the Green River," Science (N.S.), vol. 6, pp. 19-21.

.... (1907), "The Uinta Mountains," Geol. Soc. America Bull. 18, pp. 287-302.

Espach, R. H. (1951), "Unusual Res. Oils in Elk Basin and Rangely Fields," Amer. Petroleum Inst. Drilling and Production Practices 1951, pp. 224-233.

Estergren, E. F. (1945), "Rangely Field, Colorado," Petrol. Eng., vol. 17, no. 3, pp. 79-84.

Fenneman, N. M. and Gale, H. S. (1906), "Yampa Coal Field, Routt County, Colorado," U. S. Geol. Survey Bull. 297.

Ferris, B. J. (1950a), "Petroleum in Northwest Colorado," World Oil, vol. 151, no. 3, pp. 80, 82, 84, 86, 88.

.... (1950b), "Petroleum in Northwest Colorado," World Oil, vol. 151, no. 5, pp. 85-88.

Finley, W. L., and Bauer, A. D. (1926), "Coking of Oil Shales," U. S. Bur. Mines Tech. Paper 398, pp. 1-11.

Forrester, J. D. (1937), "Structure of the Uinta Mountains," Bull. Geol. Soc. America, vol. 48, pp. 631-666.

Gale, H. S. (1906), "Hahn's Peak Gold Field, Colorado," U. S. Geol. Survey Bull. 285, pp. 28-34.

.... (1907a), "Carnotite in Rio Blanco County, Colorado," U. S. Geol. Survey Bull. 315, pp. 110-117.

.... (1907b), "Coal Fields of the Danforth Hills and Grand Hogback in Northwestern Colorado," U. S. Geol. Survey Bull. 316, pp. 264-301.

.... (1908a), "Gold Placer Deposits Near Lay, Routt County, Colorado," U. S. Geol. Survey Bull. 340, pp. 84-95.

.... (1908b), "Carnotite and Associated Minerals in Western Routt County, Colorado," U. S. Geol. Survey Bull. 340, pp. 256-262.

.... (1908c), "Geology of the Rangely Oil District, Rio Blanco County, Colorado," U. S. Geol. Survey Bull. 350.

.... (1909), "Coal Fields of Northwestern Colorado and Northeastern Utah," U. S. Geol. Survey Bull. 341, pp. 283-315.

.... (1910), "Coal Fields of Northwestern Colorado and Northeastern Utah," U. S. Geol. Survey Bull. 415.

Gale, H. S., and Fenneman, N. M. (1906a), "The Yampa Coal Field, Routt County, Colorado," U. S. Geol. Survey Bull. 285, pp. 226-239.

.... (1906b), "The Yampa Coal Field, Routt County, Colorado," U. S. Geol. Survey Bull. 297, 81 pp.

Gilluly, J., and Reeside, J. B., Jr. (1926), "Jurassic Formations of Eastern Utah," (Abstract), Bull. Geol. Soc. America, vol. 37, no. 1, pp. 158, 159; Pan-Am. Geol., vol. 45, no. 2, pp. 160, 161.

Girty, G. H. (1903), "The Carboniferous Formations and Faunas of Colorado," U. S. Geol. Survey Prof. Paper 16.

Hancock, E. T. (1915), "The History of a Portion of Yampa River, Colorado, and Its Possible Bearing on That of the Green River," U. S. Geol. Survey Prof. Paper 90-K, pp. 183-189.

.... (1925), "Geology and Coal Resources of Axial and Monument Butte Quadrangles, Moffat County, Colorado," U. S. Geol. Survey Bull. 757, 134 pp.

Hayden, F. V. (1862), "The Primordial Sandstone of the Rocky Mountains in the Northwest Territories of the U. S.," Amer. Jour. Sci., vol. 83 (2s, 33), pp. 68-79.

.... (1869), "Notes on the Geology of Wyoming and Colorado Territories," Amer. Philos. Soc., vol. 10, pp. 463-478; vol. 11, pp. 25-56.

.... (1871), "Preliminary Report (Fourth Annual) of the United States Geological Survey of Wyoming and Portions of Contiguous Territories (being a second annual report of progress)," 511 pp.

.... (1873), "Sixth Annual Report of the United States Geological Survey of the Territories . . . for the Year 1872," 844 pp.

.... (1874), "Seventh Annual Report Embracing Colorado, Being a Report of Progress of the Exploration for the Year 1873: U. S. Geol. and Geog. Survey Terr. 7th Ann. Rept., 718 pp.

.... (1877), "Geological and Geographical Atlas of Colorado and Portions of Adjacent Territory," U. S. Geol. and Geog. Survey Terr. Rept.

.... (1881), "Geological and Geographical Atlas of Colorado and Portions of Adjacent Territory," U. S. Geol. and Geog. Survey Terr. Rept.

Heaton, R. L. (1929), "Relation of Accumulation to Structure in Northwestern Colorado," Structure of Typical American Oil Fields (Amer. Assoc. Petrol. Geol. Symposium), vol. 2, pp. 93-114.

.... (1933), "Ancestral Rockies and Mesozoic and Late Paleozoic Stratigraphy of Rocky Mountain Region," Bull. Amer. Assoc. Petrol. Geol., vol. 17, no. 2, pp. 109-168.

.... (1938), "Stratigraphy versus Structure in Rocky Mountain Region," Bull. Amer. Assoc. Petrol. Geol., vol. 21, no. 10, pp. 1241-1267; World Petroleum, vol. 9, no. 1, p. 61.

.... (1939), "Contribution to the Jurassic Stratigraphy of the Rocky Mountain Region," Bull. Amer. Assoc. Petrol. Geol., vol. 23, no. 8, pp. 1153-1177.

Henderson, J. R. (1954), "Airborne Radioactivity Survey of Parts of Baggs SW, Baggs SE, Sand Creek SW, and Sand Creek SE Quadrangles, Sweetwater and Carbon Counties, Wyoming," U. S. Geol. Survey Open File Rept.

Hinds, N. E. A. (1925), "The Green River Formation," Science (N.S.), vol. 62, p. 34.

─────── (1936a), "Ep-Archean and Ep-Algonkian Intervals in Western North America," Carnegie Inst. Washington Pub. 463, pp. 1-52.

─────── (1936b), "Uncompahgran and Beltian Deposits of Western North America," Carnegie Inst. Washington Pub. 463, pp. 53-136.

─────── (1939), "Pre-Cambrian Formations of Western North America," 6th Pacific Sci. Cong. Proc., pp. 289-307.

Hubbard, S., et al. (1952), "Colorado Oil Shale," U. S. Bur Mines Rept. Inv. 4872.

Imlay, R. W. (1948), "Characteristic Marine Jurassic Fossils from the Western Interior of the United States," U. S. Geol. Survey Prof. Paper 214-B, pp. 13-33.

─────── (1949), "Paleoecology of Jurassic Seas in the Western Interior of the United States," Nat. Research Council, Rept. of Comm. on a Treatise on Marine Ecology & Paleoecology, 1948-1949, no. 9, pp. 72-104.

Ingram, T. R. (1927), "Exploration Near Craig and on Moffat Dome," Oil and Gas Jour., vol. 26, no. 24, pp. 32, 93, 96.

Jensen, F. S., et al. (1954), "The Oil and Gas Fields of Colorado," Symposium, Rocky Mountain Assoc. Petrol. Geol., pp. 15-37, 39-50, 63-290.

King, C. (1870), "The Green River Coal Basin (Utah)," U. S. Geol. Expl. 40th Par. Rept., vol. 3, pp. 451-473.

─────── (1871), "Geological Exploration of the Fortieth Parallel," U. S. War Dept. Ann. Rept.

─────── (1873), "Annual Report Upon the Geological Exploration of the Fortieth Parallel from the Sierra Nevada to the Eastern Slope of the Rocky Mountains," U. S. War Dept. Ann. Rept.

─────── (1874), "Annual Report Upon the Geological Exploration of the Fortieth Parallel From the Sierra Nevada to the Eastern Slope of the Rocky Mountains," U. S. War Dept. Ann. Rept.

─────── (1875), "Annual Report Upon the Geological Exploration of the Fortieth Parallel From the Sierra Nevada to the Eastern Slope of the Rocky Mountains," U. S. War Dept. Ann. Rept.

─────── (1876a), "Annual Report Upon the Geological Exploration of the Fortieth Parallel From the Sierra Nevada to the Eastern Slope of the Rocky Mountains," U. S. War Dept. Ann. Report 1876.

─────── (1876b), "Paleozoic Subdivisions on the Fortieth Parallel," Amer. Jour. Sci., vol. 11, pp. 477-479.

─────── (1876c), "Note on the Uinta and Wasatch Ranges," Amer. Jour. Sci., 3d ser., vol. 11, p. 494.

─────── (1877), "Annual Report Upon the Geological Exploration of the Fortieth Parallel From the Sierra Nevada to the Eastern Slope of the Rocky Mountains," U. S. War Dept. Ann. Rept. 1877.

─────── (1878), "Annual Report Upon the Geological Exploration of the Fortieth Parallel From the Sierra Nevada to the Eastern Slope of the Rocky Mountains," U. S. War Dept. Ann. Rept. 1878.

Knowlton, F. H. (1908), "Description of Fossil Plants From Mesozoic and Cenozoic of North America," Smithsonian Misc. Coll., vol. 52, pp. 489-495.

─────── (1923), "Revision of the Flora of the Green River Formation with Descriptions of New Species," U. S. Geol. Survey Prof. Paper 131, pp. 133-182.

Kramer, W. B., and McMillan, R. (1939), "Geologic Map of Powder Wash Dome, Moffat County, Colorado," U. S. Geol. Survey Press Notice 66152.

Lammers, E. C. H. (1939), "Origin and Correlation of the Cloverly Conglomerate," Jour. Geology, vol. 47, no. 2, pp. 113-132.

Lochman, C. (1949), "Paleoecology of the Cambrian in Montana and Wyoming," Nat. Research Council Rept. of Committee on Treatise on Marine Ecology and Paleocology, 1948-1949, no. 9, pp. 31-71.

Marsh, O. C. (1871), "On the Geology of the Eastern Portion of the Uinta Mountains," Amer. Jour. Sci., 3d. ser., vol. 1, pp. 191-198.

─────── (1875), "Ancient Lake Basins of the Rocky Mountain Region," Amer. Jour. Sci., 3d. ser., vol. 9, pp. 49-52.

Mook, C. C. (1915), "Origin and Distribution of the Morrison Formation," Bull. Geol. Soc. America, vol. 26, pp. 315-322.

─────── (1916), "A Study of the Morrison Formation," New York Acad. Sci. Annals, vol. 27, pp. 39-191.

Morgan, A. M. (1940), "Prospective Water Well Sites in Vicinity of Rangely, Colorado," U. S. Geol. Survey Open File Rept.

Nightingale, W. T. (1930), "Geology of Vermilion Creek Gas Area in Southwest Wyoming and Northwest Colorado," Bull. Amer. Assoc. Petrol. Geol., vol. 14, pp. 1013-1040.

─────── (1935), "Geology of Natural Gas Fields in Southwestern Wyoming and Northwestern Colorado," Geology of Natural Gas (Amer. Assoc. Petrol. Geol. Symposium), pp. 341-361.

─────── (1938a), "Petroleum Geology in Colorado and Southwestern Wyoming," Bull. Amer. Assoc. Petrol. Geol., vol. 19, pp. 537-543.

─────── (1938b), "Petroleum and Natural Gas in Non-Marine Sediments of Powder Wash Field in Northwest Colorado," Bull. Amer. Assoc. Petrol. Geol., vol. 22, pp. 1020-1047.

Peterson, O. A. (1926), "The Brown's Park Formation," Science (N.S.), vol. 63, pp. 231.

─────── (1928), "The Brown's Park Formation," Carnegie Mus. Mem., vol. II, no. 2, pp. 87-121.

Pickering, W. Y., and Dorn, C. L. (1948), "Rangely Oil Field, Rio Blanco County, Colorado," Structures of Typical American Oil Fields (Amer. Assoc. Petrol. Geol. Symposium), vol. 3, pp. 132-152.

Powell, J. W. (1876), "Report on the Geology of the Eastern Portion of the Uinta Mountains and a Region of Country Adjacent Thereto," U. S. Geol. and Geog. Survey Terr., pp. 111-216.

Reeside, J. B., Jr. (1920), "Some American Jurassic Ammonites of the Genera *Quenstedticeras*, *Cardioceras* and *Amoeboceras*, Family Cardioceratidae," U. S. Geol. Survey Prof. Paper 118.

─────── (1928), "New Cretaceous Mollusks From Colorado and Utah," Washington Acad. Sci. Jour., vol. 18, no. 11, pp. 306-313.

.... (1930a), "Descriptive Geology of Green River Valley Between Green River, Wyoming, and Green River, Utah," U. S. Geol. Survey Water-Supply Paper 618, pp. 56-63.

.... (1930b), "The Cretaceous Faunas in the Section on Vermilion Creek, Moffat County, Colorado," Washington Acad. Sci. Jour., vol. 20, no. 3, pp. 35-41.

.... (1933), "Notes on the Geology of the Green River Valley Between Green River, Wyoming, and Green River, Utah," U. S. Geol. Survey Prof. Paper 132-C, pp. 35-50.

.... (1944), "Review of Upper Cretaceous Floras of the Rocky Mountain Region; Pt. 1, Stratigraphy and Paleontology of the Fox Hills and Lower Medicine Bow Formations of Southwestern Wyoming and Northwestern Colorado; Pt. 2, Flora of the Lance Formation at its Type Locality, Niobrara County, Wyoming, by Erling Dorf, 1938-42," Jour. Geology, vol. 52, no. 2, pp. 139-140.

.... (1946), "Map Showing Thickness and General Character of Cretaceous Deposits in the Western Interior of the United States," U. S. Geol. Survey Prelim. Map 10, Oil and Gas Inv. Ser.

Ritzma, H. R. (1949), "Geology Along the Southwest Flank of the Sierra Madre, Carbon County, Wyoming," unpublished Master's thesis, University of Wyoming.

Schmitt, G. T. (1953), "Marine Jurassic in the Rocky Mountains," Bull. Amer. Assoc. Petrol. Geol., vol. 37, pp. 355-393.

Schultz, A. R. (1920), "Oil Possibilities in and Around Baxter Basin in Rock Springs Uplift, Sweetwater County, Wyoming," U. S. Geol. Survey Bull. 702, 107 pp.

Sears, J. D. (1924a), "Geology and Oil and Gas Prospects of Part of Moffat County, Colorado, and Southern Sweetwater County, Wyoming," U. S. Geol. Survey Bull. 751-G, pp. 269-319.

.... (1924b), "Relations of the Brown's Park Formation and the Bishop Conglomerate and Their Role in the Origin of Green and Yampa Rivers," Bull. Geol. Soc. America, vol. 35, pp. 279-304.

Sears, J. D., and Bradley, W. H. (1924), "Wasatch and Green River Formations in Northwestern Colorado and Southern Wyoming. With Notes on Oil Shale in the Green River Formation," U. S. Geol. Survey Prof. Paper 132-F, pp. 93-107.

Simpson, G. G. (1926), "The Age of the Morrison Formation," Amer. Jour. Sci., 5th ser., vol. 12, pp. 198-216.

Stagner, W. L. (1941), "The Paleogeography of the Eastern Part of the Uinta Basin During Uinta B (Eocene) Time," Carnegie Mus. Annals, vol. 28, pp. 273-308.

Stanton, T. W. (1922), "Some Problems Connected With the Dakota Sandstone," Bull. Geol. Soc. America, vol. 33, pp. 255-272.

Stokes, W. L. (1944), "Morrison and Related Deposits in and Adjacent to the Colorado Plateau," Bull. Geol. Soc. America, vol. 55, pp. 951-992.

.... (1949), "Dinosaur National Monument, Past and Present," U. S. Gov't. Printing Office, 20 pp.

.... (1950a), "Pediment Concept Applied to Shinarump and Similar Conglomerates," Bull. Geol. Soc. America, vol. 61, pp. 91-98.

.... (1950b), "Mesozoic Stratigraphy of the Uinta Basin," Petroleum Geology of the Uinta Basin (Guidebook to the Geology of Utah, no. 5, Intermountain Assoc. Petrol. Geol.), pp. 97-99.

Stuart, R. W. (1947), "Rangely Field, Colorado," Petroleum Eng., vol. 18, no. 4, pp. 43-48.

Swain, F. M. (1949), "Early Tertiary Ostracoda From the Western Interior United States," Jour. Paleontology, vol. 23, no. 2, pp. 172-181.

Swain, F. M., and Peterson, J. A. (1951a), "Ostracoda From the 'Upper Sundance' Formation of South Dakota, Wyoming and Southern Montana," U. S. Geol. Survey Prof. Paper 243-A, pp. 1-17.

.... (1951b), "Ostracoda From the Upper Jurassic Redwater Shale Member of the Sundance Formation at the Type Locality in South Dakota," Jour. Paleontology, vol. 25, pp. 796-807.

Thomas, C. R. (1945), "Rangely Field," Oil and Gas Jour., vol. 44, no. 29, pp. 90-92, 95-96.

Thomas, C. R., McCann, F. T., and Raman, N. D. (1945), "Mesozoic and Paleozoic Stratigraphy in Northwestern Colorado and Northeastern Utah," U. S. Geol. Survey Prelim. Map 16, Oil and Gas Inv. Ser.

Thomas, H. D., and Krueger, M. L. (1946), "Late Paleozoic and Early Mesozoic Stratigraphy of Uinta Mountains, Utah," Bull. Amer. Assoc. Petrol. Geol., vol. 30, pp. 1255-1293.

Thompson, M. L. (1945), "Pennsylvanian Rocks and Fusulinids of East Utah and Northwest Colorado Correlated With Kansas Section," Kansas Univ. Geol. Survey Bull. 60, pt. 2, 84 pp.

Untermann, G. E., and Untermann, B. R. (1949), "Geology of Green and Yampa River Canyons and Vicinity, Dinosaur National Monument, Utah and Colorado," Bull. Amer. Assoc. Petrol. Geol., vol. 33, pp. 683-694.

.... (1954), "Geology of Dinosaur National Monument," Utah Geol. and Min. Survey Bull. 42.

Veatch, A. C. (1907), "Geography and Geology of a Portion of Southwestern Wyoming, with Special Reference to Coal and Oil," U. S. Geol. Survey Prof. Paper 56.

Vine, J. D., and Moore, G. W. (1952), "Reconnaissance for Uranium-Bearing Carbonaceous Rocks in Northwest Colorado, Southwestern Wyoming, and Adjacent Parts of Utah and Idaho," U. S. Geol. Survey Trace Elements Inv. Rept. 281.

Vine, J. D., and Prichard, G. E. (1954), "Uranium in Poison Basin, Carbon County, Wyoming," U. S. Geol. Survey Circ. 344.

Weeks, F. B. (1907), "Stratigraphy and Structure of the Uinta Range," Bull. Geol. Soc. America, vol. 18, pp. 427-448.

White, C. A. (1878), "Report on the Geology of a Portion of Northwestern Colorado," U. S. Geol. and Geog. Survey Terr. 10th Ann. Rept., pp. 1-60.

Williams, N. (1946a), "Rangely Field," Oil and Gas Jour., vol. 44, no. 37, pp. 70-73.

.... (1946b), "Rangely Field," Oil and Gas Jour., vol. 44, no. 40, pp. 77-78, 81.

.... (1947), "Rangely Field," Oil and Gas Jour., vol. 45, no. 30, pp. 86-87, 93.

Wilson, G. (1950), "Oak Creek Discovery," World Oil, vol. 130, no. 1, pp. 65, 66.

Woodruff, E. G. (1913), "Geology and Petroleum Resources of DeBeque Oil Field, Colorado," U. S. Geol. Survey Bull. 531, pp. 54-68.

ALPHABETICAL LIST OF ADVERTISERS

Aerial Geologic Surveys
American Stratigraphic Company
Jack Ammann Photogrammetric Engineers Inc.
Baroid Well Logging Service
Graham S. Campbell, Consultant
The Central Bank & Trust Company
Century Geophysical Corporation
Christensen Diamond Products Company
Colorado National Bank
Cortez Petroleum
George H. Gaul
General Petroleum
Geomap, Inc.
Geophoto Services
Geophysical Service, Inc.
Geoprofessional Services
Gulf Oil Corporation
Dorsey Hager, Consultant
Hatch's City Cafe
Herb J. Hawthorne, Inc.
Knox-Bergman-Shearer
Lane-Wells Company
Moench Letter Service
Don Mount, Consultant
Mountain Fuel Supply Company
Muldrow Aerial Surveys Corporation
New Drilling Company
Owanah Oil & Development Corporation
Petroleum Consultants
Petroleum Information
Powers Elevation Company
W. Don Quigley and Chas. W. Hendel, Consultants
Riley's Reproduction Service
Salt Lake Blue Print & Supply Company
Schlumberger Well Surveying Corporation
Shell Oil Company
U. S. National Bank

When **YOU** need **GEOPHYSICS** in the ROCKY MOUNTAINS...call

Experienced field supervision is another GSI guarantee of geophysical survey quality, and GSI offers your exploration program experienced field supervision **in depth**. To protect your geophysical investment, call GSI. Consult with a geophysicist skilled in the problems of the Rocky Mountain area.

...In DENVER **AMherst 60601** Robert Dyk Manager, Mid-Continent Division	...in CASPER **36762** P. M. Ralph	...in BILLINGS **20332** G. D. Cloepfil	...in MIDLAND, Tex. **20892** M. E. Trostle

GEOPHYSICAL SERVICE INC.
5900 LEMMON AVENUE • DALLAS 9, TEXAS

DAVID H. MOFFAT, JR.

Moffat the "Empire Builder," Moffat, "Conqueror of the Rockies," rose from messenger boy to multi-millionaire. In the process he speeded the development of Colorado. Often the first to enter a new mining camp and invest capital, he is said to have had interests in 125 mining companies at one time. His world-wide reputation for honesty and sound judgment enticed other investors and his interest was enough to assure the future of a new camp. Moffat brought railroads to the camps, giving them connections with processing and trade centers. His investments and promotion of these centers — mills, smelters, financial and commercial institutions, railroads and mines, development of ranching and farm land, town lots and the like — did much to assure Colorado's future.

Moffat was connected with many leading railroads in Colorado. Before his death he had become a leading figure in nine major undertakings in the state as well as personally building many branch lines.

Moffat's best known work was not completed during his lifetime. Early in 1900 Moffat began a rail line connecting Denver and Salt Lake City. The line reached Craig, Colorado, in 1913—two years after his death. Moffat had dreamed of a tunnel through the mountain to shorten the route and make it more useful. William G. Evans with others continued this project; in 1922 the Colorado legislature created the Moffat Tunnel Improvement District. Bonds were sold and work got under way late in 1923. The railroad tunnel was holed through in July of 1927. Seven months later the first regular train went through the six mile bore, and one of David Moffat's dreams became a reality for the benefit of a fast growing state.

State Historical Society of Colorado

Robert J. Knox   Denzil W. Bergman
Eugene M. Shearer

KNOX •
BERGMAN •
SHEARER

Economic, rapid and reliable surface mapping thru . . .
PHOTOGEOLOGIC EVALUATION

315 Colorado Bldg.   Ta 5-4795

DENVER, COLORADO

WESTERN STATES MAP CO.

Services:

CUSTOM DRAFTING BY EXPERIENCED DRAFTSMEN

MAPS FOR THE OIL AND URANIUM INDUSTRIES

REPRODUCTIONS OF OUT-OF-PRINT U.S.G.S. MAPS

U.S.G.S. - A.E.C. PHOTO-GEOLOGIC MAPS

DISTRIBUTOR OF THE "FAMOUS" BABBEL COUNTER

CLAIM SURVEYING

Suite 205 — State Exchange Bldg.

343 South State Street

SALT LAKE CITY, UTAH

C. W. HENDEL
and
W. DON QUIGLEY

☆

GEOLOGISTS

URANIUM & OIL

☆

Newhouse Building - Salt Lake City, Utah

FORMATION DATA
by
Analysis of Mud and Cuttings
On Location Core Analysis
Portable Gas Detectors

CALL

JAMES F. FOUTS   **D. N. GREGG**
Phone 2-7166   Phone 8-0061
Casper, Wyo.   Salt Lake City, Utah

Four Corners Area
DON L. MOORE
Phone 2-4381   Midland, Texas

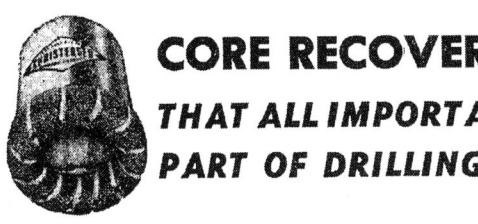

CORE RECOVERY:
THAT ALL IMPORTANT PART OF DRILLING...

Taking into consideration the type formation you're drilling, Christensen then goes to work to manufacture that custom-built bit which offers your maximum core recovery.

CHRISTENSEN
Diamond Products Company
1937 South 2nd West - Salt Lake City, Utah

DON L. MOUNT
Consulting Geologist

Telephone 84-6554

2961 Cascade Way

Salt Lake City, Utah

H. R. Ritzma

Looking half a mile down into Whirlpool Canyon from Harpers Corner, Dinosaur National Monument, Colorado.

HATCH'S CITY CAFE

DELTA, UTAH

❖

Caterer to the

I. A. P. G.

1951 - 1952 - 1953 - 1954 - 1955

Caterer to the

New Mexico Geological Society

1954 - 1955

Caterer to the

Four Corners Geological Society

1955

❖

Success on Your Field Trip

CONTROLLED REGIONAL BASE MAPS

New Mexico, Arizona,
No. Dakota, Nebraska,
Montana, S. Dakota,
Nevada, Colorado,
Kansas, Oklahoma,
Utah, Wyoming,
Idaho, Texas,
Alberta, Manitoba,
British Columbia
Saskatchewan

☆

MULDROW AERIAL SURVEYS CORP.

212 North Colorado St.

Midland, Texas

Compliments of

MARVIN F. OWENS
Vice-President Oil and Gas Department
Member R.M.A.G.

"THE OIL MAN'S BANK"

THE CENTRAL BANK & TRUST COMPANY

15th & Arapahoe   Denver

Member Federal Deposit Insurance Corp.
Federal Reserve System

H. R. Ritzma

Monument Butte (basal Mesaverde) and the Morapos rimrock, Iles Oil Field.

Secrets from the Ancient Seas of the Rockies

ARE GRADUALLY REVEALED THROUGH THE USE OF PI's TOOLS FOR SUBSURFACE STUDY.

- ELECTRIC & RADIOACTIVITY LOGS
- CORE ANALYSIS LOGS
- OIL & GAS REPORTS AND WELL HISTORY COMPLETION CARDS

Current availabilities of all materials listed weekly. Have your name placed on our mailing list to receive this information.

CONTACT ANY PI OFFICE OR WRITE TO

PETROLEUM INFORMATION, P. O. BOX 2612, DENVER 1, COLO.

THE
CORTEZ
OIL
COMPANY

DENVER

GRAHAM S. CAMPBELL

GEOLOGIST

1102 Continental Bank Building

SALT LAKE CITY, UTAH

POWELL'S PARK

The second Powell expedition (1868), designated the Rocky Mountain Scientific Exploring Expedition, left Cheyenne early in July for Denver and then turned west into the Rockies. It was an ill-sorted group of college students, clergymen, scientists, tourists and mountaineer guides that straggled over Berthoud Pass and into Middle Park. Mrs. Emma Powell and her sister, the wife of the expedition's entomologist, accompanied the group. Some of the group made the first ascent of Long's Peak. Specialists pursued their studies with vigor but often without much inspired direction.

In August, the expedition, pared of its tourist members, headed west for more serious exploring in territory uncluttered by the hordes of gold-seekers who swarmed over the Colorado Rockies. Mrs. Powell accompanied her husband and was the only woman in the party. From Hot Sulphur Springs, with Jim Bridger as guide, the expedition traversed the little used stage trail blazed by Berthoud in 1861. Mutiny, desertions and disagreements over geography slowed their progress, but they reached the fertile bottomlands along the White River below present day Meeker in mid-October. These bottoms came to be known as Powell's Park. Late October found the remnants of the expedition preparing winter quarters. This under way, Powell and most of the party headed northwest for the Union Pacific railhead at Green River City, Wyoming. Here, after a harrowing trek in the beginnings of winter, most of the expedition members disbanded, many by train to the East. Powell and one associate struck out in late November for the return trip to winter camp on the White River where Mrs. Powell and six others had remained. It was a speedy, remarkable journey.

From winter camp Powell vigorously pushed his explorations into the surrounding snowy country. His relations with the Utes who wintered in the White River bottoms were amicable, and here Powell laid the foundation for his encyclopedic knowledge of Indian ethnology that characterized his thirty years of Western exploration. And ironically, it was in Powell's Park ten years later that the Utes were provoked to the Meeker Massacre by less understanding settlers.

—H. R. Ritzma

SALT LAKE BLUE PRINT AND SUPPLY CO.

Engineering, Architectural and Artists' Supplies

☆

ALL PROCESSES FOR REPRODUCING

☆

245 South State Street

SALT LAKE CITY 1, UTAH

Telephone 4-7823

GULF Oil Corporation

☆ ☆ ☆

Exploration Department

P. O. Box 1346

First Security Bank Bldg.

Salt Lake City, Utah

H. R. Ritzma

Mesaverde (Iles) cliffs above Williams Fork Valley, Moffat County, Colorado.

GENERAL PETROLEUM CORP.

A SOCONY MOBIL CO.

Aerial Geologic Surveys

CONSULTING GEOLOGY

PHOTOGRAPHIC INTERPRETATION

Bernard M. Bench

1608 BROADWAY   DENVER, COLORADO

DORSEY HAGER
Consulting Geologist

908-909 Continental Bank Building

Telephone 5-9832

Salt Lake City 1, Utah

Complete Duplicating

and

Mailing Service

MOENCH LETTER SERVICE

DISTINCTIVE LETTER AND PRINTING WORK

53 West Broadway   Phone 22-3447

Salt Lake City, Utah

OIL MEANS MILES

The family car . . . the modern bus . . . the diesel train . . . depend on oil to speed them along mile after mile. Speed in the oil fields saves valuable rig time . . . the simultaneous recording of all curves permits Schlumberger to immediately deliver complete electrical log prints in the field.

. . . and Schlumberger means Service

Schlumberger Well Surveying Corp. • Houston, Texas

OWANAH OIL AND DEVELOPMENT CORPORATION

OIL AND GAS LEASES
and
BROKERAGE SERVICE

1163 First Security Bldg.
SALT LAKE CITY, UTAH
Phones: EL 9-6111 — EL 9-7801

408 C. A. Johnson Bldg.
DENVER, COLO.
Phone: AM 6-1727

The New Drilling Co. Inc.

DENVER, COLORADO

☆

ROTARY DRILLING

☆

1018 PATTERSON BLDG.   PHONE TABOR 5-0277

AMERICAN STRATIGRAPHIC COMPANY

☆

- Stratigraphic Studies
- Sample Distribution and Library
- Paleontological Determinations
- Research and Consulting

☆

ROCKY MOUNTAINS
and
CANADA

Billings, Mont.
Box 1444
Ph. 2-0902

Williston, N. Dak.
Gen. Del.
Ph. 3-5319

☆

Powers Elevation Company

ELEVATIONS AND LOCATIONS

P. O. Box 474
CASPER, WYOMING

☆

Sterling, Colo.
Box 981
Ph. 1675

Salt Lake City, Utah
Box 513
Ph. 84-6481

L. A. Weeks

North entrance to Irish Canyon, Weber cliffs left, Weber-Morgan contact in notch farthest right, center.

Best wishes for a successful field trip

•

MOUNTAIN FUEL SUPPLY COMPANY

Serving 68 communities in Wyoming and Utah

General and Distribution
Offices
36 South State Street
Salt Lake City, Utah

Production and Transmission
Offices
Rock Springs,
Wyoming

EXPERIENCED IN

PETROLEUM

AND

URANIUM

EXPLORATION

✯

CENTURY GEOPHYSICAL CORPORATION

TULSA, OKLA.

GEOPROFESSIONAL SERVICES, Inc.

Geologic Consultants

Contract Geology, Ore Deposits,
Oil, Gas and Non-metallics
Appraisals, Investigations, Exploration
Programs including Photo Geology

✯

Geologists

Armand J. Eardley, Ph.D. - - - - Res. Ph. 8-9448
Norman C. Williams, Ph.D. - - Res. Ph. 87-1812
F. W. Christiansen, Ph.D. - - - - Res. Ph. 7-4980
Wm. Lee Stokes, Ph.D. - - - - - - Res. Ph. 6-5614

✯

GEOPROFESSIONAL SERVICES, Inc.

Phone 4-6451

141 East 2nd South   Salt Lake City, Utah

REGIONAL BASE MAPS and AIR PHOTOGRAPHY

Accurate, up-to-date, regional base maps at scales of 1" = 8000' and 1" = 4000' are now immediately available for Montana, Wyoming, Nevada, Utah, and Colorado. Scheduled for early completion, with some areas available now, are all of North and South Dakota and portions of Nebraska, Idaho, Arizona, and New Mexico. Well data brought up to date periodically.

Also available are aerial photography and aerial mosaics of more than 200,000 square miles of this oil-rich area.

For new, free indexes of this regional base map, aerial photographic and mosaic coverage, wire, write, or call:

JACK AMMANN PHOTOGRAMMETRIC ENGINEERS, INC.

Broadway at Tenth   San Antonio 5, Texas

Branch Offices:

Manhasset, New York   Denver, Colorado

Atlanta, Georgia   Fresno, California   Houston, Texas

Yes, Sir!
Mr. Jennings is in...

You'll not only find Mr. Jennings in ... you'll also find him a "man of experience" in many fields. From roughnecking to completion, he knows the job and the problems.

He *should* know them, for he's a Registered Petroleum Engineer ... definitely the oilman's banker.

Best wishes for a successful Sixth Annual Field Conference!

Colorado National Bank

THE DOMINGUEZ-ESCALANTE EXPEDITION

First white men to explore the intermountain region were Catholic Fathers Francisco Antanasio Dominguez and Francisco Velez de Escalante accompanied by Don Juan Pedro Cisneros, Don Bernardo Miera y Pacheco, Don Joaquin Lain, Lorenzo Olivares, Lucrecio Muniz, Andres Muniz, Juan de Aguilar and Simon Lucero. Attempting to find a more practical northern route to Monterey, the above party left Santa Fe July 29, 1776, and after journeying from here northwestward, along the Utah-Colorado boundary, crossed into Utah near the present town of Jensen in Uintah County. From here they travelled westward, entering Utah Valley through Spanish Fork Canyon on September 23. Two days later the explorers encamped on the banks of Provo River, the most northerly point reached in Utah. Proceeding southwesterly, the party reached Black Rock near Sevier Lake on October 5. Alarmed at the early approach of winter, the travellers decided to return from this point to Santa Fe. The Colorado River was reached a few miles north of the Utah-Arizona Line at a point still known as The Crossing of the Fathers. Santa Fe was reached January 2, 1777. Within five months, Escalante and his companions had completed a circuit of more than eighteen hundred miles on horseback, crossing mountains and deserts practically unknown, most of the time without guides and oftentimes enduring untold hardships and privations. Although their efforts to discover a new route to Monterey were unsuccessful, these Spanish explorers did much ... by leaving to posterity the first recorded description of this intermountain country.

—L. H. Creer

UTAH OIL REPORT

225 DAVID-KEITH BLDG.
248 SOUTH MAIN STREET
P. O. Box 2606 Salt Lake City 10, Utah

... covering UTAH, IDAHO & NEVADA

Utah Oil News ...from Utah!

- Lease, Drilling & Exploration Highlights
- Field Reports on All Active Wells
- Maps of Current Activity
- State and Federal Oil & Gas Lease Applications and Assignments

PUBLISHED WEEKLY
$12.50 mo. complete

$5.00 mo. without lease applications and assignments

Compliments of

PETROLEUM CONSULTANTS

Geology and Engineering

Nellie Esperson Bldg.
CA 8-5174
Houston, Texas

722 Patterson Bldg.
AC 2-3888
Denver, Colorado

Oil saturated basal Dakota conglomerate, section 23, T. 9 N., R. 100 W., Moffat County, Colorado.

H. R. Ritzma

The bit that invites comparison!

HAWTHORNE *Blue Demon* INSERT ROCK BIT

ALL-FORMATION BIT FOR

SHOT HOLE
SLIM HOLE
STRATIGRAPHIC
CORE HOLE
and **WATER WELL DRILLING**

U S PATENTS
2,615,684
2,666,622
OTHERS PENDING

5307   P. O. Box 7366 • Houston 8, Texas

GEOMAP INC.

1014 Patterson Building

DENVER, COLORADO

☆

Geologic Maps and Base Maps
of the
Denver-Julesburg Basin

Electric Logs (Sand Section)
of the
Denver-Julesburg Basin

☆

AL 5-1196   AL 5-2749

LANE WELLS COMPANY

Technical Oilfield Service

DUALIZED...

WELL LOGGING SERVICE
Gamma Ray • Neutron • Collar Log
...all on one run!

PERFORATING
Bullet • Koneshot
...casing — open hole — tubing

DRILLABLE BRIDGING PLUGS
Cast Iron • Magnesium
...Lane-Wells or Baker

ENGINEERED PACKER SERVICE
Hookwall • Anchor • Inverted
...production — disposal — fracturing

FOR SERVICE PHONE:
Casper 2-7351   Williston 2-2651   Cody 236
Sterling, Lawrence 2-4413   Newcastle 656-W
Glendive, EMpire 5-2714   Farmington, Davis 5-2122
Kimball 244   Denver KE 4-7528   Billings 9-9984

J. R. WILLIAMS

OIL BROKERAGE SERVICE

304 Atlas Building

SALT LAKE CITY, UTAH

Phones: 5-1749 — 9-7441

Price List of I. A. P. G. and R. M. A. G. Publications

PUBLICATIONS OF THE ROCKY MOUNTAIN ASSOCIATION OF GEOLOGISTS:

GUIDEBOOK — "FIELD CONFERENCE IN CENTRAL COLORADO"	1947	Out of print
"GUIDE TO THE GEOLOGY OF CENTRAL COLORADO" (Colo. School of Mines Quart. Vol. 43, No. 2; order from Dept. of Publications, Colo. School of Mines, Golden, Colo.)	1948	$ 3.00
GUIDEBOOK — "FIELD CONFERENCE IN NORTHWESTERN COLORADO"	1953	Out of print
GUIDEBOOK — "FIELD CONFERENCE" 1st trip Denver to Colorado Springs and return. 2nd trip Denver to Canon City and return	1954	$ 3.00
"OIL AND GAS FIELDS OF COLORADO"	1954	$15.00
GUIDEBOOK — "JOINT FIELD CONFERENCE, GEOLOGY OF NORTHWEST COLORADO" — Intermountain Association of Petroleum Geologists and Rocky Mountain Association of Geologists. (Available August 24, 1955)	1955	$ 7.50
GUIDEBOOK — "ONE-DAY FIELD CONFERENCE, GEOLOGY OF THE FRONT RANGE, WEST OF DENVER." (Available May 20, 1955)	1955	$ 3.50

Order From:

PETROLEUM INFORMATION

P. O. Box 2612   Denver, Colorado

PUBLICATIONS OF THE INTERMOUNTAIN ASSOCIATION OF PETROLEUM GEOLOGISTS

GUIDEBOOK — FIRST ANNUAL FIELD CONFERENCE — 1950 "Petroleum Geology of the Uintah Basin," Guidebook to the Geology of Utah, No. 5. (Out of print — to be superseded by U. G. & M. S. symposium volume.)	
GUIDEBOOK — SECOND ANNUAL FIELD CONFERENCE — 1951 "Geology of the Canyon, House and Confusion Ranges, Millard County, Utah," Guidebook to the Geology of Utah, No. 6	$ 4.00
GUIDEBOOK — THIRD ANNUAL FIELD CONFERENCE — 1952 Cedar City, Utah to Las Vegas, Nevada, Guidebook to the Geology of Utah, No. 7	$ 4.00
GUIDEBOOK — FOURTH ANNUAL FIELD CONFERENCE — 1953 Guide to the Geology of Northern Utah and Southeastern Idaho	$ 7.50
GUIDEBOOK — FIFTH ANNUAL FIELD CONFERENCE — 1954 Geology of Portions of the High Plateaus and Adjacent Canyon Lands, Central and South-Central Utah	$ 7.50
GUIDEBOOK — SIXTH ANNUAL FIELD CONFERENCE — 1955 (Held jointly with Rocky Mountain Association of Geologists, Denver) Guidebook to the Geology of Northwest Colorado (Available August 24, 1955)	$ 7.50
RS-34 — Lecture notes by the I.A.P.G. on "Symposium: Oil Well Logging, Testing and Completion," February 23, 1953 by R. B. Downing and J. H. Terry	$ 1.50

Order From:

UTAH GEOLOGICAL AND MINERALOGICAL SURVEY

Mines Building, University of Utah, Salt Lake City, Utah

# Always Superior

REPRODUCTIONS
"World's Largest" ELECTRICAL LOG FILE

CONFIDENTIAL — CERTIFIED

PHOTO COPIES
Rapid Pick Up and Delivery

BLUE PRINTS **DIRECT PRINTS**

FILMS
ENLARGED or REDUCED
K & E SUPPLIES and EQUIPMENT

Main Office and Plant:
1540 GLENARM, DENVER, COLO.

Plant at:
145 SO. STATE ST., SALT LAKE CITY, UTAH

Plant at:
2924 2nd AVE. NO., BILLINGS, MONT.

Sales Office:
132 NO. WOLCOTT, CASPER, WYO.

ACKNOWLEDGMENTS

Assistance in Reproductions:
RILEY'S REPRODUCTION SERVICE

Drafting Time and Facilities Contributed:
EL PASO (Salt Lake)
GENERAL PETROLEUM (Salt Lake)
GULF (Salt Lake)
LION OIL (Denver)
PHILLIPS (Salt Lake)
SHELL (Salt Lake)
SINCLAIR (Salt Lake)
STANDARD OF CALIFORNIA (Salt Lake)

Sound Truck:
LANE-WELLS COMPANY

Catering:
HATCH'S CITY CAFE, Hatch Farnsworth

Guided Tour:
RANGELY FIELD INSTALLATIONS
THE CALIFORNIA COMPANY

Lithographed and Printed by
QUALITY PRESS
52 Exchange Place
Salt Lake City, Utah

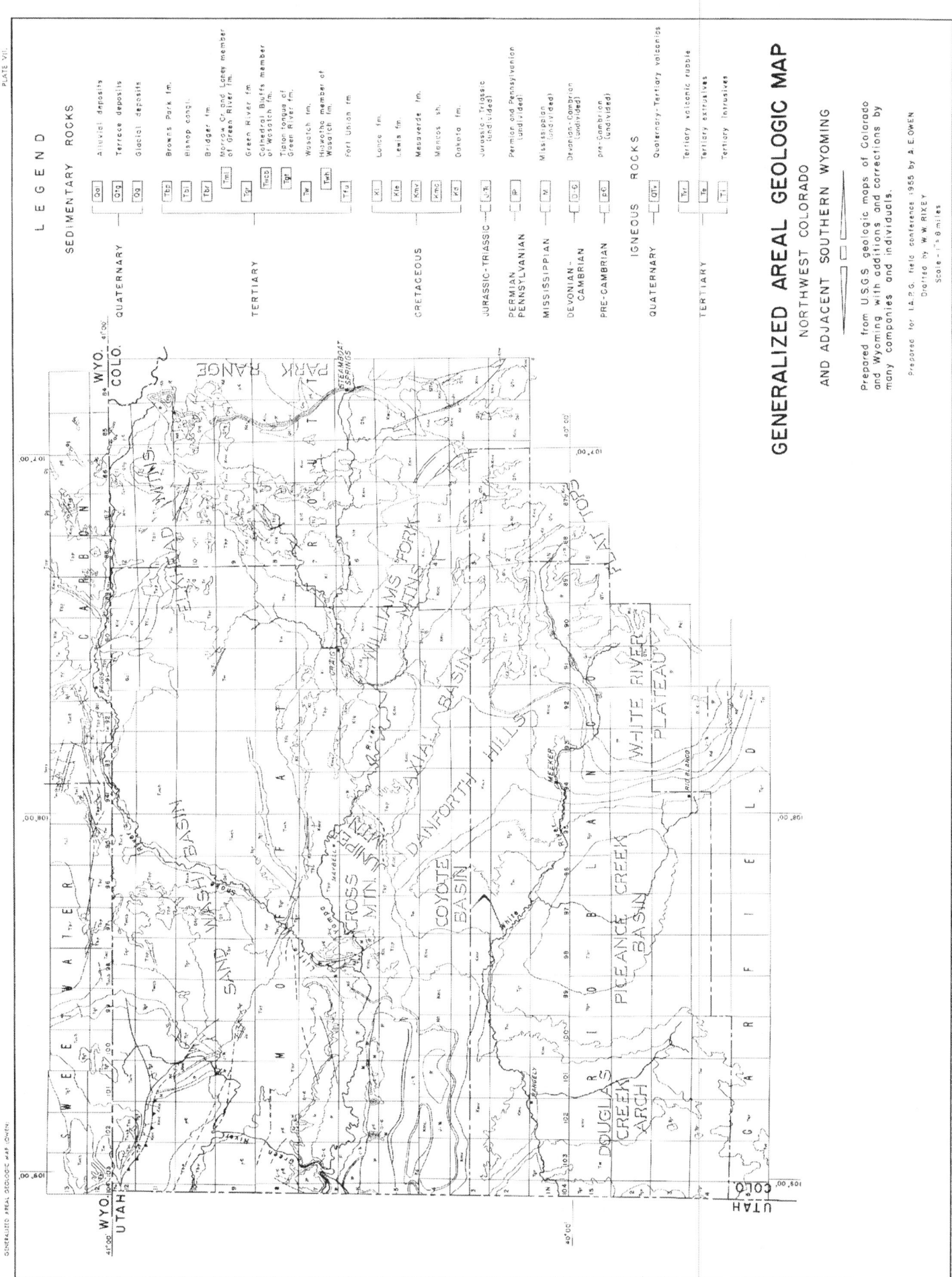

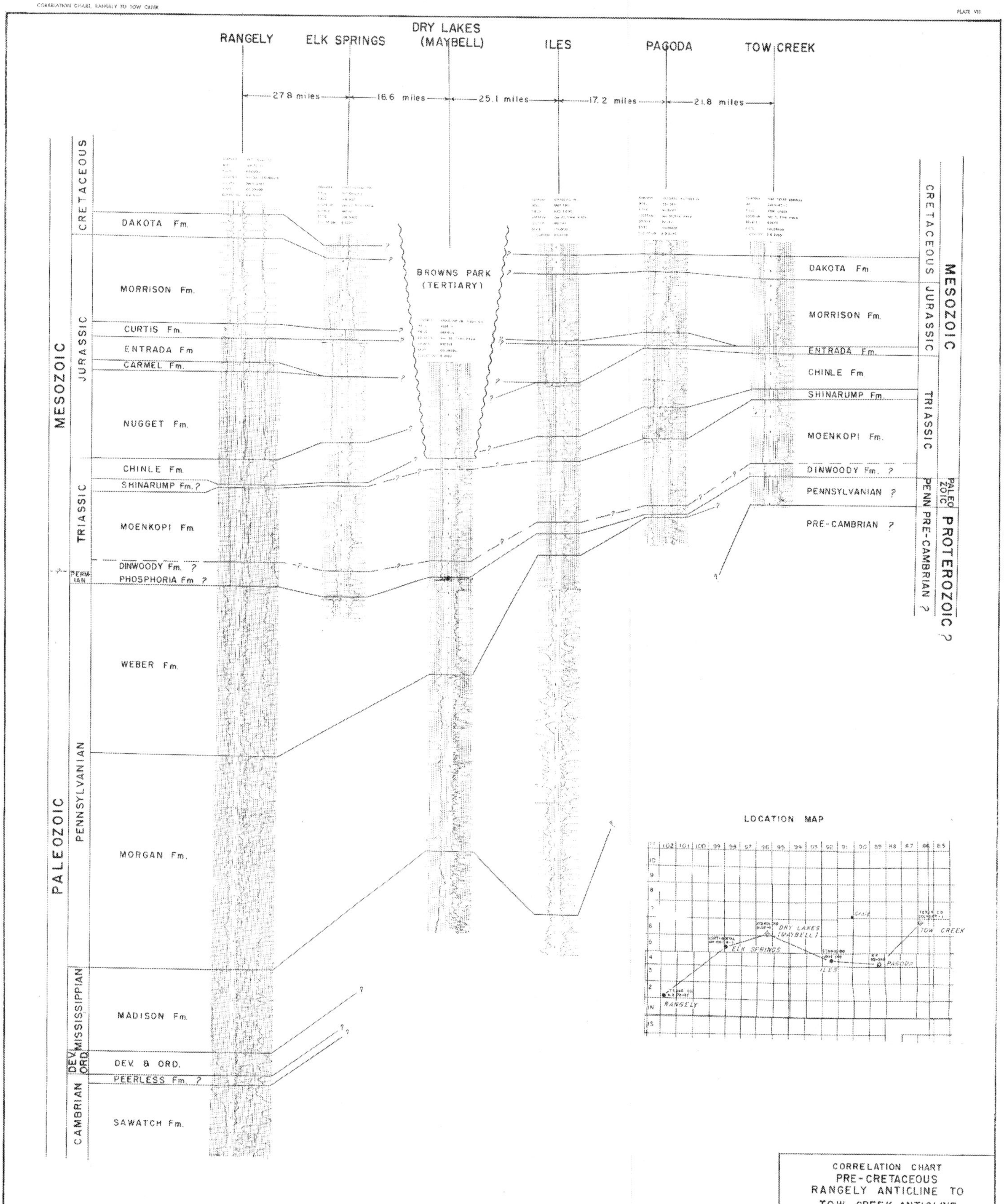
RANGELY
ELK SPRINGS
DRY LAKES (MAYBELL)
ILES
PAGODA
TOW CREEK
27.8 miles
16.6 miles
25.1 miles
17.2 miles
21.8 miles
MESOZOIC
PALEOZOIC
CRETACEOUS
JURASSIC
TRIASSIC
PERMIAN
PENNSYLVANIAN
MISSISSIPPIAN
DEV. ORD.
CAMBRIAN
DAKOTA Fm.
MORRISON Fm.
CURTIS Fm.
ENTRADA Fm.
CARMEL Fm.
NUGGET Fm.
CHINLE Fm.
SHINARUMP Fm. ?
MOENKOPI Fm.
DINWOODY Fm. ?
PHOSPHORIA Fm. ?
WEBER Fm.
MORGAN Fm.
MADISON Fm.
DEV. & ORD.
PEERLESS Fm. ?
SAWATCH Fm.
BROWNS PARK (TERTIARY)
DAKOTA Fm
MORRISON Fm.
ENTRADA Fm.
CHINLE Fm.
SHINARUMP Fm.
MOENKOPI Fm.
DINWOODY Fm. ?
PENNSYLVANIAN ?
PRE-CAMBRIAN ?
CRETACEOUS
JURASSIC
TRIASSIC
PALEOZOIC
PENN
PRE-CAMBRIAN ?
MESOZOIC
PROTEROZOIC ?
LOCATION MAP
DRY LAKES (MAYBELL)
ELK SPRINGS
ILES
PAGODA
TOW CREEK
RANGELY
Craig
CORRELATION CHART
PRE-CRETACEOUS
RANGELY ANTICLINE TO
TOW CREEK ANTICLINE
NORTHWEST COLORADO
Scale: Vertical 1"=300' Horizontal: Not to scale

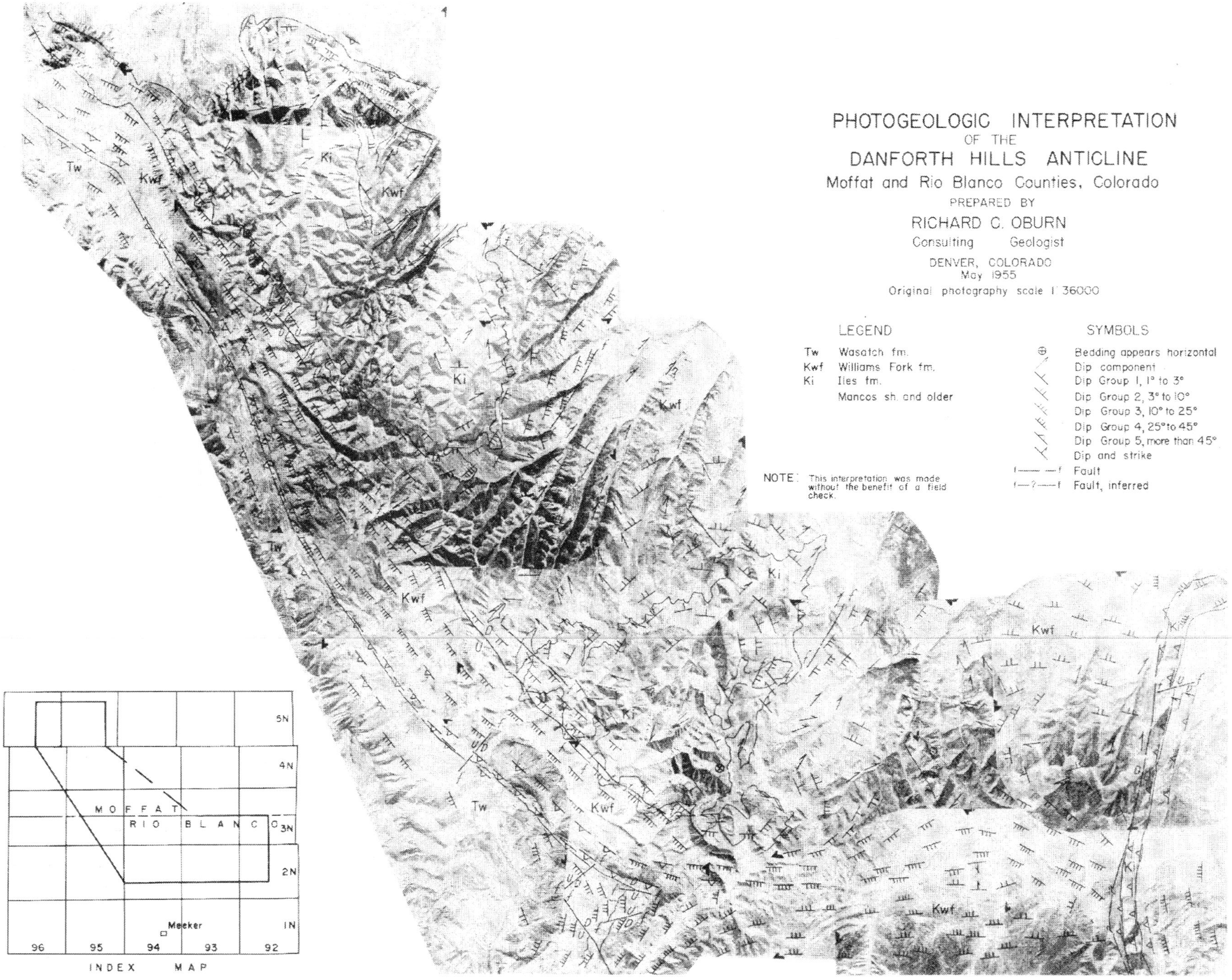

PLATE IX.

PHOTOGEOLOGIC INTERPRETATION
OF THE
DANFORTH HILLS ANTICLINE
Moffat and Rio Blanco Counties, Colorado

PREPARED BY

RICHARD C. OBURN
Consulting Geologist
DENVER, COLORADO
May 1955

Original photography scale 1:36000

LEGEND

- Tw — Wasatch fm.
- Kwf — Williams Fork fm.
- Ki — Iles fm.
- — Mancos sh and older

SYMBOLS

- Bedding appears horizontal
- Dip component
- Dip Group 1, 1° to 3°
- Dip Group 2, 3° to 10°
- Dip Group 3, 10° to 25°
- Dip Group 4, 25° to 45°
- Dip Group 5, more than 45°
- Dip and strike
- Fault
- Fault, inferred

NOTE: This interpretation was made without the benefit of a field check.

INDEX MAP

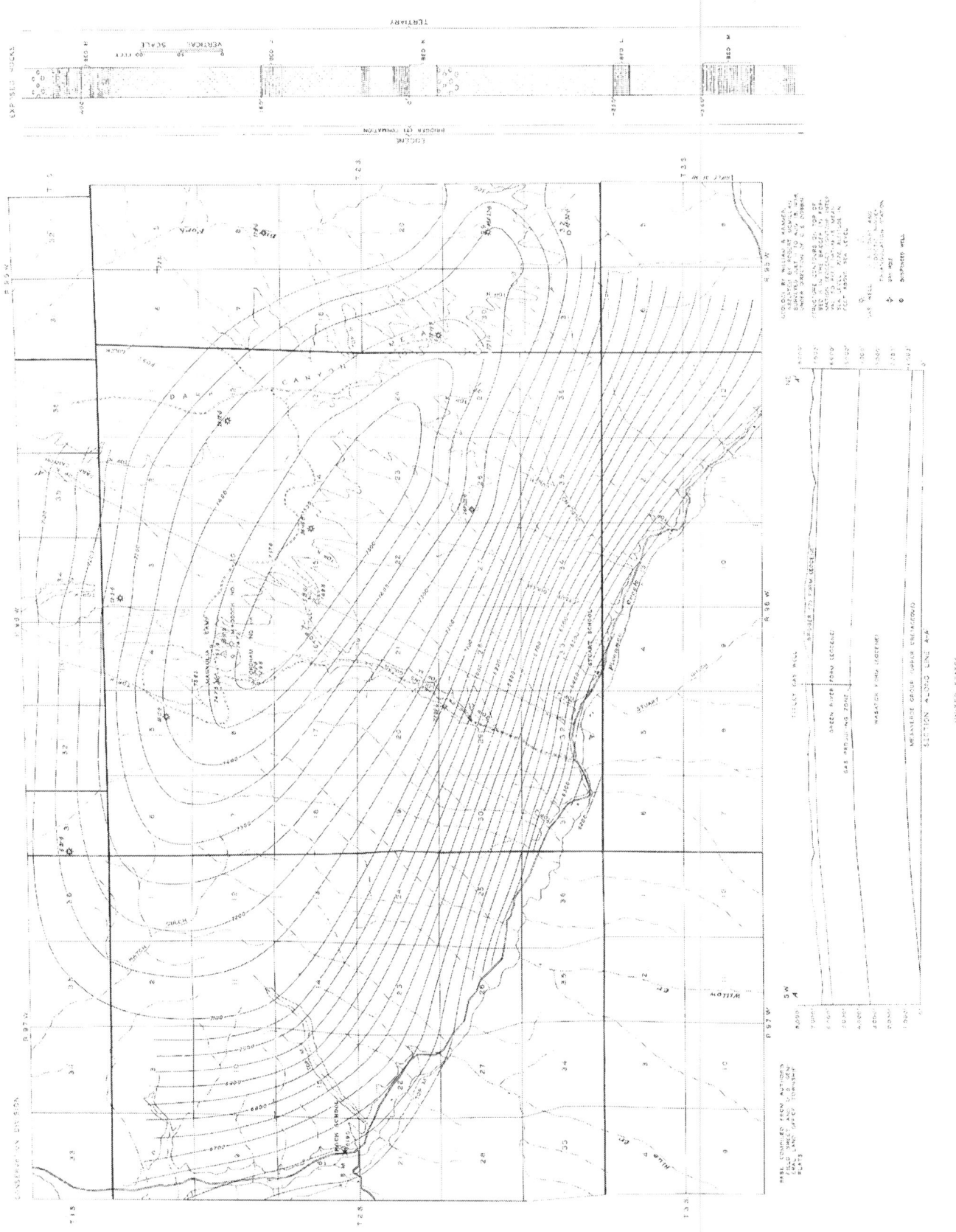